Genetic Models of Down Syndrome

Bing Ye • Roger H. Reeves
Editors

Genetic Models of Down Syndrome

Springer

Editors
Bing Ye
Life Sciences Institute
University of Michigan–Ann Arbor
Ann Arbor, MI, USA

Roger H. Reeves
Department of Physiology
Johns Hopkins University School
of Medicine
Baltimore, MD, USA

ISBN 978-3-031-78610-5   ISBN 978-3-031-78611-2 (eBook)
https://doi.org/10.1007/978-3-031-78611-2

This Springer imprint is published by the registered company Springer Nature Switzerland AG
The registered company address is: Gewerbestrasse 11, 6330 Cham, Switzerland

Preface

Down syndrome (DS) is a common condition, affecting about 1 in every 700 live births in the United States, for example. Individuals with DS often have a higher risk of various health issues but are simultaneously protected from several other medical conditions. Thus, DS presents both challenges in addressing co-occurring medical conditions and opportunities for understanding the protections it offers against other conditions.

Genetic models are crucial for studying the etiology of medical conditions. Our understanding of DS has primarily relied on animal and human cell models. DS is caused by trisomy of human chromosome 21 (HSA21), causing hundreds of protein coding and non-protein coding genes to be misexpressed. This overexpression has cascading effects throughout the genome as hundreds or thousands of genes throughout the genome respond to the changes induced by the overrepresentation of HSA21 sequences.

The consequences of trisomy 21 are ubiquitous and complex at the level of transcription, with correspondingly widespread effects on phenotype during development and throughout life. This complexity necessitates the use of equally complex experimental systems to understand the mechanisms underlying the medical conditions associated with DS. Consequently, genetic models of DS are uniquely different from those for other medical conditions. For example, rodent models of DS differ significantly from those used for single-gene conditions. Additionally, traditional genetic models such as *Drosophila* and *C. elegans* are used differently in DS studies compared to their use in other medical conditions, emphasizing their utility in screens and in multigenic interactions of individually non-pathogenic genes when misexpressed.

While there is extensive research on DS, a comprehensive guide on its genetic models is absent. This book aims to bridge that gap by providing an in-depth examination of the most commonly used models from invertebrate to "simpler" vertebrate, mammalian, and human pluripotent cell models. Strengths and weaknesses of each model in the study of DS are highlighted, offering researchers and students an understanding of the capabilities of each model and the current state of the field.

Additionally, practical guidance is provided for using model organisms in the study of medical conditions in humans and animals.

The intended audience for this book includes biomedical researchers interested in using model systems to study DS; graduate and undergraduate students who want to learn how model organisms can be applied to study human diseases; medical students learning about DS and other chromosomal disorders; veterinary students interested in academic careers; parents and caregivers in the DS community who would like a deeper understanding of research methods and possibilities; researchers from other fields such as Alzheimer disease, cancer, obesity, and diabetes.

Mouse models have been paramount in DS research, leading our book to begin with a comprehensive review by Randall Roper and co-authors on existing mouse models of DS (Chapter "Mouse Models of Down Syndrome"). This is followed by discussions on newer mouse models using Cre/LoxP technology by Eugene Yu and co-authors (Chapter "Impact of Cre/LoxP-Mediated Chromosome Engineering Technology on Down Syndrome Research"), and humanized rodent models by Roger H. Reeves (Chapter "Humanized Rodent Models of Down Syndrome"). The subsequent chapters trace the evolutionary complexity through different models: *C. elegans* (Chapter "Simple Worm, Complex Disorder: Studying Down syndrome Mechanisms in the Unassuming Model *C. elegans*" by Sophia Sanchez and Jonathan Pierce), *Drosophila* (Chapter "Using *Drosophila* to Study the Mechanisms That Underlie Down Syndrome" by Bing Ye), zebrafish (Chapter "Using Zebrafish to Study the Mechanisms That Underlie Down Syndrome" by Anna Moyer and Summer Thyme), and human cell and organoid models (Chapter "Human Models of Down Syndrome" by Anita Bhattacharyya, Yan Liu, and co-authors).

Each chapter underwent rigorous review by at least one senior author of a different model chapter and two external experts. We are very grateful of the external reviewers: Karen Chang, Jean Delabar, Mara Dierssen, Lizzy Fisher, Xuejun Li, Jie Lv, Bill Mobley, Cecilia Moens, Stefan Pinter, Gary Smith, Kelly Sullivan, Eduardo Torres, and three anonymous reviewers. Their rigorous feedback has shaped this book into a comprehensive and effective resource that we believe will be pivotal for DS research and education.

Ann Arbor, MI, USA Bing Ye
August, 2024

Contents

Mouse Models of Down Syndrome 1
Andrew Folz, Kourtney Sloan, and Randall J. Roper

Impact of Cre/LoxP-Mediated Chromosome Engineering Technology on Down Syndrome Research 45
Zhuo Xing, Avrium Douglas, Annie Pao, Tiffany Glass, Nadine Connor, and Y. Eugene Yu

Humanized Rodent Models of Down Syndrome 71
Roger H. Reeves

Simple Worm, Complex Disorder: Studying Down Syndrome Mechanisms in the Unassuming Model *C. elegans* 87
Sophia M. Sanchez and Jonathan T. Pierce

Using *Drosophila* to Study the Mechanisms That Underlie Down Syndrome 151
Bing Ye

Using Zebrafish to Study the Mechanisms That Underlie Down Syndrome 187
Anna J. Moyer and Summer B. Thyme

Human Models of Down Syndrome 233
Megan Jandy, Hao Hu, Yan Liu, and Anita Bhattacharyya

Afterword 269

Index 273

Mouse Models of Down Syndrome

Andrew Folz, Kourtney Sloan, and Randall J. Roper

Abstract Understanding the relationship between triplicated genes and Down syndrome (DS) phenotypes, including cellular and molecular mechanisms leading to these traits, has been facilitated through the use of DS mouse models. Mouse models provide an indispensable resource to understand geneotype–phenotype relationships in Trisomy 21 (Ts21). Experiments utilizing DS mouse models have produced key results addressing the molecular pathophysiology of DS. DS mouse models are useful in better understanding Ts21 and its resultant attributes, as well as predicting potential treatments. Although they do not exactly recapitulate human genetics or phenotypes, mouse models share functional mechanisms with humans. An overall best DS mouse model for Ts21 and every DS-related phenotype does not exist; instead, an array of different models are available with unique features to test hypotheses and understand more about DS. DS mouse models include different genetic constructs to create dosage imbalance to model Ts21. Phenotypic variability is apparent in individuals with Ts21 and is also observed in DS mouse models; this variation must be accounted for when utilizing DS mouse models. Because of differing genetic composition and Ts21-related phenotypes in DS mouse models, the ability to represent traits accurately and to predict how results from animal models may represent the human condition may vary between models. Therefore, it is important to understand DS mouse models, their individual benefits, and limitations, and select those models that have the highest probability for success in representing a particular DS-related condition. Potential confounding factors, including presence or absence of an extra chromosome, differences between sexes, genetic background, and phenotypic modifiers should be considered as DS mouse models are selected to test the chosen hypothesis. The following discussion of these issues will be useful in selecting DS mouse models, maximizing the knowledge generated using these mice, and better understanding the genotype–phenotype relationship in DS.

A. Folz · K. Sloan · R. J. Roper (✉)
Department of Biology, Indiana University Indianapolis, Indianapolis, IN, USA
e-mail: rjroper@iu.edu

B. Ye, R. H. Reeves (eds.), *Genetic Models of Down Syndrome*,
https://doi.org/10.1007/978-3-031-78611-2_1

Keywords Trisomy 21 · Genetic background · Dosage imbalance · Phenotype variability · Environmental effects · Rodent models · Segmental trisomy · Segmental duplication · Translocation · Transchromosomic

1 Down Syndrome Mouse Models Provide a Resource to Understand the Impact of Trisomy 21

Trisomy 21 (Ts21) causes biological alterations that may be understood only in the context of development and function of an appropriate genetic modification to a living organism. Laboratory mice provide a robustly characterized system for studies relating the effects of the extra chromosomal material to traits associated with Down syndrome (DS). The comparative genetics of *Mus musculus* (Mmu) and *Homo sapiens* (Hsa) at the DNA sequence level form a basis for utilizing DS mouse models. Comparing phenotypes across species is more challenging, especially in the domain of cognitive effects. Nonetheless, many genotype–phenotype comparisons between humans with Ts21 and DS mouse models have been formulated (Antonarakis et al. 2020).

Mouse models offer a system that allows precisely defined genetics, access to all tissues for analysis of physiological, anatomical, and behavioral outcomes at all stages of development, and these changes occur in a relatively short lifespan (Fisher and Bannerman 2019). Yet only humans have Ts21 and the resultant DS phenotypes; while mouse models provide an entrée into analysis of the molecular, cellular, and genetic mechanisms, they do not express human phenotypes nor have the exact genetic or phenotypic alterations. Different methodologies used in creating DS mouse models may have enabling or limiting factors for addressing a specific hypothesis. Therefore, a mouse model must be examined and selected for its ability to answer a particular question or hypothesis. To understand the genotype–phenotype relationship between Ts21 and DS and to identify dosage-sensitive genes of major effect in a given aspect of the pathophysiology of DS, details of the model must be understood. There is no perfect DS mouse model, and different models may provide a unique perspective to help identify candidate genes linked to molecular mechanisms of DS phenotypes (Herault et al. 2017). When selecting and using DS mouse models, the strengths and weaknesses of each model must be critically analyzed. The specific enabling and limiting characteristics of DS models should be detailed in scientific publications and presentations, especially when presenting results derived from tested hypotheses.

2 The Genetic Bases of DS Mouse Models

While humans have 22 autosome pairs and 2 sex chromosomes, mice have 19 autosome pairs and 2 sex chromosomes. Human chromosome 21 (Hsa21) is the smallest of the autosomes and is one of the least gene-dense chromosomes with 221 protein-coding genes (PCG), 447 noncoding genes, and 185 pseudogenes (Ensembl Human GRCh38.p14). This limited gene number is believed to contribute to the increased survival rate in those with Ts21 when compared to other autosomal aneuploidies (Muniz Moreno et al. 2020). DS is a contiguous gene syndrome, where multiple syntenic genes are found in three copies, and there is no naturally occurring comparable trisomy in mice as the genes found on Hsa21 are found on three different mouse chromosomes, mouse chromosome (Mmu)10 (from *Prmt2* to *Pdxk*, 2.1 Mb, 37 genes), Mmu16 (from *Lipi* to *Zbtb21*, 22.5 Mb, 117 genes), and Mmu17 (from *Umodl1* to *Rrp1b*,1.1 Mb, 19 genes). Gene order and orientation are largely conserved within these segments (Herault et al. 2017; Guedj et al. 2023).

Some human PCG and other genetic regions on Hsa21, however, are not represented on homologous mouse chromosomes, and conversely, mice have genetic elements encoded in these homologous regions that are not found on Hsa21. At least four protein-coding genes on Hsa21 are not found in mice and at least four protein-coding genes on Hsa21 homologous regions are not found in humans (Sturgeon and Gardiner 2011; Gupta et al. 2016; Fisher and Bannerman 2019). For example, in mouse, the integrin beta-2-like (*Itgb2l*) gene is not in the Hsa21 region and a human gene, prostate-, ovary-, testis-, and placenta-expressed ankyrin domain family member D (*POTED*), is not found in mouse (Herault et al. 2017). The function or regulation of homologous genes in humans and mice may also differ. For instance, differences in amino acid sequences that result from the amyloid beta precursor protein (*APP*) gene in humans and *App* in mice alter phenotypic manifestation (Wiseman et al. 2015; Xu et al. 2015). In humans, triplicated *APP* results in early onset Alzheimer's disease (onset at 40–50 years), but in mice, (which live about 2 years), triplicated *App* does not result in a similar phenotype. Mice without a functional copy of superoxide dismutase 1 (*Sod1*) survive but no humans without a functional *SOD1* have been reported (Fisher and Bannerman 2019). These phenotypic differences may be caused by the genes themselves or potentially important interacting effects on Hsa21 may not be seen in the orthologous regions of Hsa21 on Mmu10, 16, and 17, and therefore there is no equivalent phenotype in humans or DS mouse models (Gupta et al. 2016).

Phenotypes observed in DS are likely the result of the direct impact of dosage-sensitive genes, the presence of an extra chromosome, and/or the global effect of an extra chromosome and genes on genome-wide gene expression (Herault et al. 2017). Additionally, noncoding genes and epigenetic factors also interact with the extra genomic information. The effects of these genetic changes also impact transcriptional changes affecting gene expression. To model Ts21 and DS effectively, there must be an adequate meeting of the criteria of construct, face, and predictive validity in the model system (Rueda et al. 2012). Construct validity is a similarity etiology of the human and animal disorder or genetic representation, including the triplication of Hsa21 homologous genes. Face validity is how well the animal model represents molecular, cellular, physiological, behavioral, and other phenotypes. Predictive validity is new knowledge obtained in a DS animal model that portends what will be found in humans with DS.

When evaluating the genetic construct of DS mouse models, one of the issues is whether it is preferable to utilize mice with smaller numbers of triplicated genes or if it is better to have more triplicated genes in a mouse model (Herault et al. 2017). Trisomic segments with smaller numbers of genes may help to localize triplicated genes of major effect contributing to a phenotype. Yet, using models with only a few triplicated genes may eliminate or modify a trait that only occurs when multiple triplicated genes are present, either because two or more triplicated genes are needed for the phenotype or other influences of the triplicated chromosome are necessary (Lana-Elola et al. 2016; Sloan et al. 2023). Additionally, the extra chromosome itself has been shown to play an important role in gene expression and the DS-associated phenotypes (Xing et al. 2023).

3 Phenotypic Variability in DS Mouse Models

There is a high degree of phenotypic variability in individuals with DS. Some of this phenotypic variability may be attributable to genetic influences. Both trisomic and non-trisomic genes, including allelic differences in these genes, may affect the incidence and severity of DS phenotypes (Roper and Reeves 2006; Roper et al. 2020b). When there are three copies of a gene, one or more of these alleles may cause molecular alterations that affect gene expression or function. The sensitizing effect of trisomy may cause allelic differences to affect a phenotype whereas on a non-trisomic background, these allelic differences would be benign. Examples of phenotypic variability in humans that have been influenced by allelic differences include atrioventricular septal defects, B lymphoblastic acute leukemia, and acute megakaryoblastic leukemia [reviewed in (Roper et al. 2020b)].

There is considerable phenotypic variability among DS mouse models. However, there has been little effort to assess the degree of variability within a model strain, still less to compare variability in presentation between different models, and few comparisons with variability of the human condition modeled. Unfortunately, quantitative expression of a given phenotype across labs and even within labs is problematic. Inbred strains of mice have the same alleles at (virtually) every gene and by breeding with a different inbred strain, have the potential to identify modifier genes throughout the genome. Some DS mouse models can be bred onto an inbred background (generally those with genomic duplications) while others cannot (usually those with trisomy, i.e., an extra freely segregating chromosome). DS mouse models with a non-inbred background may have increased variability in some phenotypes. A few DS mouse models, including Ts65Dn, are maintained as an advanced intercross line and allelic differences even in a single litter may contribute to phenotypic variability. Genetic drift within each model strain, possibly related to unintended selection for better husbandry properties, may also contribute to inadvertent changes in other phenotypes (Costa et al. 2010; Shaw et al. 2020). However, significant variability is also seen in models maintained on an inbred background. Even inbred mice may develop allelic differences (mutations) when maintained in a closed breeding colony for many generations. For this reason, the designation of mice maintained in a different colony for more than ten generations is modified to indicate a substrain (e.g., C67BL6/J vs C57BL6/N). Phenotypic differences between substrains have been defined [for example (Bryant et al. 2008; Simon et al. 2013)]. When describing the DS model mice and resultant experiments in presentations and publications, it is important to identify those that have been maintained in a closed colony for ten generations or more in consideration of possible genetic drift. Ts65Dn mice maintained as an advanced intercross on a B6N.C3H background in comparison to a B6J.C3H background displayed fewer and/or attenuated cardiac deficits (Li et al. 2016). Phenotypic variability has also been associated with the presence of a freely segregating chromosome in addition to gene dosage effects (Villar et al. 2005; Xing et al. 2023).

Phenotypic variability is not solely relegated to allelic differences. Differences in the environment, such as housing conditions and feed, mouse husbandry, pathogen exposure, environmental enrichment, and epigenetic modification may lead to variability in phenotypes. Even developmental noise and stochastic gene expression, though difficult to identify and control, may contribute to phenotypic variability.

4 Choosing an Appropriate DS Mouse Model

Choosing an appropriate mouse model to test a specific hypothesis depends on the triplication of gene(s) (construct validity), the presence of the phenotype of interest (face validity) in a DS mouse model, and the hypothesis that the phenotype will have predictive validity for humans with DS. Some DS mouse models may exhibit traits at different stages of development as demonstrated by differences in fetal development and perinatal neurological abnormalities between Dp(16)1Yey, Ts65Dn, and Ts1Cje mice (Aziz et al. 2018; Adams et al. 2021). Additionally, expression of triplicated genes may differ between DS mouse models (Aziz et al. 2018; Guedj et al. 2023). Temporal expression of altered phenotypes and gene expression must also be considered when picking an appropriate mouse model; where and when a gene or trait is expressed may significantly differ between models.

How mice are generated and how many mice may be generated is important. Generally, when an extra chromosome is present in a mouse model (Ts65Dn, TcMAC21, Tc1), males are less fertile and trisomic offspring must be generated through the trisomic female mouse. For Ts65Dn mice, rare fertile male mice were selectively bred to produce a substrain of males with increased fertility (Moore et al. 2010). Cryopreserved sperm from rare fertile males may be used to create large numbers of progeny simultaneously through in vitro fertilization, though because of the limited genetic variability of these sperm, a genetic bottleneck effect may be introduced. Husbandry for DS models may be complicated further as the ratio of trisomic to non-trisomic offspring is often less than the expected Mendelian ratios, due to prenatal and postnatal lethality. This should be considered when planning experiments with DS mouse models (Moore 2006; Roper et al. 2006). The numbers of mice that can be generated are affected by reductions in both fecundity and fertility in most DS mouse models and it is important to design studies that are sufficiently powered to test a hypothesis with appropriate effect size (Thomas et al. 2020).

Different mouse models may display a known phenotype of interest, may not have the desired trait, or may never have been tested for a particular phenotype. Known phenotypes of DS mouse models may be found by examining phenotype databases (e.g., Mouse Genome Informatics [MGI], https://www.informatics.jax.org/marker/MGI:2178111), or through literature searches. Some phenotypes may have subtle differences between models including the severity of phenotype (Olson et al. 2004b) and the developmental stage at which a phenotype occurs (Aziz et al. 2018). Two DS mouse models may have markedly different phenotypes due to genetic construct, genetic background, environmental influences, or differences in protocols (Roper et al. 2020b). Because of genetic and phenotypic differences between different DS mouse models, a novel phenotype should be established in at least two or more DS mouse models to ensure the validity of the DS associated trait.

Mouse models offer a system that can be manipulated and utilized to understand the genotype–phenotype relationship in humans with DS. An understanding of genetic, molecular, and cellular mechanisms of a particular phenotype predicts that similar mechanisms will be found in humans with DS and may be used to improve a particular phenotype in individuals with Ts21. Examples of predictive validity of DS mouse models includes a finding of fewer cerebellar granule cells and decrease in motor neurons first in DS model mice and then in humans with DS (Baxter et al. 2000; Watson-Scales et al. 2018).

DS mouse models may be divided into different categories based on how they inherit extra genomic material including segmental trisomy, translocation, duplication, and transchromosomic (Table 1 and Figs. 1 and 2). Like in most individuals with DS (Antonarakis et al. 2020), aneuploidy or the addition of an extra chromosome is found in segmental trisomic and transchromosomic models, and this may affect gene expression and ultimately phenotypes independent of gene dosage effects (Xing et al. 2023). When choosing a model, genetic factors should be considered including aneuploidy (containing an extra centromere), number of genes at dosage imbalance, number of nonhomologous genes that are at dosage imbalance, origin of the extra genetic material, and mosaicism (the expression of the extra genetic material in only a subset of cells) (Kazuki et al. 2020).

5 Segmental Trisomy Mouse Models

Trisomy is an aneuploidy that refers to an additional freely segregating copy of a chromosome and represents a majority (95%) of individuals with DS. Most trisomic mouse models of DS are considered segmentally trisomic as they do not contain an entire extra chromosome; rather, they only contain a portion, or segment of genes on a freely segregating chromosome, and have an extra centromere.

Table 1 DS mouse models

MGI approved name	Lab or common name	Type of DS model	Short description	Genetic background	Chromosomal position (MGI)	Number of Hsa21 orthologous genes at dosage imbalance (GRCm39)	First publication	Availability
Segmental trisomy models								
Ts(17 < 16>)65Dn (001924) https://www.informatics.jax.org/marker/MGI:2178111	Ts65Dn (001924)	Radiation-induced translocation	Trisomy for a marker chromosome containing a segment of *Mrpl39* to the telomere of Mmu16 homologous to Hsa21, translocated to the centromeric end of Mmu17.	B6EiC3Sn	Breakpoint is Chr 16: 84,351,351 bp and Chr 17 9,426,822 bp	101	Davisson MT et al. (1993) https://pubmed.ncbi.nlm.nih.gov/8115398/	https://www.jax.org/strain/001924
Ts(17 < 16>)65Dn (005252) https://www.informatics.jax.org/marker/MGI:2178111	Ts65Dn (005252)	Congenic made from radiation-induced translocation	Trisomy for a marker chromosome containing a segment of *Mrpl39* to the telomere of Mmu16 homologous to Hsa21, translocated to the centromeric end of Mmu17. These mice are homozygous for the wild-type allele of *Pde6b* to eliminate possible retinal degeneration and have slightly different phenotypes than Ts65Dn 001924	B6EiC3Sn. BLiA	Breakpoint is Chr 16: 84,351,351 bp and Chr 17 9,426,822 bp	101	Costa AC et al. (2010) https://pubmed.ncbi.nlm.nih.gov/19720087/	https://www.jax.org/strain/005252
Ts(17^{16})66Yah https://www.informatics.jax.org/allele/MGI:6459295	Ts66Yah	CRISPR/Cas9 derivative of Ts65Dn	This mouse strain is derived from the Ts65Dn 01924 strain and removes the non Hsa21 homologous *Pisd-ps2* to *Pde10a* region derived from Mmu17 in the Ts65Dn mouse using CRISPR/Cas9	B6C3B		101	Duchon A et al. (2022) https://pubmed.ncbi.nlm.nih.gov/36374158/	https://www.jax.org/strain/036600

(continued)

Table 1 (continued)

MGI approved name	Lab or common name	Type of DS model	Short description	Genetic background	Chromosomal position (MGI)	Number of Hsa21 orthologous genes at dosage imbalance (GRCm39)	First publication	Availability
Translocation models								
Ts(16C-tel)1Cje https://www.informatics.jax.org/marker/MGI:3623021	Ts1Cje	Translocation	Mice are trisomic for *Sod1* to *Mx1* and 94 genes on Mmu16 that translocated to Mmu12, deleting **7** genes from Mmu12. *Sod1* is not functionally triplicated because the *Sod1* gene in the translocated segment is inactivated by the insertion of the neomycin resistance sequence. These mice are monosomic for 7 genes in Mmu12 (Duchon A, et al. 2011 https://pubmed.ncbi.nlm.nih.gov/21953411/	B6EiC3Sn	Translocation breakpoint between positions 119,278,499 and 119,289,499, in the *Dnahc11* gene, between exons 35 and 41.	64	Sago H et al. (1998) https://pubmed.ncbi.nlm.nih.gov/9600952/	https://www.jax.org/strain/004861
Rb(12.Ts17^{16}65Dn)2Cje https://www.informatics.jax.org/allele/MGI:3531161	Ts2Cje	Spontaneous Robertsonian translocation	Spontaneous fusion between the small marker chromosome in Ts65Dn mice (carrying *Mrpl39* to the distal telomere of Mmu16) and Mmu12, forming a Robertsonian fusion chromosome	B6EiC3Sn		101	Villar AJ et al. (2005) https://pubmed.ncbi.nlm.nih.gov/15859352/	https://www.jax.org/strain/004850

Tc(HSA21)1TybEmcf https://www.informatics.jax.org/marker/MGI:3814702	Tc1	Transchromosomic	Contains freely segregating copy of Hsa21, with ~75% of Hsa21 in mouse. Mice are mosaic for the extra chromosome. The copy of Hsa21 has internal deletions and rearrangements (Gribble SM et al., 2013) https://pubmed.ncbi.nlm.nih.gov/23596509/.	B6129S	NA	227* (Gribble et al. 2013 https://www.ncbi.nlm.nih.gov/pmc/articles/PMC3626651/) *includes duplications and anticipated single copy genes)	O'Doherty A et al. (2005) https://pubmed.ncbi.nlm.nih.gov/16179473/	https://www.jax.org/strain/010801
Tc(HSA21,CAG-EGFP)1Yakaz https://www.informatics.jax.org/allele/MGI:6386757	TcMAC21	Transchromosomic	Contains freely segregating copy of Hsa21 with 93% of Hsa21 represented in the mouse.	B6D2	NA	581 (Kazuki Y et al. 2020 https://pubmed.ncbi.nlm.nih.gov/32597754/)	Kazuki Y, et al. (2020) https://pubmed.ncbi.nlm.nih.gov/32597754/	https://www.jax.org/strain/035561
Duplication models								
Dp(16Cbr1-Fam3b)1Rhr https://www.informatics.jax.org/allele/MGI:3487283	Ts1Rhr	Duplication	Cre-mediated chromosomal rearrangement on Mmu16 between loxP sites in regions proximal to *Cbr1* and distal to *Fam3b* results in duplication of the genes, excluding *Cbr1* and including *Fam3b*, in the region that is syntenic with the distal part of Hsa21q22. Also produced the reciprocal deletion as Ms1Rhr.	B6.129S6	Chr16:93404725-97306136	28	Olson LE, et al. (2004a) https://pubmed.ncbi.nlm.nih.gov/15499018/	https://www.jax.org/strain/005383
Dp(16Lipi-Zbtb21)1Yey https://www.informatics.jax.org/marker/MGI:3714521	Dp(16)1Yey	Duplication	Cre-mediated chromosomal rearrangement on Mmu16 between loxP sites in regions proximal to D930038D03Rik (*Lipi*) and distal to *Zfp295* (*Zbtb21*) resulted in the duplication of the genes.	B6.129S7	Chr16:75155590-97794739	117	Li Z et al. (2007) https://pubmed.ncbi.nlm.nih.gov/17412756/	https://www.jax.org/strain/013530

(continued)

Table 1 (continued)

MGI approved name	Lab or common name	Type of DS model	Short description	Genetic background	Chromosomal position (MGI)	Number of Hsa21 orthologous genes at dosage imbalance (GRCm39)	First publication	Availability
Dp(16Tiam1-Kcnj6)6Yey https://www.informatics.jax.org/marker/MGI:5305761	Dp(16)2Yey	Duplication	Cre-mediated chromosomal rearrangement on Mmu16 between loxP sites in regions proximal to *Tiam1* and distal to *Kcnj6* resulted in the duplication of the genes, including *Tiam1* and *Kcnj6*, in the region that is syntenic with the part of Hsa21q22.	B6;129S7	Chr16:89583999-94798555	52	Liu C et al. (2011) https://pubmed.ncbi.nlm.nih.gov/24362460/	https://www.jax.org/strain/018852
Dp(10Prmt2-Pdxk)2Yey https://www.informatics.jax.org/marker/MGI:4461397	Dp(10)1Yey	Duplication	Cre-mediated chromosomal rearrangement on Mmu10 between loxP sites in regions proximal to *Prmt2* and distal to *Pdxk* results in duplication of the genes, including *Prmt2* and *Pdxk*, in the region that is syntenic with the distal part of Hsa21q22.	B6;129S7	Chr10:76043060-78300782	37	Yu T et al. (2010a) https://pubmed.ncbi.nlm.nih.gov/20442137/	https://www.jax.org/strain/013529
Dp(17Abcg1-Rrp1b)3Yey https://www.informatics.jax.org/marker/MGI:4461398	Dp(17)1Yey	Duplication	Cre-mediated chromosomal rearrangement on Mmu17 between loxP sites in regions proximal to *Abcg1* and distal to *Rrp1b* results in duplication of the genes, including *Abcg1* and *Rrp1b*, in the region that is syntenic with the proximal part of Hsa21q22.	B6;129S7	Chr17:31276672-32281839	18	Yu T et al. (2010a) https://pubmed.ncbi.nlm.nih.gov/20442137/	https://www.jax.org/strain/013531

Dp(16Tiam1-Il10rb)8Yey https://www.informatics.jax.org/marker/MGI:5562362	Dp(16)3Yey	Duplication	Cre-mediated chromosomal rearrangement on Mmu16 between loxP sites in regions proximal to *Tiam1* and distal to *Il10rb* resulted in the duplication of the genes, including *Tiam1* and *Il10rb*, in the region that is syntenic with the part of Hsa21q22.	129Sv	Not available	17	Liu C et al. (2011) https://pubmed.ncbi.nlm.nih.gov/24362460/	https://www.jax.org/strain/024906
Dp(16Ifnar1-Kcnj6)10Yey https://www.informatics.jax.org/marker/MGI:5562361	Dp(16)4Yey	Duplication	Cre-mediated chromosomal rearrangement on Mmu16 between loxP sites in regions proximal to *Ifnar1* and distal to *Kcnj6* resulted in the duplication of the genes, including *Ifnar1* and *Kcnj6*, in the region that is syntenic with the part of Hsa21q22.	129Sv	Not available	35	Liu C et al. (2011) https://pubmed.ncbi.nlm.nih.gov/24362460/	https://www.jax.org/strain/024908
Dp(17Abcg1-Cbs)1Yah Dp(17Abcg1-Cbs)1Yah	Ts1Yah	Duplication	Cre-mediated chromosomal rearrangement on Mmu17 between loxP inserted in *Abcg1* and downstream of *Cbs* (before *U2af1*) region resulted in the deletion of *Abcg1* and the duplication of the segment homologous and syntenic to the Hsa21q22	B6;129P2	Chr17:31276672-31856173	13	Lopes Pereira P, et al. (2009) https://pubmed.ncbi.nlm.nih.gov/19783846/	https://www.infrafrontier.eu/emma/strain-search/straindetails/?q=1987
Dp(16Hspa13-app)2Yah Dp(16Hspa13-App)2Yah	Ts2Yah	Duplication	Cre-mediated chromosomal rearrangement on Mmu16 between loxP inserted in *Hspa13* and *App* resulted in the inactivation of both flanking genes and the duplication of the segment homologous to Hsa21q11-q21	B6;129P2	Chr16:75552078-84970840	19	Brault V et al. (2015) https://pubmed.ncbi.nlm.nih.gov/25803843/	https://www.infrafrontier.eu/emma/strain-search/straindetails/?q=2101

(continued)

Table 1 (continued)

MGI approved name	Lab or common name	Type of DS model	Short description	Genetic background	Chromosomal position (MGI)	Number of Hsa21 orthologous genes at dosage imbalance (GRCm39)	First publication	Availability
Dp(10Col6a1-Cstb)3Yah	Ts3Yah	Duplication	Cre-mediated chromosomal rearrangement on Hsa16 between loxP inserted in *Cstb* and *Col6a1* resulted in the inactivation of both flanking genes and the duplication of the segment homologous to Hsa21	B6		25	*Unpublished*	Herault's lab
Dp(10Prmt2-Cstb)4Yah	Ts4Yah	Duplication	Cre-mediated chromosomal rearrangement on Mmu16 between loxP inserted in *Cstb* and *Prmt2* resulted in the inactivation of both flanking genes and the duplication of the segment homologous to Hsa21	B6		36	*Unpublished*	Herault's lab
Dp(16App-Runx1)5Yah	Ts5Yah	Duplication	Cre-mediated chromosomal rearrangement on Hsa16 between loxP inserted in *App* and *Runx1* resulted in the inactivation of both flanking genes and the duplication of the segment homologous to chromosome 21	B6	Chr16:84954440-92826149	60	*Unpublished*	Herault's lab
Dp(16Lipi-Zbtb21)1TybEmcf https://www.informatics.jax.org/allele/MGI:5703798	Dp1Tyb	Duplication	Cre-mediated chromosomal rearrangement on Mmu16 to duplicate the region between two loxP sites inserted proximal to *Lipi* and distal to *Zbtb21*.	B6.129P2	Chr16:74930370-97982380	117	Lana-Elola E. et al. (2016) https://pubmed.ncbi.nlm.nih.gov/26765563/	https://www.infrafrontier.eu/emma/strain-search/straindetails/?q=10565 https://www.jax.org/strain/037183

Dp(16Mis18a-Runx1)2TybEmcf Dp(16Mis18a-Runx1)2TybEmcf	Dp2Tyb	Duplication	Cre-mediated chromosomal rearrangement on Mmu16 to duplicate the region between two loxP sites inserted proximal to *Mis18a* and distal to *Runx1*.	B6.129P2	Chr16:90563769-93062456	32	Lana-Elola E. et al. (2016) https://pubmed.ncbi.nlm.nih.gov/26765563/	https://www.infrafrontier.eu/emma/strain-search/straindetails/?q=10558
Dp(16Mir802-Zbtb21)3TybEmcf Dp(16Mir802-Zbtb21)3TybEmcf	Dp3Tyb	Duplication	Cre-mediated chromosomal rearrangement on Mmu16 to duplicate the region between two loxP sites inserted proximal to *Mir802* and distal to *Zbtb21*.	B6.129P2	Chr16:93054020-97982380	37	Lana-Elola, E. et al. (2016) https://pubmed.ncbi.nlm.nih.gov/26765563/	https://www.infrafrontier.eu/emma/strain-search/straindetails/?q=10559
Dp(16Mir802-Dscr3)4TybEmcf Dp(16Mir802-Vps26c)4TybEmcf	Dp4Tyb	Duplication	Cre-mediated chromosomal rearrangement on Mmu16 to duplicate the region between two loxP sites inserted proximal to *Mir802* and distal to *Dscr3*.	B6.129P2	Chr16:93054020-94546849	14	Lana-Elola, E. et al. (2016) https://pubmed.ncbi.nlm.nih.gov/26765563/	https://www.infrafrontier.eu/emma/strain-search/straindetails/?q=10560
Dp(16Dyrk1a-B3galt5)5TybEmcf Dp(16Dyrk1a-B3galt5)5TybEmcf	Dp5Tyb	Duplication	Cre-mediated chromosomal rearrangement on Mmu16 to duplicate the region between two loxP sites inserted proximal to *Dyrk1a* and distal to *B3galt5*.	B6.129P2	Chr16:94538615-96331804	12	Lana-Elola, E. et al. (2016) https://pubmed.ncbi.nlm.nih.gov/26765563/	https://www.infrafrontier.eu/emma/strain-search/straindetails/?q=10561
Dp(16Igsf5-Zbtb21)6TybEmcf Dp(16Igsf5-Zbtb21)6TybEmcf	Dp6Tyb	Duplication	Cre-mediated chromosomal rearrangement on Mmu16 to duplicate the region between two loxP sites inserted proximal to *Igsf5* and distal to *Zbtb21*.	B6.129P2	Chr16:96327324-97982380	11	Lana-Elola, E. et al. (2016) https://pubmed.ncbi.nlm.nih.gov/26765563/	https://www.infrafrontier.eu/emma/strain-search/straindetails/?q=10562
Dp(16Mis18a-Il10rb)7TybEmcf	Dp7Tyb	Duplication	Cre-mediated chromosomal rearrangement on Mmu16 to duplicate the region between two loxP sites inserted proximal to *Mis18a* and distal to *Il10rb*.	B6.129P2	Chr16:90563769-91443378	13	Unpublished	Tybulewicz Lab

(continued)

Table 1 (continued)

MGI approved name	Lab or common name	Type of DS model	Short description	Genetic background	Chromosomal position (MGI)	Number of Hsa21 orthologous genes at dosage imbalance (GRCm39)	First publication	Availability
Dp(16Ifnar1-Runx1)8TybEmcf	Dp8Tyb	Duplication	Cre-mediated chromosomal rearrangement on Mmu16 to duplicate the region between two loxP sites inserted proximal to *Ifnar1* and distal to *Runx1*.	B6.129P2	Chr16:91436064-93062456	19	Unpublished	Tybulewicz Lab
Dp(16Lipi-hunk)9TybEmcf Dp(16Lipi-Hunk)9TybEmcf	Dp9Tyb	Duplication	Cre-mediated chromosomal rearrangement on Mmu16 to duplicate the region between two loxP sites inserted proximal to *Lipi* and distal to *hunk*.	B6.129P2	Chr16:74930370-90577148	48	Lana-Elola, E. et al. (2016) https://pubmed.ncbi.nlm.nih.gov/26765563/	https://www.infrafrontier.eu/emma/strain-search/straindetails/?q=10565
Deletion models								
Del(16Cbr1-Fam3b)1Rhr Del(16Cbr1-Fam3b)1Rhr	Ms1Rhr	Deletion	Cre-mediated chromosomal rearrangement on Mmu16 between loxP sites in regions proximal to *Cbr1* and distal to *Fam3b* results in deletion of the genes, including *Cbr1* and *Fam3b*, in the region that is syntenic with the distal part of Hsa21q22	B6C3	Chr16:93404725-97306136	29	Olson LE et al. (2004a) https://pubmed.ncbi.nlm.nih.gov/15499018/	https://www.jax.org/strain/005654
Del(10Prmt2-Pdxk)4Yey Del(10Prmt2-Pdxk)4Yey	Df(10)1Yey	Deletion	Cre-mediated chromosomal rearrangement on Mmu10 between loxP sites in regions proximal to *Prmt2* and distal to *Pdxk* resulted in deletion of the genes, including *Prmt2* and *Pdxk*, in the region that is syntenic with the distal part of Hsa21q22.	Mixed 129S7 and 129S1 background.	Chr10:76043060-78300782	37	Yu T, et al. (2010a) https://pubmed.ncbi.nlm.nih.gov/20512340/	https://www.jax.org/strain/017439

Del(17Abcg1-Rrp1b)5Yey Del(17Abcg1-Rrp1b)5Yey	Df(17)1Yey	Deletion	Cre-mediated chromosomal rearrangement on Mmu17 between loxP sites in regions proximal to *Abcg1* and distal to *Rrp1b* resulted in the deletion of the genes, including *Abcg1* and *Rrp1b*, in the region that is syntenic with the proximal part of Hsa21q22.	Mixed 129S7 and 129S1 background.	Chr17:31276672-32281839	18	Yu T, et al. (2010a) https://pubmed.ncbi.nlm.nih.gov/20512340/	https://www.jax.org/strain/017440
Del(16Tiam1-Kcnj6)7Yey Del(16Tiam1-Kcnj6)7Yey	Df(16)2Yey	Deletion	Cre-mediated chromosomal rearrangement on Mmu16 between loxP sites in regions proximal to *Tiam1* and distal to *Kcnj6* resulted in the deletion of the genes, including *Tiam1* and *Kcnj6*, in the region that is syntenic with the part of Hsa21q22.	Mixed 129S7 and 129S1 background.	Chr16:89583999-94798555	52	Liu C et al. (2011) https://pubmed.ncbi.nlm.nih.gov/21442329/	https://www.jax.org/strain/018853
Del(16Tiam1-Il10rb)9Yey Del(16Tiam1-Il10rb)9Yey	Df(16)3Yey	Deletion	Cre-mediated chromosomal rearrangement on Mmu16 between loxP sites in regions proximal to *Tiam1* and distal to *Il10rb* resulted in the deletion of the genes, including *Tiam1* and *Il10rb*, in the region that is syntenic with the part of Hsa21q22.	Mixed 129S7 and 129S1 background.	Not available	17	Liu C et al. (2011) https://pubmed.ncbi.nlm.nih.gov/21442329/	https://www.jax.org/strain/024907
Del(16Ifnar1-Kcnj6)11Yey Del(16Ifnar1-Kcnj6)11Yey	Df(16)4Yey	Deletion	Cre-mediated chromosomal rearrangement on Mmu16 between loxP sites in regions proximal to *Ifnar1* and distal to *Kcnj6* resulted in the deletion of the genes, including *Ifnar1* and *Kcnj6*, in the region that is syntenic with the part of Hsa21q22.	Mixed 129S7 and 129S1 background.	Not available	35	Liu C et al. (2011) https://pubmed.ncbi.nlm.nih.gov/21442329/	https://www.jax.org/strain/024908

(continued)

Table 1 (continued)

MGI approved name	Lab or common name	Type of DS model	Short description	Genetic background	Chromosomal position (MGI)	Number of Hsa21 orthologous genes at dosage imbalance (GRCm39)	First publication	Availability
Del(16Setd4-Kcnj6)12Yey Del(16Setd4-Kcnj6)12Yey	Df(16)5Yey	Deletion	Cre-mediated chromosomal rearrangement on Mmu16 between loxP sites in regions proximal to *Setd4* and distal to *Kcnj6* resulted in the deletion of the genes, including *Setd4* and *Kcnj6*, in the region that is syntenic with the part of Hsa21q22.	B6;129S7	Not available	15	Jiang X et al. (2015) https://pubmed.ncbi.nlm.nih.gov/26374847/	https://www.jax.org/strain/028289 (In combination with Dp(16)1Yey)
Del(16Kcnj15-Mx2)13Yey Del(16Kcnj15-Mx2)13Yey	Df(16)6Yey	Deletion	Cre-mediated chromosomal rearrangement on Mmu16 between loxP sites in regions proximal to *Kcnj15* and distal to *Mx2* resulted in the deletion of the genes, including *Kcnj15* and *Mx2*, in the region that is syntenic with the part of Hsa21q22.	B6.129S7	Not available	16	Jiang X et al. (2015) https://pubmed.ncbi.nlm.nih.gov/26374847/	https://www.jax.org/strain/028292
Del(16Dyrk1a-Kcnj6)14Yey Del(16Dyrk1a-Kcnj6)14Yey	Df(16)7Yey	Deletion	Cre-mediated chromosomal rearrangement on Mmu16 between loxP sites in regions proximal to *Dyrk1a* and distal to *Kcnj6* resulted in the deletion of the genes, including *Dyrk1a* and *Kcnj6*, in the region that is syntenic with the part of Hsa21q22.	B6.129S7	Not available	2	Jiang X, et al. (2015) https://pubmed.ncbi.nlm.nih.gov/26374847/	https://www.jax.org/strain/028290

Del(16Lipi-Ncam2)8Yey/J https://www.informatics.jax.org/allele/MGI:7509496	Df(16)8Yey	Deletion	Cre-mediated chromosomal rearrangement on Mmu16 between loxP sites in regions proximal to *Lipi* and distal to *Ncam2* resulted in the deletion of the genes, including *Lipi* and *Ncam2*, in the region that is syntenic with the part of Hsa21q22.	129S7	Not available	15	Xing X et al. (2023) https://pubmed.ncbi.nlm.nih.gov/37014740/	https://www.jax.org/strain/038567
Del(17Scaf8-Pde10a)2Yey/J https://www.informatics.jax.org/allele/MGI:7509492	Df(17)2Yey	Deletion	Cre-mediated chromosomal rearrangement on Mmu17 between loxP sites in regions proximal to *Scaf8* and distal to *Pde10a* resulted in the deletion of the genes, including *Scaf8* and *Pde10a*.	B6;129S7	Not available	0	Xing X et al. (2023) https://pubmed.ncbi.nlm.nih.gov/37014740/	https://www.jax.org/strain/038566
Del(10Prmt2-Col6a1)1Yah Del(10Prmt2-Col6a1)1Yah	Ms1Yah	Deletion	Cre-mediated chromosomal rearrangement on Mmu16 between loxP inserted in *Col6a1* and *Prmt2* resulted in the inactivation of both flanking genes and the deletion of the segment homologous to Hsa21q22	B6.129P2	Chr10:76043060-76561878	12	Besson V et al. (2007) https://pubmed.ncbi.nlm.nih.gov/17591625/	https://www.infrafrontier.eu/emma/strain-search/straindetails/?q=1808
Del(17Abcg1-Cbs)2Yah Del(17Abcg1-Cbs)2Yah	Ms2Yah	Deletion	Cre-mediated chromosomal rearrangement on Mmu16 between loxP inserted in *Abcg1* and *Cbs* resulted in the inactivation of both flanking genes and the deletion of the segment homologous to Hsa21q22	B6;129P2	Chr17:31276672-31856173	13	Lopes Pereira P et al. (2009) https://pubmed.ncbi.nlm.nih.gov/19783846/	https://www.infrafrontier.eu/emma/strain-search/straindetails/?q=1812

(continued)

Table 1 (continued)

MGI approved name	Lab or common name	Type of DS model	Short description	Genetic background	Chromosomal position (MGI)	Number of Hsa21 orthologous genes at dosage imbalance (GRCm39)	First publication	Availability
Del(16Hspa13-app)3Yah Del(16Hspa13-App)3Yah	Ms3Yah	Deletion	Cre-mediated chromosomal rearrangement on Mmu16 between loxP inserted in *Hspa13* and *App* resulted in the inactivation of both flanking genes and the deletion of the segment homologous to Hsa21q22	B6;129P2	Chr16:75552078-84970840	19	Brault V et al. (2015) https://pubmed.ncbi.nlm.nih.gov/25803843/	https://www.infrafrontier.eu/emma/strain-search/straindetails/?q=2100
Del(10Prmt2-Cstb)4Yah https://www.informatics.jax.org/allele/MGI:4868385	Ms4Yah	Deletion	Cre-mediated chromosomal rearrangement on Mmu16 between loxP inserted in *Cstb* and *Prmt2* resulted in the inactivation of both flanking genes and the deletion of the segment homologous to Hsa21q22	B6;129P2	Chr10:76043060-78263456	36	Duchon A, et al. (2008) https://pubmed.ncbi.nlm.nih.gov/18757940/	https://www.infrafrontier.eu/search?keyword=EM:02099
Del(16App-Runx1)5Yah Del(16App-Runx1)5Yah	Ms5Yah	Deletion	Cre-mediated chromosomal rearrangement on Mmu16 between loxP inserted in *App* and *Runx1* resulted in the inactivation of both flanking genes and the deletion of the segment homologous to Hsa21q22	B6J	Chr16:84749554-92622962	60	Raveau M et al. (2012) https://pubmed.ncbi.nlm.nih.gov/22693452/	Herault's lab

All coordinates from Mouse GRCm39 and Human GRCm38 unless otherwise noted

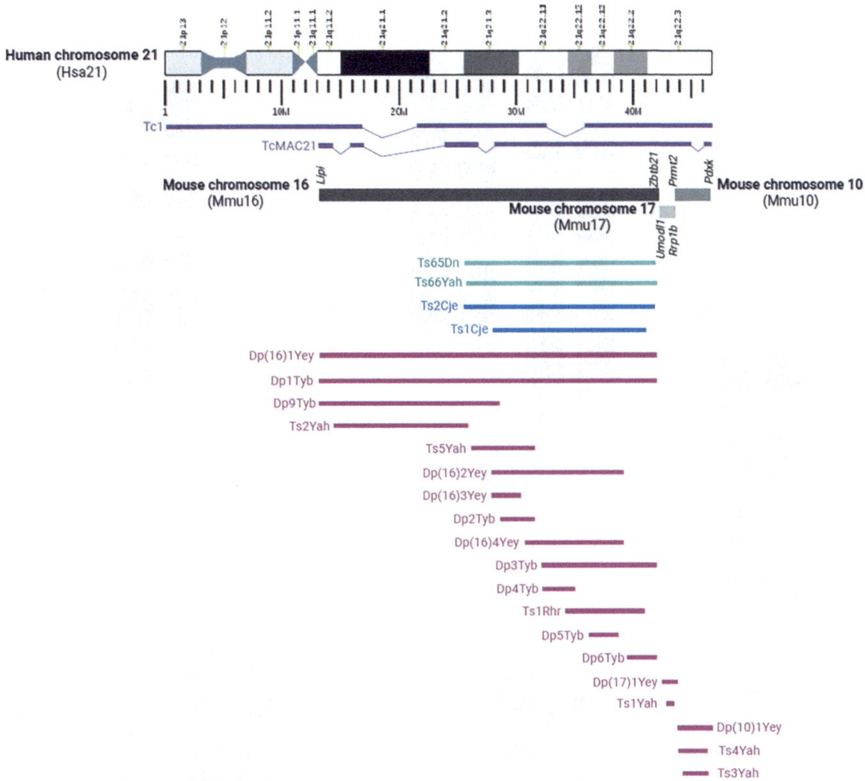

Fig. 1 Chromosomal regions covered by DS mouse models. Transchromosomic, segmental trisomy, translocation, and segmental duplication models are shown in purple, teal, dark blue, and pink, respectively. Made using BioRender.com

The earliest attempt to model DS utilized the Trisomy 16 mouse. These mice were made by breeding mice that were doubly heterozygous for two metacentric Robertsonian translocation chromosomes; 10–20% of the offspring received two metacentric chromosomes and displayed "trisomy" of Mmu16 (Miyabara et al. 1982). These mice showed hypoplasia, anasarca, slight developmental delay, and cardiovascular anomalies but did not survive past birth (Galdzicki et al. 2001). Only about 20% of the triplicated genes in the Trisomy 16 mice were homologous to Hsa21 and thus do not provide a good genetic construct to model DS.

5.1 *Ts65Dn DS Mouse Model*

The B6EiC3Sn *a/A*-Ts(17^{16})65Dn/J (Ts65Dn) mouse (Jackson Laboratory strain 001924) is the oldest and most widely utilized DS mouse model to date. It was produced by Muriel Davisson at the Jackson Laboratory by irradiating the testes of

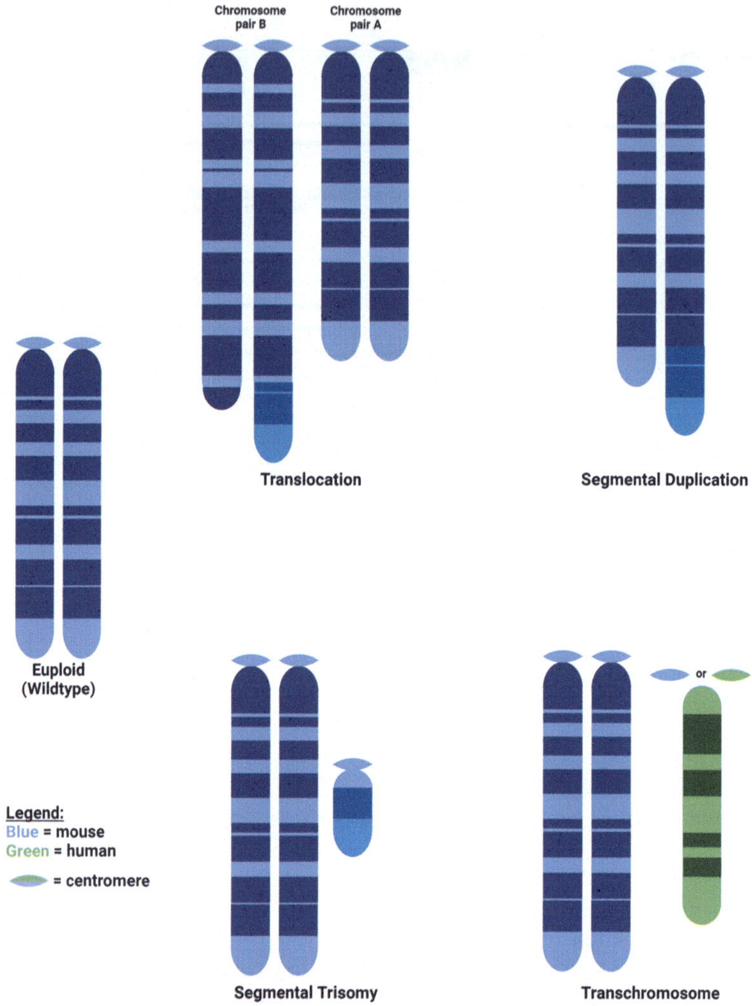

Fig. 2 Schematic representations of different genetic constructs of DS mouse models. Made using BioRender.com

DBA/2J male mice which were subsequently crossed with C57BL/6J female mice. The resulting F1 mice were screened for chromosomal abnormalities to identify reciprocal translocations (Davisson et al. 1993). The F1 carrier was crossed repeatedly with C57BL/6J x C3H/HeJ F1 (B6C3F1) hybrid mice (Davisson et al. 1990) because that background provided a higher baseline litter size than other inbred background strains (Davisson et al. 1993). In Ts65Dn mice, about 15% of the distal end of Mmu16 from *Mrpl139* to *Zbtb21* is translocated to the centromeric end of Mmu17, resulting in a freely segregating chromosome that contains 101 genes homologous to Hsa21 and ~35 nonhomologous PCG from Mmu17 (Duchon et al.

2011; Rueda et al. 2012). The translocation breakpoint between Mmu16 and Mmu17 has facilitated the usage of PCR primers to determine ploidy in the offspring of Ts65Dn and B6C3F1 mice (Reinholdt et al. 2011).

With the freely segregating extra chromosome, a cross between Ts65Dn and B6C3F1 mice results in 50%, 44%, 40%, and 36% trisomic offspring at embryonic day (E)9.5, birth, postnatal day (P)6, and weaning, respectively. This decreasing number of surviving trisomic pups indicates prenatal loss and resorption of trisomic embryos before birth and the poor mothering skills of Ts65Dn mothers postnatally (Roper et al. 2006). Ts65Dn mice were the first DS mouse model to survive postnatally with sufficient genetic construct to allow for the characterization of a variety of DS-related phenotypes including hippocampal abnormalities (Reeves et al. 1995; Insausti et al. 1998), Alzheimer's disease symptoms (Hamlett et al. 2016), muscle traits (Glass and Connor 2016), appendicular skeletal phenotypes (Blazek et al. 2011), and craniofacial changes (Richtsmeier et al. 2000). Ts65Dn mice have also been used to predict a cellular cerebellar phenotype in individuals with DS (Baxter et al. 2000).

Because of a freely segregating chromosome, extensively characterized phenotypes, and historical value, the Ts65Dn model may be considered when characterizing DS phenotypes. The limited quantity of triplicated homologous genes and the nonhomologous triplicated Mmu17 genes may circumscribe its value as a DS mouse model (Duchon et al. 2022). Many features identified in Ts65Dn mice have been found in other DS models, although the expressivity (severity) of a given phenotype may vary between models. With any genotype–phenotype correlation for DS, phenotypes must be validated in more than one DS mouse model.

Phenotypic drift across generations has been documented in the Ts65Dn DS mouse model (Shaw et al. 2020). In addition, the creation of a substrain of Ts65Dn mice (Jackson Laboratory strain 005252) to avoid having some progeny that are blind has added complexity to phenotypic comparisons in Ts65Dn mice. The original Ts65Dn strain (Jackson Laboratory strain 001924) is maintained on a B6C3 background that segregates the mutant C3H *Pde6b*^{rd1} allele (the wildtype *Pdeb6* allele is important for retinal functions), and ~25% of offspring homozygous for the C3H *Pde6b*^{rd1} allele exhibit blindness by 2–3 months of age. These mice must be individually genotyped (or crossed with known homozygous wildtype mice) to avoid blindness. The derived 005252 substrain of Ts65Dn mice is homozygous for the B6 alleles of a region estimated at 20–40 Mb that contains *Pdeb6* and does not produce progeny that are blind. Blindness may be avoided in Ts65Dn 005252 mice by breeding to B6C3F1 mice that do not segregate the mutant C3H *Pde6b*^{rd1} allele (B6EiC3Sn.BLiAF1). In the first report of Ts65Dn 005252 mice, differences in body weights between trisomic and littermate euploid mice were noted at 4–5 months. Significant differences were seen between Ts65Dn 001924 and littermate control mice for mean body length between nose and tip of tail and between nose and base of tail, but only between nose and tip of tail in Ts65Dn 005252 and littermate control mice (Costa et al. 2010). Some differences were also observed between Ts65Dn 001924 and Ts65Dn 005252 mice in different aspects of the

Morris water maze, an important test of learning and memory associated with hippocampal changes in DS.

Further differences were noted between Ts65Dn 001924 and 005252 lines when developmental characteristics were quantified in these mice. Ts65Dn 001924 mice generally have reduced body weight up to 5 months of age (Roper et al. 2006). At P15 and P30, significant differences in body weight were observed between Ts65Dn 005252 and littermate control mice that were not detected at P60 (Shaw et al. 2020). Likewise, in a separate study, there were no differences in weight between Ts65Dn 005252 and littermate control mice at 4 or 7–8 months (Tallino et al. 2022). Both studies included some groups with only 3–4 mice, which, because of the variability in weight, may have skewed these results. Therefore, monitoring weight is recommended when using DS model mice as a simple marker for detecting potential genetic drift within a colony. In summary, it appears that the Ts65Dn 001924 and Ts65Dn 005252 lines are distinct lines, and which line is used to understand DS genotype–phenotype relationships should be noted in research results.

5.2 Ts66Yah DS Mouse Model

The Ts(17^{16})66Yah mouse model eliminates the non-Hsa21-homologous Mmu17 region on the Ts65Dn marker chromosome that is homologous to a region on Hsa6 (*SCAF8* to *PDE10A*) (Muniz Moreno et al. 2020; Duchon et al. 2022). These nonhomologous genes were removed from embryos derived from in vitro fertilization of C3H/HeN.BALB/c-*Pde6b*$^{+}$ F1 (B6C3BF1) oocytes and male Ts65Dn sperm via CRISPR/Cas9. The embryos were implanted in CD1 females, and one male pup with the expected deletion was found and used to start the Ts66Yah colony by breeding with two C57BL/6 N female mice. As with selectively bred Ts65Dn mice, a subset of male Ts66Yah mice is fertile. Male Ts66Yah are reported to have a similar transmission of the marker chromosome to their offspring as female Ts66Yah mice, and transmission is similar on either a hybrid B6C3BF1 or inbred C57BL/6J background (around 30%). Ts66Yah colonies have been maintained on a hybrid B6C3B background for comparison to previous Ts65Dn results.

To date, some behavioral, craniofacial, cardiac, aging, and embryonic growth phenotypes, along with some congenital anomalies, have been characterized in Ts66Yah mice (Duchon et al. 2022; Guedj et al. 2023; Adams et al. 2024; Lanzillotta et al. 2024). Adult Ts66Yah mice lack the hyperactivity phenotype sometimes seen in Ts65Dn mice while retaining some cognitive deficits; however, which aspects of cognition are affected varies between studies and research methodologies used (Duchon et al. 2022; Guedj et al. 2023; Lanzillotta et al. 2024). Morphological differences were present in specific Ts66Yah brain structures and overall skull morphology like those of Ts65Dn mice; however, deficits were less pronounced in Ts66Yah as compared to Ts65Dn mice. The Ts66Yah mouse model has improved genetic construct validity compared to Ts65Dn mice, and these mice may have less severe phenotypes than Ts65Dn mice, possibly due to fewer genes in three copies or the effect of triplicated non-Hsa21- homologous genes on the observed Ts65Dn phenotypes.

6 Translocation DS Mouse Models

About 3% of individuals with DS present with a Robertsonian translocation, where the q arm of Hsa21 is triplicated but attaches to another acrocentric autosome (often Hsa14, but also 13, 15, or 22). It is also possible for the translocation to occur on Hsa21, creating a 21q21q translocation. Some mice have been found to have translocations of portions of Hsa21 homologous genes, resulting spontaneously or as an unintended consequence of gene targeting. These mouse models are considered translocation mouse models of DS, which are triplicated for a portion of Hsa21 homologous genes.

6.1 Ts1Cje DS Mouse Model

The B6EiC3Sn-Ts(16C-tel)1Cje/DnJ (Ts1Cje) mouse model was developed during *Sod1* gene targeting. Ts1Cje has a translocation between Mmu16 proximal to *Sod1* to the very distal end of Mmu12, with the breakpoint of Mmu16 between *App* and *Sod1* (Sago et al. 1998; Duchon et al. 2011). Ts1Cje is triplicated for only the segment of Mmu16 spanning *Sod1* to *Mx1* but does not include a functioning *Sod1*, resulting in ~64 genes at dosage imbalance in this model (Davisson et al. 2007). The translocation also produced a deletion of the of the telomeric part of Mmu12 leading to a monosomy of seven genes (*Dnach11*, *Sp4*, *Sp8*, *Abcb5*, *Itgb8*, *Macc1*, and *Tmem196*) in Ts1Cje mice (Duchon et al. 2011). Monosomy is much less tolerated than is trisomy (no complete autosomal monosomy survives to term in humans), and decades of work with single gene transgenic and "gene knockout" mice demonstrates that monosomy is more likely to produce a visible phenotype than is trisomy. Nonetheless, Ts1Cje mice display a number of phenotypes related to those seen in other DS models and have been especially useful when examining numbers of triplicated genes or specific triplicated genes (Salehi et al. 2006). In most cases, Ts1Cje has been shown to have more septal defects than Ts65Dn mice (Olson and Mohan 2011; Li et al. 2016; Xing et al. 2023). Some studies suggest that DS-like phenotypes are generally attenuated in Ts1Cje mice (Olson et al. 2004b; Xing et al. 2023), possibly due to the reduced number of triplicated genes, single copy Mmu12 genes, and/or the lack of a freely segregating chromosome.

6.2 *Ts2Cje DS Mouse Model*

The B6EiC3Sn-Rb(12.Ts17^{16}65Dn)2Cje/CjeDnJ (Ts2Cje) mouse model was identified when a female offspring from the Ts65Dn breeding colony was identified with the freely segregating 17^{16} marker chromosome attached to Mmu12, resulting in a Robertsonian fusion (Villar et al. 2005). The founder female was bred to a B6C3F1 male; litter sizes were comparable to those from Ts65Dn mothers (6 on average), but the Ts2Cje mothers had a higher rate of translocation transmission (45%) than Ts65Dn mothers. Ts2Cje males were also fertile, though they had a significantly lower rate of trisomy transmission than Ts2Cje females. Similar to Ts65Dn mice, Ts2Cje mice had steady state RNA transcript levels that were very close to 1.5 for *App*, *Sod1*, *Sim2*, *Dyrk1a*, and *Mx1*. Although Ts2Cje mice appear to have increased transmission of the genes at dosage imbalance that are homologous to Hsa21, and males are fertile, they have not been widely used to characterize DS-like phenotypes. Like Ts65Dn mice, Ts2Cje mice still have nonhomologous Mmu17 genes segregating within the strain.

7 Segmental Duplication Mouse Models

Early transgenic models utilized human yeast artificial chromosome (YAC) insertions to duplicate segments of Mmu16 in mice (Smith et al. 1997). With the advent of Cre-*loxP*-mediated chromosomal engineering, precise duplications and deletions could be performed by chromosomal rearrangement to generate DS mouse models. This is done by the introduction of *loxP* sites upstream and downstream of the desired segment in mouse embryonic stem (ES) cells to create a double-targeted cell, then transferring a Cre recombinase-expressing vector into the double-targeted cell using electroporation (Ramirez-Solis et al. 1995). Cre will recognize the *loxP* sites and induce recombination. When the two *loxP* sites are on separate homologues (in *trans*), and are oriented in the same direction, recombination results in one of the two chromosomes possessing a duplication of the targeted region and the other chromosome possessing a deletion of the targeted region (Herault et al. 2017).

Segmental duplication models utilizing Cre-*loxP* recombination are not trisomic, as they do not contain an extra chromosome (no aneuploidy); however, they are useful in studying gene dosage effects related to DS (Xing et al. 2023). These triplication model mice are also generally more amenable to inbreeding than trisomic mice. Male and female triplicated mice are both generally fertile, allowing for use of female mice in studies because they do not need to be reserved for colony maintenance, and supporting studies of potential parent of origin effects.

7.1 Mmu16 Duplication DS Mouse Models

The B6.129S7-Dp(16*Lipi-Zbtb21*)1Yey/+ (Dp(16)1Yey) mouse model was developed to duplicate the entire Hsa21 homologous region of Mmu16 (~117 genes) (Li et al. 2007). AB2.2 ES cells, derived from the 129S7/SvEvBrd-Hprt1$^{b\text{-}m2}$ strain, were used to generate the chromosomal rearrangement, then injected into C57BL/6J-*Tyr*$^{c\text{-}Brd}$ blastocysts to create chimeric mice. The chimeric mice were then mated to C57BL/6J mice. This mouse model has been maintained on both C57BL/6J × 129S7/SvEvBrd F1 and congenic C57BL/6J backgrounds. Transmission of the segmental duplication to offspring was around 38% at weaning.

The B6J.129P2-Dp(16*Lipi-Zbtb21*)1TybEmcf/Nimr (Dp1Tyb) mouse model was also developed to duplicate the entire Hsa21 homologous region of Mmu16 (Lana-Elola et al. 2016), like the Dp(16)1Yey mouse model. HM-1 ES cells, derived from the 129P2/OlaHsd-*Hprt*$^{b\text{-}m3}$ strain, were used for chromosomal rearrangement, then injected into C57BL/6J-*Tyr*$^{c\text{-}Brd}$ blastocysts to create chimeric mice. The chimeric mice were then mated to C57BL/6JNimr mice, which were used to maintain the colony. Similar to segmental trisomy models of DS, the transmission of the segmental duplication to offspring at weaning was around 33% at the time of development, but a later study showed rates of 46% and 61% of male and female duplicated Dp1Tyb mouse survival, respectively, at weaning (Lana-Elola et al. 2021). Hydrocephalus was also noted in this model, which rarely occurs in humans with DS (Lana-Elola et al. 2016). Increased metabolism (opposite of what is experienced by humans with DS) was indicated as well, but weight is not significantly altered as in other DS mouse models (see TcMAC21), as Dp1Tyb mice normalize in weight by 5-weeks-old (Lana-Elola et al. 2021).

Both of these Dp16 (Dp(16)1Yey and Dp1Tyb) models contain triplication of the same number of Hsa21 orthologous genes (117); however, the upstream and downstream sites of the duplication are in different intergenic regions, which may affect the number of non-protein-coding genes and gene regulation (personal communication, Eugene Yu and Victor Tybulewicz). Despite this and other differences during their development, similar phenotypes are present in these models, including skeletal deficits (Thomas et al. 2020; Lamantia et al. 2024) and congenital heart defects, which may be the cause of decreased viability (about 62% of Dp1Tyb embryonic day (E)14.5 embryos and 37% of Dp(16)1Yey E18.5 embryos). Dp16 models have more triplicated Hsa21 homologous genes than Ts65Dn, Ts1Cje, or Ts66Yah mice and do not have non-Hsa21 homologous genes at dosage imbalance. Both males and females are fertile and can be utilized to transmit the Mmu16 with the duplication. Dp16 mice do not have a freely segregating chromosome which may affect DS-like phenotypes. Dp(16)1Yey mice develop neurological abnormalities at later postnatal stages than Ts65Dn mice (Aziz et al. 2018).

Other mouse models have segmental duplications of smaller regions of genes on Mmu16, including B6.129S6-Dp(16*Cbr1-Fam3b*)1Rhr (Ts1Rhr or Dp1Rhr) and 13 other strains with various segmental duplications (Table 1 and Fig. 1). These models have been developed through Cre/lox-mediated engineering to understand minimum gene dosage for certain phenotypes or identify regions of genes responsible for a particular phenotype. The Ts1Rhr model was the first of these to be published, and it was designed to test the hypothesis that genes in the Down syndrome critical region (DSCR) alone were sufficient to cause all DS-like phenotypes (Olson et al. 2004a). The DSCR hypothesis was postulated because some individuals with DS had chromosomal rearrangements of subsets of triplicated Hsa21 genes, and presented with DS phenotypes, suggesting that a small region of Hsa21 genes was sufficient to cause DS. Ts1Rhr mice have a segmental duplication of the most precise description of the DSCR, a region of 28 genes homologous to Hsa21 from *Cbr1* (nonfunctional; two copies in these mice) to *Gm9242* (mouse-specific pseudogene) (Sloan et al. 2023). A number of phenotypes have been examined in Ts1Rhr mice, including craniofacial, brain, cardiac, neurological, and skeletal deficits. While some DS-related neurological and behavioral phenotypes had deficits in the Ts1Rhr mouse model (Belichenko et al. 2009; Haas et al. 2013), and some single genes in this genetic region have been shown to have a major impact on DS phenotypes, the triplication of the DSCR in Ts1Rhr mice was not sufficient to produce some common DS phenotypes, and largely disproved the DSCR hypothesis (Olson et al. 2004a; Olson et al. 2007; Olson and Mohan 2011; Lana-Elola et al. 2016; Sloan et al. 2023). The DSCR hypothesis was also refuted by studies of individuals with partial Hsa21 trisomy, where several regions of Hsa21 were associated with DS phenotypes (Korbel et al. 2009; Lyle et al. 2009)

DS models with different duplicated segments of Mmu16 have been utilized, including as part of a DS mouse mapping panel (Lana-Elola et al. 2016), to determine triplicated regions and/or genes that contribute to DS-like traits. If there is not a trait present with fewer triplicated genes, it may be that these duplication models do not have sufficient triplicated genes to cause a particular DS phenotype, may not have the necessary triplicated genes that interact, or do not have a freely segregating chromosome. Used as part of a mouse mapping panel, these DS mouse models may uncover triplicated genes that contribute to or prevent DS-like phenotypes (Sloan et al. 2023; Lana-Elola et al. 2024).

7.2 *Mmu10 Duplication DS Mouse Models*

While a majority of Hsa21 orthologous genes are on Mmu16, there are also 37 orthologous genes on Mmu10 from *Prmt2* to *Pdxk*. Three segmental duplication models for Mmu10 genes have been generated: B6;129S7-Dp(10*Prmt2-Pdxk*)2Yey (Dp(10)1Yey), B6-Dp(10*Cstb-Col6a1*)3Yah, and B6-Dp(10*Cstb-Prmt2*)4Yah. Both the Dp(10)Yah models consist of slightly smaller segments of Mmu10 genes and have not yet been published.

The Dp(10)1Yey mouse was developed using a similar protocol as Dp(16)1Yey, using AB2.2 ES cells for chromosomal rearrangement and C57BL/6J-$Tyr^{c\text{-}Brd}$ blastocysts to generate chimeric mice (Yu et al. 2010a). Chimeric males were then bred with

C57BL/6J female mice. This model was created with the intention of breeding with Dp(16)1Yey and Dp(17)1Yey to generate triple "trisomic" mice (see below), but some have investigated this model to evaluate the contribution of Mmu10 genes alone. Dp(10)1Yey mice survived at a similar rate as control mice and did not have congenital heart defects or otitis media (middle ear infection and inflammation) (Yu et al. 2010a; Liu et al. 2011; Bhutta et al. 2013). Some data indicated there may be hippocampus-related cognitive deficits in Dp(10)1Yey mice, but early data showed no differences (Yu et al. 2010b; Chang et al. 2020; Muza et al. 2023). There may also be cerebellar dysmorphology and alterations in primary ciliogenesis; however, behavioral assays investigating cerebellar function indicated Dp(10)1Yey mice were improved compared to wildtype controls (Watson-Scales et al. 2018; Jewett et al. 2023).

7.3 Mmu17 Duplication DS Mouse Models

The fewest number of Hsa21 orthologous genes (19) are found on Mmu17 from *Umodl1* to *Rrp1b*. Two segmental duplication models for Mmu17 genes have been generated: B6;129P2-Dp(17*Abcg1-Cbs*)1Yah (Ts1Yah) and Dp(17*Abcg1-Rrp1b*)3Yey (Dp(17)1Yey). The Ts1Yah model has a segmental duplication of 13 of the 19 Hsa21 orthologous genes (Pereira et al. 2009). The chromosomal rearrangement was done in HM-1 ES cells (like Dp1Tyb), then injected into C57BL/6J blastocysts to generate chimeric mice. Chimeric mice were bred to C57BL/6J mice. While this model had no deficits associated with the cerebellum (motor coordination, motor learning, balance), no altered muscle tone, nor anxiety-like behavior, some cognitive deficits related to short-term learning and memory and habituation were noted. Interestingly, there was improvement in visual-spatial memory with correlated increases in hippocampal long-term potentiation (LTP).

The Dp(17)1Yey model was developed in the same manner as the Dp(10)1Yey model (Yu et al. 2010a). Again, this model was initially intended to be bred with the Dp(16)1Yey and Dp(10)1Yey models to create the triple "trisomic" model. When assessed alone, Dp(17)1Yey mice survived at a similar rate as control mice and did not display cognitive deficits (hippocampus-, cerebellum-, or otherwise mediated), congenital heart defects, nor otitis media (Yu et al. 2010a, b; Liu et al. 2011; Bhutta et al. 2013; Watson-Scales et al. 2018; Chang et al. 2020). One phenotype that has been observed is an increase in hippocampal LTP, similar to the Ts1Yah model (Yu et al. 2010b).

7.4 Triple Transgenic/Trisomic Model

The triple "trisomic" (there is no extra chromosome in this mouse) model is made when Dp(16)1Yey, Dp(10)1Yey, and Dp(17)1Yey are bred together so mice contain all three segmental duplications, effectively meaning all Hsa21 orthologous mouse genes are triplicated in this mouse model (Yu et al. 2010a). The expected Mendelian ratio for obtaining triple "trisomic" mice is 12.5% due to the multiple crossings required to get all duplications in one mouse (Belichenko et al. 2015). This number is usually lower because about 26% of triple "trisomic" mice die shortly after birth, potentially due to congenital heart defects (Yu et al. 2010a). Of the surviving perinatal mice, 74% were reported to survive to weaning age.

Hydrocephalus was discovered in 6.5% of triple "trisomic" mice, which was clinically observable by 6–8 weeks of age and led to mortality at 8–10 weeks of age (Yu et al. 2010a). Triple "trisomic" mice have been shown to have extensive cognitive deficits, decreased body weight, hyperactive behaviors, decreased grip strength, splenomegaly (enlarged spleen), megakaryocytosis, anemia, macrocytosis (enlarged red blood cells), and thrombocytosis (high platelet count) (Yu et al. 2010a; Belichenko et al. 2015; Liu et al. 2018).

Due to the complexity of breeding the triple "trisomic" mouse, three different genotyping protocols must be performed to know if a mouse contains the segmental duplications and if so, which one(s). To streamline this process and reduce the number of genotyping protocols necessary, a coat-color method of determining which mice would need to be genotyped was devised (Li et al. 2021). Each of the segmental duplication mouse strains were backcrossed to albino C57BL/6J-*Tyr*$^{c\text{-Brd}}$ in order to produce a cream coat color (due to *Ty* minigene) in Dp(16)1Yey and Dp(10)1Yey mice and a white coat color in Dp(17)1Yey mice. When Dp(16)1Yey and Dp(10)1Yey mice with a cream coat color are bred together, the offspring with both segmental duplications now has a darker tan coat color. This tan Dp(16)1Yey;Dp(10)1Yey mouse can be bred with an albino Dp(17)1Yey mouse to produce all three coat colors, but only those with coat colors that match the parents need to be genotyped, because other mice will only have a single or double segmental duplication. Additionally, the target mice (wildtype and triple "trisomic") can be identified using only the Dp(17)1Yey genotyping protocol and mouse coat color. All mice with a tan coat color would either be a triple "trisomic" or Dp(16)1Yey;Dp(10)1Yey (parent genotype), so mice with a tan coat color and a Dp(17)1Yey segmental duplication genotype would be the triple "trisomic" mice. On the other hand, albino mice would either be Dp(17)1Yey (parent genotype) or wildtype, so they can also be distinguished using the Dp(17)1Yey genotyping protocol. Triple "trisomic" mice identified using this method appeared to have the same cognitive deficits as were previously reported, despite changes in the background of these mice; however, more phenotypes need to be confirmed.

8 Deletion Mouse Models

Due to the use of Cre/lox engineering to create DS mouse models, mice are often created with deletions of the same genetic elements that are duplicated (Table 1). These mice, on their own, are not DS mouse models. If a gene(s) within these deletion models is dosage sensitive, haploinsufficiency of this genomic sequence may cause a phenotype, perhaps even related to or opposite from traits seen with triplication of these genes (Aldridge et al. 2007; Besson et al. 2007; Olson et al. 2007; Olson and Mohan 2011; Sahun et al. 2014; Arbogast et al. 2015; Brault et al. 2015). By breeding mice with deleted chromosomal material with DS mouse models with larger segments of gene triplication, mice can be made that are triplicated for a number of genes, but the genes corresponding to the deletion now have a normal copy number of those genes in that mouse. If a triplicated gene or region is suspected of causing or contributing to DS-like phenotypes, DS mouse models with a normal copy number of some genes would have an attenuated or absent phenotype (Olson et al. 2004a, b; Garcia-Cerro et al. 2014, 2017, 2018; Sloan et al. 2023; Lana-Elola et al. 2024). Recently, the Del(17*Scaf8-Pde10a*)2Yey/J (Df(17)2Yey/+) mouse model was generated to normalize the Ts65Dn non-Hsa21 homologous genomic region by breeding this model with the Ts65Dn mice (Xing et al. 2023).

9 Transchromosomic Mouse Models

Transchromosomic models are made when a chromosome from one species is expressed in another species. In the case of transchromosomic DS mouse models, an Hsa21 chromosome is introduced into a mouse. These models are considered trisomic/aneuploid due to the presence of an extra centromere with the triplicated genetic content. These models may exhibit the most similar construct to humans with DS since they express the greatest percentage of Hsa21 genes and are aneuploid; however, there are concerns about how a human chromosome functions in a mouse, including how the human genes are regulated, transcribed, and translated and how well the translated proteins can function. Transcriptional regulation of the human and mouse chromosome in comparable tissue of transchromosomic mice appears similar and depends mostly on genetic sequence, though epigenetics, cellular environment, and other transcriptional factors may affect species-specific gene expression (Wilson et al. 2008).

9.1 Tc1 Mouse Model

The B6129S-Tc(HSA21)1TybEmcf/J (Tc1) mouse model was developed by utilizing irradiated microcell-mediated chromosome transfer from line 739 derived from human cell line HT1080 into the MPI VI cell line derived from a female 129S2 mouse. An ES cell with this extra Hsa21 was injected into C57BL/6 blastocysts to create transchromosomic chimeras that were then mated with C57BL/6J mice (O'Doherty et al. 2005). A successful germline transmission was achieved from one female chimera, who had one litter of two pups, one male and one female (O'Doherty et al. 2005), which both retained the freely segregating Hsa21. When the Tc1 mice were inbred, and the number of backcrosses increased, the transmission of the human chromosome diminished to the point where the colonies could no longer be maintained (O'Doherty et al. 2005). With a stable transmission frequency of ~40% through the female germline, only the F1(C57BL/6Jx129S8) colony could be maintained (O'Doherty et al. 2005). The extra chromosomal material in Tc1 mice contains 42 Mb, or approximately 90%, of a freely segregating Hsa21. The genes contained in the transchromosome were found in two groups, one bounded by *CXADR* and *D21S1922* and the other bounded by *IFNAR1* and *RUNX1* (O'Doherty et al. 2005). With more advanced oligonucleotide assays, the deletion areas were redefined, and new deletion, duplications, and rearrangements were revealed with 15 genes structurally rearranged, 45 genes partially or fully deleted, 13 genes fully duplicated, and 4 genes partially duplicated, leaving 209 Hsa21 genes at the anticipated single copy number (Gribble et al. 2013).

Tc1 mice display deficiencies related to DS including cognitive, craniofacial, and cardiac phenotypes. When examining different tissues of Tc1 mice, there was mosaicism of Hsa21. In the brain, there was around a 66% retention of Hsa21 while the spleen had 24% retention of Hsa21 (O'Doherty et al. 2005). Mosaicism is seen in some individuals with DS and generally attenuates deficiencies in these individuals. Due to the mosaicism, each Tc1 mouse may present different phenotypes and intensity of phenotypes. This may make it difficult to make strong statistical comparisons of the data. Additionally, human genes from Hsa21 are likely transcribed by proteins encoded in mouse and there could be interacting proteins from human and mice (or the lack thereof) in this model (Muniz Moreno et al. 2020).

9.2 TcMAC21 DS Mouse Model

The Tc(Hsa21,CAG-EGFP)1Yakaz/J (TcMAC21) mouse model is a transchromosomic model that contains a freely segregating Hsa21q-MAC hybrid chromosome that has no detectable mosaicism (Kazuki et al. 2020). The hybrid chromosome was designed to contain ~13 Mbp with the MAC1 sequence including the mouse chromosome 11 centromere plus 2.5 kbp of the peri-centromeric region, an HS4 insulator, CAG-EGFP, an HS4 insulator, PGK-neo selection cassette and a *loxP* site. Following that was ~26Mbp Hsa21q sequences that contained all annotated Hsa21q PCGs from *POTED* to *PRMT2* and all except for one non-protein-coding gene (*AP001461.1*). The hybrid chromosome was introduced into embryonic stem cells in a (C57BL/6NCrlj x CBA/JNCrlj) F1-derived TT2F female using microcell-mediated chromosome transfer. Two clones that were determined to have the Hsa21q-MAC freely segregating chromosome were used to produce chimeras. GFP-positive female chimeras were crossed with ICR males to establish the mouse line. The peri-centromeric GFP allows for identification using UV light. The luminescence from the GFP is more noticeable before fur growth. Using the GFP-tag along with real-time PCR, there was no evidence of mosaicism, unlike what has been seen in the Tc1 model (Kazuki et al. 2020). TcMAC21 are currently maintained by breeding female TcMAC21 to B6D2F1/J male mice.

There was variable expression of the intact Hsa21 protein coding genes, when combined with the expression of the Mmu genes, approximately a third had less than a 1.3 fold increase compared to euploid levels, a third had 1.3–1.7 fold overexpression, and a third were highly overexpressed with over a 1.7 fold increase (Kazuki et al. 2020). Additionally, there are four deletions identified in mice carrying the transchromosome, resulting in the loss of 14 protein-coding genes and 105 non-protein-coding genes, leaving 199 protein-coding and 382 non-protein-coding genes present.

B6/D2.TcMAC21 females crossed with B6 males produced 15 litters with an average of 6 pups per litter. 48% of offspring were TcMAC21, and the TcMAC21 dams took better care of their pups as compared to Ts65Dn dams as there was no observed loss of pups from birth to weaning (Kazuki et al. 2020). According to the Jackson Laboratory, there is high variation in fecundity in their TcMAC21 colony, with an average of ~35% of TcMAC21 females being nonproductive in each breeding cycle. At birth, there was no difference in weight between euploid and TcMAC21 mice. As they age, the TcMAC21 mice show a consistent delay in growth rate, reaching approximately a 25% reduction in weight compared to euploid mice (Kazuki et al. 2020).

TcMAC21 mice present with a number of phenotypes that occur in both individuals with DS along with other DS mouse models, including small size/short stature, altered brain morphology with cerebellar hypoplasia, altered craniofacial morphology, congenital heart defects, elevated levels of the *APP* gene, and deficits in learning and memory. TcMAC21 mice also present with hematological abnormalities that resemble the increased leukemia incidence in children with DS, along with increased sensitivity to X-ray damage to bone marrow cells (Kazuki et al. 2020).

Despite these similarities, TcMAC21 mice also present with hypermetabolism, which is the opposite of what humans with DS experience (Sarver et al. 2023). The delayed growth rate and weight reduction after birth in TcMAC21 mice was not found to be caused by reductions in IGF-1 or growth hormone nor reduced caloric intake, both of which were elevated in TcMAC21 mice, suggesting hypermetabolism could be the cause. Even when fed a high fat diet, TcMAC21 mice remained low in weight compared to euploid mice. There was also no alteration in nutrient absorption due to dysfunction of the gut. It is thought that hypermetabolism is caused by the skeletal muscle's increased mitochondrial respiration, which is occurring to meet energy demands due to the mouse's hyperactivity and futile cycling. There is also evidence that transchromosomic genes may not have the ability to perturb disomic mouse chromosome gene expression similar to segmental trisomy models with three chromosomes from the same species, which may lead to less severe or absent phenotypes (Moyer et al. 2023).

The TcMAC21 mice provide a unique DS mouse model with an extra chromosome including a high percentage of Hsa21 genes, and no significant mosaicism. Because of its recent development, there remain questions as to how well it replicates DS-like phenotypes. Additionally, reduced fecundity in this model may limit its use. Phenotypes that have been characterized have been done mostly in older mice. More work needs to be done to characterize multiple phenotypes of TcMAC21 mice at various ages.

10 Genotyping of DS Mouse Models

Because of variability in DS mouse models, mice should be genotyped to identify the genetic change (construct validity). This is most often done by PCR, and protocols can be found on websites distributing the mice or in the papers that originally described the mice. Visual phenotyping or genotyping by amplifying one or a few genes by qPCR is not sufficient to accurately genotype DS mouse models. There is considerable overlap between DS mouse model and control littermate phenotypes including size or weight (Roper et al. 2006), and triplicated genes may or may not be overexpressed depending on age, sex, and tissue-specific factors (Hawley et al. 2023). More recently, TcMAC21 mice may be genotyped by the presence or absence of GFP included in the genetic construct of this model (Kazuki et al. 2020).

11 Genetic Backgrounds of DS Mouse Models

Mouse models should be obtained from reliable resources including the Jackson Laboratory (JAX) or the European Mouse Mutant Archive (EMMA). DS mouse models are most often found on an inbred (B6) or an advanced intercross (B6 and C3H or B6 and DBA) genetic background. Inbred strains of mice facilitate genetic studies because these strains have homozygous alleles at all genes throughout the genome and eliminate genetic variability when using mice of the same inbred strain (Silver 2001). Due to issues with viability described above, DS mouse models are found with different genetic backgrounds and may not be inbred (Table 1). Generally, DS mice with a segmental trisomy or a transchromosome are maintained on a mixed genetic background, but these mice on a mixed background may be more susceptible to genetic drift. Because ~10 generations of backcrossing are required for an inbred genetic line, mice that have recently been developed contain a mixed genetic background. Because the nomenclature of mouse strains is often difficult to understand, Table 2 explains some of the mouse nomenclature related to DS mouse models. Additional explanation is found at the Jackson Laboratory Website (https://www.jax.org/). These genetic backgrounds may play a role in the phenotype, with different genetic backgrounds giving different phenotypes with the same DS mouse model (Deitz and Roper 2011). When comparing traits between different DS mouse models, it is important to maintain a consistent genetic background. To avoid genetic drift, it is important to replenish founder mice in breeding colonies before genetic drift takes place.

When the genetic background is altered, phenotypes will often change in both occurrence and severity (Olson et al. 2004b; Deitz and Roper 2011). When crossing DS models with lines on a different genetic background, for example to add or eliminate trisomic genes, nonhomozygous passenger or hitchhiker allelic differences will be introduced (Armstrong et al. 2006; Lusis et al. 2007; Su et al. 2009). Genetic background should be noted when breeding DS model mice to mice with

Table 2 DS model genetic background explanations

Symbol	Description
B6	Found on a C57BL/6J background
B6C3	Advanced intercross between B6 and C3H/HeJ (on average, 50% B6 and 50% C3H); C3H mice carry retinal degeneration alleles for *Pde6b* that, when found in two copies, cause blindness in offspring
B6C3B	Advanced intercross between B6 and C3HB (on average, 50% B6 and 50% C3HB); C3HB mice do not carry retinal degeneration alleles for *Pde6b*, and all offspring are sighted mice.
B6EiC3Sn	Advanced intercross between C57BL/6JEiJ and C3H/HeSnJ (on average, ~50% B6 and ~50% C3H); C3H mice carry retinal degeneration alleles for *Pde6b* that, when found in two copies, cause blindness in offspring. C57BL/6JEiJ and C3H/HeSnJ are specific lines that are close to but not exactly equal to C57BL/6J and C3H/HeJ lines
B6EiC3Sn.BLiA	Advanced intercross between C57BL/6JEiJ and C3H/HeSn.BliA/J (on average, ~50% B6 and ~50% C3H); C3H/HeSn.BliA/J mice do not carry retinal degeneration alleles for Pde6b, and all offspring are sighted mice.
B6129S	Advanced intercross between B6 and 129S (usually ~50% B6 and ~50% 129S)
B6D2	Advanced intercross between B6 and DBA2/J (usually ~50% B6 and ~50% DBA);
B6.129S7	Mouse is considered "congenic" or fully on the inbred B6 background even though it started on a 129S7 genetic background; Notice the period "." between B6 and 129S7, this means congenic on the inbred B6 background. The 129S7 is the donor strain and B6 is the recipient strain.
B6;129S7	Mixed background between B6 and 129S7. Mouse was generated on a 129S7 background and has been bred one or more times to a B6 background, but not enough times to be considered congenic or a complete B6 background; Notice the semicolon ";" between B6 and 129S7, this means that the mouse is still on a mixed background

other genetic variants (e.g., deletions of triplicated genes or a knockout of a certain gene). Breeding mice on an identical genetic background is preferred, but if this cannot be done, the breeding scheme and approximation of the resultant genetic background should be noted. Changes genetic backgrounds have also been shown in response to context fear conditioning in DS mouse models (Ahmed et al. 2021).

12 Sex Differences in DS Mouse Models

Differences between sexes have been noted for many DS traits, including behavioral and skeletal phenotypes. Behavioral development in humans with Ts21 differs between the sexes (Aoki et al. 2018; Channell et al. 2019). Girls as compared with boys with Ts21 are more often found in the above-average group on intelligence and adaptive function (Marchal et al. 2016). The developmental age of 8-year-old girls with DS was significantly higher than boys (van Gameren-Oosterom et al. 2011), and male as compared to female adolescents with Ts21 show more frequent behavioral problems (van Gameren-Oosterom et al. 2013). In adulthood, women with DS show higher cognitive abilities compared to men, and profound intellectual disability was twice as frequent in men as women with DS (Maatta et al. 2006). Males as compared to females with Ts21 exhibit greater bone mineral density (BMD) and bone area in the upper and lower limbs (Baptista et al. 2005). Males with DS begin losing BMD in the femur much earlier than females with DS, suggesting a protective effect of the Hsa21 trisomy in the female biological sex in terms of maintaining BMD (Carfi et al. 2017; Costa et al. 2017, 2018; Tang et al. 2019). When many individuals with DS were compared in specific age ranges, males with DS as compared to controls had reduced femoral neck BMD beginning in their 20 s but females with DS did not show reduced femoral neck BMD until their 40 s (Carfi et al. 2017).

In DS mouse models, sex had a significant modulatory role in behavioral changes in environmental enrichment of Ts65Dn mice (Martinez-Cue et al. 2002). Female but not male Ts65Dn mice showed higher anxiety measures and altered defensive behaviors than control mice (Martinez-Cue et al. 2006). Differences between sexes in skeletal abnormalities seen in humans with DS have been recapitulated in DS mouse models (LaCombe and Roper 2020; Thomas et al. 2020; Lamantia et al. 2024). The differences between sexes have been used to better define how trisomy causes DS-associated bone and brain phenotypes (Hawley et al. 2023; Sloan et al. 2023). Therefore, it is imperative to utilize both sexes when designing experiments with mouse models of DS. Sufficient animals of both sexes should be used so that there is adequate power to analyze potential differences in male and female animals in separate cohorts (Drobniak et al. 2024).

13 Parent of Origin Influences in DS Mouse Models

Parent of origin influences, or which sex the triplicated genetic information is inherited from, may alter phenotypes of DS mouse models. The most accurate representation of how T21 is transmitted in most humans is when the dam is non-trisomic. Male fertility, however, is often compromised in DS mouse models containing an extra freely segregating chromosome, and in these models, trisomic females are most often used to generate DS mouse model offspring and maintain DS mouse colonies. This places both trisomic and euploid offspring in the uterus of a trisomic mother. Some experiments have shown that DS mouse model offspring from trisomic mothers have different phenotypes than DS mouse model offspring from non-trisomic mothers. For example, in the Ts66Yah mouse model, male but not female Ts66Yah mice have reduced body rotations, a proxy measurement for motor development and maturation, in the first week of life only when the trisomy is inherited from the mother. Additionally, there were more significant differences in patterns of ultrasonic vocalization in both male and female Ts66Yah mice with maternal trisomy (Guedj et al. 2023). Unless otherwise shown to have no influence, parent of origin of mice with triplicated genes should be accounted for, optimally by passing the triplicated genes through the same parent in every experiment.

14 Environmental Influences on Phenotypes of DS Mouse Models

Even mice with the same triplicated genes may show phenotypic differences between labs and even within the same lab. Some behavioral traits that were evaluated had similar strain differences when tested using the exact same protocols in different labs, while other traits showed less reliable differences when tested in different labs (Crabbe et al. 1999; Wahlsten et al. 2003). Additionally, slight differences in protocols or how mice are maintained in the laboratory setting may increase variation in traits and differences to appear in the same DS mouse model in different labs (Lanzillotta et al. 2024). Care should be given to similarities and differences in protocols, including how mice are housed, type of bedding, when testing is done and how that may affect sleep and circadian rhythms, amount or nature of handling and pretraining, how the stimulus is given, the motivation, and how long the motivation lasts (Fisher and Bannerman 2019). Additionally, social interaction among mice, including how many mice are housed together in a single cage, the number of females per litter pre-weaning, the number of trisomic mice in each litter, and whether or to what extent that cage has environmental enrichment is important (Roper et al. 2020a; Sierra et al. 2021). Additionally, diet, microbiome, and the immune status of where mice are housed may affect the variability of a trait and the outcomes of a study.

15 Information to Include When Publishing DS Mouse Models

When using DS mouse models, ARRIVE guidelines, including numbers of animals in the experimental and control groups, precise details of how the experiments were carried out, how investigators were blinded to genotypes, and steps taken to reduce subjective bias to minimize variability between animals, should be included (Percie du Sert et al. 2020; Roper et al. 2020a). In addition to the ARRIVE guidelines, specific information including the source and strain of mice (e.g., Ts65Dn 001924 or 005252), how long the DS mouse models have been continuously bred in the colony, genetic background and breeding modifications, sex of mice, age and weight of mice in the experiments, parental source of trisomy, exact number of mice used in each experiment, and defined primary and secondary outcomes of each experiment should be included (Roper et al. 2020a). Additionally, when presenting or publishing results from DS mouse models, it is advisable to delineate positive aspects and limitations of each model.

16 Using Multiple DS Mouse Models in an Experiment

As previously discussed, there are many factors that influence mouse models of DS, such as genetic background, environmental influences, parent of origin influences, and genetic drift, to name a few. Due to this, if different mouse models are used in the same experiment, one might get varied results when comparing those different models. Housing the different models in the same facility and ensuring there are no differences in the experimental protocol with the different mouse models could reduce some of this variability. It is not possible to fully eliminate variability between mouse models, so it is important to recognize the advantages and disadvantages of each model and to use multiple models when examining a phenotype, especially a novel phenotype.

Each model will have caveats that need to be considered before starting an experiment and acknowledged when analyzing and reporting the data. The Ts65Dn model has a freely segregating extra chromosome, but it has a number of genes in triplicate that are not triplicated in those with DS (Rueda et al. 2012). The duplication models allow for more refined investigation of specific genes or gene sets, but they do not have an extra segregating centromere or chromosome. The transchromosomic models utilize Hsa21 genes in a mouse model, but it is still unclear if mice are molecularly capable of fully working with the human genome and if there are any off-target effects of this combined chromosomal material. It is also important to recognize that the genetic background of the different models can also affect the phenotypes, as discussed earlier.

As there are many DS mouse models with different triplicated regions, phenotypes, and potential predictive validities, it is important to consider using multiple mouse models and to keep in mind the different caveats when examining phenotypes. When characterizing a new phenotype, *multiple mouse models of DS should be used* as phenotypes could have varied penetrance and expressivity (presentation and severity). When characterizing a previously identified phenotype in a new mouse model, it is important to recognize the caveats of both models and to compare them directly, not just with previously published results. For example, a lack of a phenotype in a new model could be due to its different construct compared to the previous model. It is also important to consider the nature of the question and the hypothesis that is being tested and whether a particular DS mouse model is appropriate for the particular inquiry.

17 Conclusions

DS mouse models offer an invaluable resource for understanding the effects of triplicated genes on DS-associated phenotypes. The DS community is fortunate to have many different mouse models with unique genetic compositions. There is no perfect or best DS mouse model. Each model has strengths and weaknesses, and acknowledging these features is important when working with different mouse models. Choosing and working with appropriate model(s) for a specific phenotype is imperative to test specific hypotheses. One model may be best to test a certain hypothesis, whereas another model may have the genetic construct and phenotype preferred for another hypothesis. As with individuals with Ts21, phenotypic variability is high within DS mouse models and experiments must be designed to control for as many confounding factors as possible and with enough power to detect potential differences in traits between DS and control mice, as well as male and female mice. When establishing a novel trait in DS mouse models it is important to replicate that trait in at least two DS mouse models. This may involve examining the trait in multiple mouse models to understand how that trait varies between models with different genetic constructs. Testing a specific hypothesis in multiple DS mouse models is suggested while understanding the genotypic and phenotypic caveats for each model (Gupta et al. 2016).

Acknowledgements We are grateful for Drs. Roger H. Reeves, Eugene Yu, Mara Dierssen, and Jean Delabar who provided excellent comments and suggestions that improved this chapter. Table 1 was made with input from the Trisomy 21 Research Society (T21RS) Preclinical Committee with significant contributions from Elizabeth Fisher, Yann Herault, Eugene Yu, and Victor Tybulewicz. RJR is supported by funds from National Institutes of Health, Eunice Kennedy Shriver National Institute for Child Health and Human Development HD090603 and the National Institute of Arthritis and Musculoskeletal and Skin Diseases AR078663.

References

Adams, A. D., V. Hoffmann, L. Koehly, F. Guedj and D. W. Bianchi, 2021 Novel insights from fetal and placental phenotyping in 3 mouse models of Down syndrome. Am J Obstet Gynecol 225: 296 e291–296 e213.

Adams, A. D., J. Lin, D. W. Bianchi, L. Bishop, T. Sato *et al.*, 2024 Embryonic statistical analyses reveal 2 growth phenotypes in mouse models of Down syndrome. Am J Obstet Gynecol 230: 258 e251–258 e211.

Ahmed, M. M., A. Block, N. Busquet and K. J. Gardiner, 2021 Context Fear Conditioning in Down Syndrome Mouse Models: Effects of Trisomic Gene Content, Age, Sex and Genetic Background. Genes (Basel) 12.

Aldridge, K., R. H. Reeves, L. E. Olson and J. T. Richtsmeier, 2007 Differential effects of trisomy on brain shape and volume in related aneuploid mouse models. Am J Med Genet A 143: 1060–1070.

Antonarakis, S. E., B. G. Skotko, M. S. Rafii, A. Strydom, S. E. Pape *et al.*, 2020 Down syndrome. Nat Rev Dis Primers 6: 9.

Aoki, S., Y. Yamauchi and K. Hashimoto, 2018 Developmental trend of children with Down's syndrome – How do sex and neonatal conditions influence their developmental patterns? Brain Dev 40: 181-187.

Arbogast, T., M. Raveau, C. Chevalier, V. Nalesso, D. Dembele *et al.*, 2015 Deletion of the App-Runx1 region in mice models human partial monosomy 21. Dis Model Mech 8: 623–634.

Armstrong, N. J., T. C. Brodnicki and T. P. Speed, 2006 Mind the gap: analysis of marker-assisted breeding strategies for inbred mouse strains. Mamm Genome 17: 273–287.

Aziz, N. M., F. Guedj, J. L. A. Pennings, J. L. Olmos-Serrano, A. Siegel *et al.*, 2018 Lifespan analysis of brain development, gene expression and behavioral phenotypes in the Ts1Cje, Ts65Dn and Dp(16)1/Yey mouse models of Down syndrome. Dis Model Mech 11.

Baptista, F., A. Varela and L. B. Sardinha, 2005 Bone mineral mass in males and females with and without Down syndrome. Osteoporos Int 16: 380–388.

Baxter, L. L., T. H. Moran, J. T. Richtsmeier, J. Troncoso and R. H. Reeves, 2000 Discovery and genetic localization of Down syndrome cerebellar phenotypes using the Ts65Dn mouse. Hum Mol Genet 9: 195–202.

Belichenko, N. P., P. V. Belichenko, A. M. Kleschevnikov, A. Salehi, R. H. Reeves *et al.*, 2009 The "Down syndrome critical region" is sufficient in the mouse model to confer behavioral, neurophysiological, and synaptic phenotypes characteristic of Down syndrome. J Neurosci 29: 5938–5948.

Belichenko, P. V., A. M. Kleschevnikov, A. Becker, G. E. Wagner, L. V. Lysenko *et al.*, 2015 Down Syndrome Cognitive Phenotypes Modeled in Mice Trisomic for All HSA 21 Homologues. PLoS One 10: e0134861.

Besson, V., V. Brault, A. Duchon, D. Togbe, J. C. Bizot *et al.*, 2007 Modeling the monosomy for the telomeric part of human chromosome 21 reveals haploinsufficient genes modulating the inflammatory and airway responses. Hum Mol Genet 16: 2040–2052.

Bhutta, M. F., M. T. Cheeseman, Y. Herault, Y. E. Yu and S. D. Brown, 2013 Surveying the Down syndrome mouse model resource identifies critical regions responsible for chronic otitis media. Mamm Genome 24: 439–445.

Blazek, J. D., A. Gaddy, R. Meyer, R. J. Roper and J. Li, 2011 Disruption of bone development and homeostasis by trisomy in Ts65Dn Down syndrome mice. Bone 48: 275–280.

Brault, V., A. Duchon, C. Romestaing, I. Sahun, S. Pothion *et al.*, 2015 Opposite phenotypes of muscle strength and locomotor function in mouse models of partial trisomy and monosomy 21 for the proximal Hspa13-App region. PLoS Genet 11: e1005062.

Bryant, C. D., N. N. Zhang, G. Sokoloff, M. S. Fanselow, H. S. Ennes *et al.*, 2008 Behavioral differences among C57BL/6 substrains: implications for transgenic and knockout studies. J Neurogenet 22: 315–331.

Carfi, A., R. Liperoti, D. Fusco, S. Giovannini, V. Brandi *et al.*, 2017 Bone mineral density in adults with Down syndrome. Osteoporos Int 28: 2929–2934.

Chang, P., D. Bush, S. Schorge, M. Good, T. Canonica *et al.*, 2020 Altered Hippocampal-Prefrontal Neural Dynamics in Mouse Models of Down Syndrome. Cell Rep 30: 1152–1163 e1154.

Channell, M. M., L. J. Hahn, T. C. Rosser, D. Hamilton, M. A. Frank-Crawford *et al.*, 2019 Characteristics Associated with Autism Spectrum Disorder Risk in Individuals with Down Syndrome. J Autism Dev Disord 49: 3543–3556.

Costa, A. C., M. R. Stasko, C. Schmidt and M. T. Davisson, 2010 Behavioral validation of the Ts65Dn mouse model for Down syndrome of a genetic background free of the retinal degeneration mutation Pde6b(rd1). Behav Brain Res 206: 52–62.

Costa, R., R. De Miguel, C. Garcia, D. R. de Asua, S. Castaneda *et al.*, 2017 Bone Mass Assessment in a Cohort of Adults With Down Syndrome: A Cross-Sectional Study. Intellect Dev Disabil 55: 315–324.

Costa, R., A. Gullon, R. De Miguel, D. R. de Asua, A. Bautista *et al.*, 2018 Bone Mineral Density Distribution Curves in Spanish Adults With Down Syndrome. J Clin Densitom 21: 493–500.

Crabbe, J. C., D. Wahlsten and B. C. Dudek, 1999 Genetics of mouse behavior: interactions with laboratory environment. Science 284: 1670–1672.

Davisson, M., E. Akeson, C. Schmidt, B. Harris, J. Farley *et al.*, 2007 Impact of trisomy on fertility and meiosis in male mice. Hum Reprod 22: 468–476.

Davisson, M. T., C. Schmidt and E. C. Akeson, 1990 Segmental trisomy of murine chromosome 16: a new model system for studying Down syndrome. Prog Clin Biol Res 360: 263–280.

Davisson, M. T., C. Schmidt, R. H. Reeves, N. G. Irving, E. C. Akeson *et al.*, 1993 Segmental trisomy as a mouse model for Down syndrome. Prog Clin Biol Res 384: 117–133.

Deitz, S. L., and R. J. Roper, 2011 Trisomic and allelic differences influence phenotypic variability during development of Down syndrome mice. Genetics 189: 1487–1495.

Drobniak, S. M., M. Lagisz, Y. Yang and S. Nakagawa, 2024 Realism and robustness require increased sample size when studying both sexes. PLoS Biol 22: e3002456.

Duchon, A., M. Del Mar Muniz Moreno, C. Chevalier, V. Nalesso, P. Andre *et al.*, 2022 Ts66Yah, a mouse model of Down syndrome with improved construct and face validity. Dis Model Mech 15.

Duchon, A., M. Raveau, C. Chevalier, V. Nalesso, A. J. Sharp *et al.*, 2011 Identification of the translocation breakpoints in the Ts65Dn and Ts1Cje mouse lines: relevance for modeling Down syndrome. Mamm Genome 22: 674–684.

Fisher, E. M. C., and D. M. Bannerman, 2019 Mouse models of neurodegeneration: Know your question, know your mouse. Sci Transl Med 11.

Galdzicki, Z., R. Siarey, R. Pearce, J. Stoll and S. I. Rapoport, 2001 On the cause of mental retardation in Down syndrome: extrapolation from full and segmental trisomy 16 mouse models. Brain Res Brain Res Rev 35: 115–145.

Garcia-Cerro, S., P. Martinez, V. Vidal, A. Corrales, J. Florez *et al.*, 2014 Overexpression of Dyrk1A is implicated in several cognitive, electrophysiological and neuromorphological alterations found in a mouse model of Down syndrome. PLoS One 9: e106572.

Garcia-Cerro, S., N. Rueda, V. Vidal, S. Lantigua and C. Martinez-Cue, 2017 Normalizing the gene dosage of Dyrk1A in a mouse model of Down syndrome rescues several Alzheimer's disease phenotypes. Neurobiol Dis 106: 76–88.

Garcia-Cerro, S., V. Vidal, S. Lantigua, M. T. Berciano, M. Lafarga *et al.*, 2018 Cerebellar alterations in a model of Down syndrome: The role of the Dyrk1A gene. Neurobiol Dis 110: 206–217.

Glass, T. J., and N. P. Connor, 2016 Digastric Muscle Phenotypes of the Ts65Dn Mouse Model of Down Syndrome. PLoS One 11: e0158008.

Gribble, S. M., F. K. Wiseman, S. Clayton, E. Prigmore, E. Langley *et al.*, 2013 Massively parallel sequencing reveals the complex structure of an irradiated human chromosome on a mouse background in the Tc1 model of Down syndrome. PLoS One 8: e60482.

Guedj, F., E. Kane, L. A. Bishop, J. L. A. Pennings, Y. Herault *et al.*, 2023 The Impact of Mmu17 Non-Hsa21 Orthologous Genes in the Ts65Dn Mouse Model of Down Syndrome: The Gold Standard Refuted. Biol Psychiatry 94: 84–97.

Gupta, M., A. R. Dhanasekaran and K. J. Gardiner, 2016 Mouse models of Down syndrome: gene content and consequences. Mamm Genome 27: 538–555.

Haas, M. A., D. Bell, A. Slender, E. Lana-Elola, S. Watson-Scales *et al.*, 2013 Alterations to dendritic spine morphology, but not dendrite patterning, of cortical projection neurons in Tc1 and Ts1Rhr mouse models of Down syndrome. PLoS One 8: e78561.

Hamlett, E. D., H. A. Boger, A. Ledreux, C. M. Kelley, E. J. Mufson *et al.*, 2016 Cognitive Impairment, Neuroimaging, and Alzheimer Neuropathology in Mouse Models of Down Syndrome. Curr Alzheimer Res 13: 35–52.

Hawley, L. E., M. Stringer, A. J. Deal, A. Folz, C. R. Goodlett *et al.*, 2023 Sex-specific developmental alterations in DYRK1A expression in the brain of a Down syndrome mouse model. Neurobiol Dis 190: 106359.

Herault, Y., J. M. Delabar, E. M. C. Fisher, V. L. J. Tybulewicz, E. Yu *et al.*, 2017 Rodent models in Down syndrome research: impact and future opportunities. Dis Model Mech 10: 1165–1186.

Insausti, A. M., M. Megias, D. Crespo, L. M. Cruz-Orive, M. Dierssen *et al.*, 1998 Hippocampal volume and neuronal number in Ts65Dn mice: a murine model of Down syndrome. Neurosci Lett 253: 175–178.

Jewett, C. E., B. L. McCurdy, E. T. O'Toole, A. J. Stemm-Wolf, K. S. Given *et al.*, 2023 Trisomy 21 induces pericentrosomal crowding delaying primary ciliogenesis and mouse cerebellar development. Elife 12.

Kazuki, Y., F. J. Gao, Y. Li, A. J. Moyer, B. Devenney *et al.*, 2020 A non-mosaic transchromosomic mouse model of down syndrome carrying the long arm of human chromosome 21. Elife 9.

Korbel, J. O., T. Tirosh-Wagner, A. E. Urban, X. N. Chen, M. Kasowski *et al.*, 2009 The genetic architecture of Down syndrome phenotypes revealed by high-resolution analysis of human segmental trisomies. Proc Natl Acad Sci U S A 106**:** 12031–12036.

LaCombe, J. M., and R. J. Roper, 2020 Skeletal dynamics of Down syndrome: A developing perspective. Bone 133**:** 115215.

Lamantia, J., K. Sloan, J. M. Wallace and R. J. Roper, 2024 Compromised femoral and lumbovertebral bone in the Dp(16)1Yey Down syndrome mouse model. Bone 181**:** 117046.

Lana-Elola, E., R. Aoidi, M. Llorian, D. Gibbins, C. Buechsenschuetz *et al.*, 2024 Increased dosage of DYRK1A leads to congenital heart defects in a mouse model of Down syndrome. Sci Transl Med 16**:** eadd6883.

Lana-Elola, E., H. Cater, S. Watson-Scales, S. Greenaway, J. Müller-Winkler *et al.*, 2021 Comprehensive phenotypic analysis of the Dp1Tyb mouse strain reveals a broad range of Down Syndrome-related phenotypes. bioRxiv**:** 2021.2002.2011.430828.

Lana-Elola, E., S. Watson-Scales, A. Slender, D. Gibbins, A. Martineau *et al.*, 2016 Genetic dissection of Down syndrome-associated congenital heart defects using a new mouse mapping panel. Elife 5.

Lanzillotta, C., M. R. Baniowska, F. Prestia, C. Sette, V. Nalesso *et al.*, 2024 Shaping down syndrome brain cognitive and molecular changes due to aging using adult animals from the Ts66Yah murine model. Neurobiol Dis 196**:** 106523.

Li, H., S. Edie, D. Klinedinst, J. S. Jeong, S. Blackshaw *et al.*, 2016 Penetrance of Congenital Heart Disease in a Mouse Model of Down Syndrome Depends on a Trisomic Potentiator of a Disomic Modifier. Genetics 203**:** 763–770.

Li, Y., Z. Xing, T. Yu, A. Pao, M. Daadi *et al.*, 2021 Coat Color-Facilitated Efficient Generation and Analysis of a Mouse Model of Down Syndrome Triplicated for All Human Chromosome 21 Orthologous Regions. Genes (Basel) 12.

Li, Z., T. Yu, M. Morishima, A. Pao, J. Laduca *et al.*, 2007 Duplication of the entire 22.9 Mb human chromosome 21 syntenic region on mouse chromosome 16 causes cardiovascular and gastrointestinal abnormalities. Hum Mol Genet 16**:** 1359–1366.

Liu, C., M. Morishima, T. Yu, S. Matsui, L. Zhang *et al.*, 2011 Genetic analysis of Down syndrome-associated heart defects in mice. Hum Genet 130**:** 623–632.

Liu, C., T. Yu, Z. Xing, X. Jiang, Y. Li *et al.*, 2018 Triplications of human chromosome 21 orthologous regions in mice result in expansion of megakaryocyte-erythroid progenitors and reduction of granulocyte-macrophage progenitors. Oncotarget 9**:** 4773–4786.

Lusis, A. J., J. Yu and S. S. Wang, 2007 The problem of passenger genes in transgenic mice. Arterioscler Thromb Vasc Biol 27**:** 2100–2103.

Lyle, R., F. Bena, S. Gagos, C. Gehrig, G. Lopez *et al.*, 2009 Genotype-phenotype correlations in Down syndrome identified by array CGH in 30 cases of partial trisomy and partial monosomy chromosome 21. Eur J Hum Genet 17**:** 454–466.

Maatta, T., T. Tervo-Maatta, A. Taanila, M. Kaski and M. Iivanainen, 2006 Mental health, behaviour and intellectual abilities of people with Down syndrome. Downs Syndr Res Pract 11**:** 37–43.

Marchal, J. P., H. Maurice-Stam, B. A. Houtzager, S. L. Rutgers van Rozenburg-Marres, K. J. Oostrom *et al.*, 2016 Growing up with Down syndrome: Development from 6 months to 10.7 years. Res Dev Disabil 59**:** 437–450.

Martinez-Cue, C., C. Baamonde, M. Lumbreras, J. Paz, M. T. Davisson *et al.*, 2002 Differential effects of environmental enrichment on behavior and learning of male and female Ts65Dn mice, a model for Down syndrome. Behav Brain Res 134**:** 185–200.

Martinez-Cue, C., N. Rueda, E. Garcia and J. Florez, 2006 Anxiety and panic responses to a predator in male and female Ts65Dn mice, a model for Down syndrome. Genes Brain Behav 5: 413–422.

Miyabara, S., A. Gropp and H. Winking, 1982 Trisomy 16 in the mouse fetus associated with generalized edema and cardiovascular and urinary tract anomalies. Teratology 25: 369–380.

Moore, C. S., 2006 Postnatal lethality and cardiac anomalies in the Ts65Dn Down syndrome mouse model. Mamm Genome 17: 1005–1012.

Moore, C. S., C. Hawkins, A. Franca, A. Lawler, B. Devenney *et al.*, 2010 Increased male reproductive success in Ts65Dn "Down syndrome" mice. Mamm Genome 21: 543–549.

Moyer, A. J., F. X. Fernandez, Y. Li, D. K. Klinedinst, L. D. Florea *et al.*, 2023 Overexpression screen of chromosome 21 genes reveals modulators of Sonic hedgehog signaling relevant to Down syndrome. Dis Model Mech 16.

Muniz Moreno, M. D. M., V. Brault, M. C. Birling, G. Pavlovic and Y. Herault, 2020 Modeling Down syndrome in animals from the early stage to the 4.0 models and next. Prog Brain Res 251: 91–143.

Muza, P. M., D. Bush, M. Perez-Gonzalez, I. Zouhair, K. Cleverley *et al.*, 2023 Cognitive impairments in a Down syndrome model with abnormal hippocampal and prefrontal dynamics and cytoarchitecture. iScience 26: 106073.

O'Doherty, A., S. Ruf, C. Mulligan, V. Hildreth, M. L. Errington *et al.*, 2005 An aneuploid mouse strain carrying human chromosome 21 with down syndrome phenotypes. Science 309: 2033–2037.

Olson, L. E., and S. Mohan, 2011 Bone density phenotypes in mice aneuploid for the Down syndrome critical region. Am J Med Genet A 155: 2436–2445.

Olson, L. E., J. T. Richtsmeier, J. Leszl and R. H. Reeves, 2004a A chromosome 21 critical region does not cause specific Down syndrome phenotypes. Science 306: 687–690.

Olson, L. E., R. J. Roper, L. L. Baxter, E. J. Carlson, C. J. Epstein *et al.*, 2004b Down syndrome mouse models Ts65Dn, Ts1Cje, and Ms1Cje/Ts65Dn exhibit variable severity of cerebellar phenotypes. Dev Dyn 230: 581–589.

Olson, L. E., R. J. Roper, C. L. Sengstaken, E. A. Peterson, V. Aquino *et al.*, 2007 Trisomy for the Down syndrome "critical region" is necessary but not sufficient for brain phenotypes of trisomic mice. Hum Mol Genet.

Percie du Sert, N., V. Hurst, A. Ahluwalia, S. Alam, M. T. Avey *et al.*, 2020 The ARRIVE guidelines 2.0: Updated guidelines for reporting animal research. PLoS Biol 18: e3000410.

Pereira, P. L., L. Magnol, I. Sahun, V. Brault, A. Duchon *et al.*, 2009 A new mouse model for the trisomy of the Abcg1-U2af1 region reveals the complexity of the combinatorial genetic code of down syndrome. Hum Mol Genet 18: 4756–4769.

Ramirez-Solis, R., P. Liu and A. Bradley, 1995 Chromosome engineering in mice. Nature 378: 720–724.

Reeves, R. H., N. G. Irving, T. H. Moran, A. Wohn, C. Kitt *et al.*, 1995 A mouse model for Down syndrome exhibits learning and behaviour deficits. Nat Genet 11: 177–184.

Reinholdt, L. G., Y. Ding, G. J. Gilbert, A. Czechanski, J. P. Solzak *et al.*, 2011 Molecular characterization of the translocation breakpoints in the Down syndrome mouse model Ts65Dn. Mamm Genome 22: 685–691.

Richtsmeier, J., L. Baxter and R. Reeves, 2000 Parallels of craniofacial maldevelopment in Down Syndrome and Ts65Dn mice. Developmental Dynamics 217: 137–145.

Roper, R. J., C. R. Goodlett, M. Martinez de Lagran and M. Dierssen, 2020a Behavioral Phenotyping for Down Syndrome in Mice. Curr Protoc Mouse Biol 10: e79.

Roper, R. J., L. Hawley and C. R. Goodlett, 2020b Influence of allelic differences in Down syndrome. Prog Brain Res 251: 29–54.

Roper, R. J., and R. H. Reeves, 2006 Understanding the basis for Down syndrome phenotypes. PLoS Genet 2: e50.

Roper, R. J., H. K. St John, J. Philip, A. Lawler and R. H. Reeves, 2006 Perinatal loss of Ts65Dn Down syndrome mice. Genetics 172: 437–443.

Rueda, N., J. Florez and C. Martinez-Cue, 2012 Mouse models of Down syndrome as a tool to unravel the causes of mental disabilities. Neural Plast 2012: 584071.

Sago, H., E. J. Carlson, D. J. Smith, J. Kilbridge, E. M. Rubin *et al.*, 1998 Ts1Cje, a partial trisomy 16 mouse model for Down syndrome, exhibits learning and behavioral abnormalities. Proc Natl Acad Sci U S A 95: 6256–6261.

Sahun, I., D. Marechal, P. L. Pereira, V. Nalesso, A. Gruart *et al.*, 2014 Cognition and hippocampal plasticity in the mouse is altered by monosomy of a genomic region implicated in Down syndrome. Genetics 197: 899–912.

Salehi, A., J. D. Delcroix, P. V. Belichenko, K. Zhan, C. Wu *et al.*, 2006 Increased App expression in a mouse model of Down's syndrome disrupts NGF transport and causes cholinergic neuron degeneration. Neuron 51: 29–42.

Sarver, D. C., C. Xu, S. Rodriguez, S. Aja, A. E. Jaffe *et al.*, 2023 Hypermetabolism in mice carrying a near-complete human chromosome 21. Elife 12.

Shaw, P. R., J. A. Klein, N. M. Aziz and T. F. Haydar, 2020 Longitudinal neuroanatomical and behavioral analyses show phenotypic drift and variability in the Ts65Dn mouse model of Down syndrome. Dis Model Mech 13.

Sierra, C., I. De Toma, L. L. Cascio, E. Vegas and M. Dierssen, 2021 Social Factors Influence Behavior in the Novel Object Recognition Task in a Mouse Model of Down Syndrome. Front Behav Neurosci 15: 772734.

Silver, L., 2001 Inbred Strain, pp. 1015–1016 in *Encyclopedia of Genetics*, edited by S. Brenner and J. H. Miller. Academic Press.

Simon, M. M., S. Greenaway, J. K. White, H. Fuchs, V. Gailus-Durner *et al.*, 2013 A comparative phenotypic and genomic analysis of C57BL/6J and C57BL/6N mouse strains. Genome Biol 14: R82.

Sloan, K., J. Thomas, M. Blackwell, D. Voisard, E. Lana-Elola *et al.*, 2023 Genetic dissection of triplicated chromosome 21 orthologs yields varying skeletal traits in Down syndrome model mice. Dis Model Mech 16.

Smith, D. J., M. E. Stevens, S. P. Sudanagunta, R. T. Bronson, M. Makhinson *et al.*, 1997 Functional screening of 2 Mb of human chromosome 21q22.2 in transgenic mice implicates minibrain in learning defects associated with Down syndrome. Nat Genet 16: 28–36.

Sturgeon, X., and K. J. Gardiner, 2011 Transcript catalogs of human chromosome 21 and orthologous chimpanzee and mouse regions. Mamm Genome 22: 261–271.

Su, Z., X. Wang, S. W. Tsaih, A. Zhang, A. Cox *et al.*, 2009 Genetic basis of HDL variation in 129/SvImJ and C57BL/6J mice: importance of testing candidate genes in targeted mutant mice. J Lipid Res 50: 116–125.

Tallino, S., W. Winslow, S. K. Bartholomew and R. Velazquez, 2022 Temporal and brain region-specific elevations of soluble Amyloid-beta(40–42) in the Ts65Dn mouse model of Down syndrome and Alzheimer's disease. Aging Cell 21: e13590.

Tang, J. Y. M., H. Luo, G. H. Y. Wong, M. M. Y. Lau, G. M. Joe *et al.*, 2019 Bone mineral density from early to middle adulthood in persons with Down syndrome. J Intellect Disabil Res 63: 936–946.

Thomas, J. R., J. LaCombe, R. Long, E. Lana-Elola, S. Watson-Scales *et al.*, 2020 Interaction of sexual dimorphism and gene dosage imbalance in skeletal deficits associated with Down syndrome. Bone 136: 115367.

van Gameren-Oosterom, H. B., M. Fekkes, S. E. Buitendijk, A. D. Mohangoo, J. Bruil *et al.*, 2011 Development, problem behavior, and quality of life in a population based sample of eight-year-old children with Down syndrome. PLoS One 6: e21879.

van Gameren-Oosterom, H. B., M. Fekkes, J. P. van Wouwe, S. B. Detmar, A. M. Oudesluys-Murphy *et al.*, 2013 Problem behavior of individuals with Down syndrome in a nationwide cohort assessed in late adolescence. J Pediatr 163: 1396–1401.

Villar, A. J., P. V. Belichenko, A. M. Gillespie, H. M. Kozy, W. C. Mobley *et al.*, 2005 Identification and characterization of a new Down syndrome model, Ts[Rb(12.1716)]2Cje, resulting from a spontaneous Robertsonian fusion between T(171)65Dn and mouse chromosome 12. Mamm Genome 16: 79–90.

Wahlsten, D., P. Metten, T. J. Phillips, S. L. Boehm, 2nd, S. Burkhart-Kasch *et al.*, 2003 Different data from different labs: lessons from studies of gene-environment interaction. J Neurobiol 54: 283–311.

Watson-Scales, S., B. Kalmar, E. Lana-Elola, D. Gibbins, F. La Russa *et al.*, 2018 Analysis of motor dysfunction in Down Syndrome reveals motor neuron degeneration. PLoS Genet 14: e1007383.

Wilson, M. D., N. L. Barbosa-Morais, D. Schmidt, C. M. Conboy, L. Vanes *et al.*, 2008 Species-specific transcription in mice carrying human chromosome 21. Science 322: 434–438.

Wiseman, F. K., T. Al-Janabi, J. Hardy, A. Karmiloff-Smith, D. Nizetic *et al.*, 2015 A genetic cause of Alzheimer disease: mechanistic insights from Down syndrome. Nat Rev Neurosci 16: 564–574.

Xing, Z., Y. Li, E. Cortes-Gomez, X. Jiang, S. Gao *et al.*, 2023 Dissection of a Down syndrome-associated trisomy to separate the gene dosage-dependent and -independent effects of an extra chromosome. Hum Mol Genet 32: 2205–2218.

Xu, G., Y. Ran, S. E. Fromholt, C. Fu, A. T. Yachnis *et al.*, 2015 Murine Abeta over-production produces diffuse and compact Alzheimer-type amyloid deposits. Acta Neuropathol Commun 3: 72.

Yu, T., Z. Li, Z. Jia, S. J. Clapcote, C. Liu *et al.*, 2010a A mouse model of Down syndrome trisomic for all human chromosome 21 syntenic regions. Hum Mol Genet 19: 2780–2791.

Yu, T., C. Liu, P. Belichenko, S. J. Clapcote, S. Li *et al.*, 2010b Effects of individual segmental trisomies of human chromosome 21 syntenic regions on hippocampal long-term potentiation and cognitive behaviors in mice. Brain Res 1366: 162–171.

Impact of Cre/LoxP-Mediated Chromosome Engineering Technology on Down Syndrome Research

Zhuo Xing, Avrium Douglas, Annie Pao, Tiffany Glass, Nadine Connor, and Y. Eugene Yu

Abstract Down syndrome (DS; trisomy 21) significantly impacts the human population, affecting up to six million people, and understanding the underlying mechanisms is crucial for developing effective interventions to enhance health and quality of life. Animal models, especially mouse models, have been instrumental in this area of research. Over the past two decades, Cre/loxP-mediated chromosome engineering technology has transformed DS research in mice. This technology allows for the creation of predetermined chromosomal duplications and deletions in regions syntenic to human chromosome 21. In this chapter, we discuss how this technology has led to the development of novel DS models, enabled genetic dissection of syntenic regions to establish precise genotype–phenotype relationships, and revealed the effects of an extra chromosome beyond gene dosage. These advances pave the way for a new era of discoveries in understanding this significant medical condition.

Authors Zhuo Xing and Avrium Douglas have equally contributed to this chapter.

Z. Xing · A. Douglas · A. Pao
The Children's Guild Foundation Down Syndrome Research Program, Department of Cancer Genetics and Genomics, Roswell Park Comprehensive Cancer Center, Buffalo, NY, USA

T. Glass
Department of Surgery, Division of Otolaryngology, University of Wisconsin-Madison, Madison, WI, USA

N. Connor
Department of Surgery, Division of Otolaryngology, University of Wisconsin-Madison, Madison, WI, USA

Department of Communication Sciences and Disorders, University of Wisconsin-Madison, Madison, WI, USA

Y. E. Yu (✉)
The Children's Guild Foundation Down Syndrome Research Program, Department of Cancer Genetics and Genomics, Roswell Park Comprehensive Cancer Center, Buffalo, NY, USA

Genetics, Genomics and Bioinformatics Program, State University of New York at Buffalo, Buffalo, NY, USA
e-mail: yuejin.yu@roswellpark.org

B. Ye, R. H. Reeves (eds.), *Genetic Models of Down Syndrome*,
https://doi.org/10.1007/978-3-031-78611-2_2

Keywords Down syndrome · Mouse models · Cre/loxP-mediated chromosome engineering · Gene dosage · Extra chromosome

1 Cre/LoxP-Mediated Chromosome Engineering in Mice

In 1995, the scientific community witnessed the emergence of a groundbreaking technology, Cre/loxP-mediated mouse chromosome engineering. Two independent teams reported the Cre/loxP-mediated engineering of chromosomal translocations in mouse embryonic stem (ES) cells (Smith et al. 1995; Van Deursen et al. 1995), while Allan Bradley's laboratory reported the engineering of deletions, duplications, and inversions both in mouse ES cells and subsequently in mice carrying those chromosomal rearrangements (Ramirez-Solis et al. 1995). The Cre/loxP site-specific recombinase system was initially established using plasmid DNA in bacteria (Fig. 1), with results from three laboratories demonstrating the successful application of the system to a mammalian genome at a multi-megabase level, which can reach up to 60 Mb (Fig. 2) (Zheng et al. 1999a).

Successful chromosomal engineering necessitates two gene targeting steps and Cre-mediated recombination (Figs. 2 and 3) (Yu and Bradley 2001). To execute the task, two complementary targeting vectors containing multiple genetic cassettes are required. To enhance efficiency, two complementary *HPRT* (hypoxanthine guanine phosphoribosyltransferase gene) phage libraries were developed to contain all necessary genetic cassettes (Fig. 3a) (Zheng et al. 1999b). These phage libraries were later converted into plasmid libraries to establish the Mouse Insertion and Chromosome Engineering Resource (MICER). MICER was developed by indexing over 150,000 clones from the 5′*HPRT* and 3′*HPRT* insertional targeting vector libraries (Zheng et al. 1999b; Adams et al. 2004). Each clone contains a drug resistance gene as a positive selection marker, 5′ or 3′ portion of an *HPRT* minigene, a loxP site, and a coat color marker (Fig. 4). An individual MICER targeting vector is established by sequencing the end sequences of the mouse genomic insert using the sequencing primers located external to the cloning site. All MICER vectors with mouse genomic inserts are displayed in the latest versions of Ensembl Mouse Genome Browser (www.ensembl.org) (Adams et al. 2004). The establishment of this resource greatly improved access to the complicated process of chromosome engineering. However, because MICER-based chromosome engineering requires multiple steps of genetic manipulations in the same ES cells, the technology remains challenging. The alternative procedures have also been reported by Yann Herault and his collaborators to accomplish Cre/loxP-mediated recombination step in vivo via targeted meiotic recombination (TAMERE) (Herault et al. 1998; Raveau et al. 2012) or targeted asymmetric sister chromatin event of recombination (TASCER) (Duchon et al. 2008), which do not require to use MICER vectors.

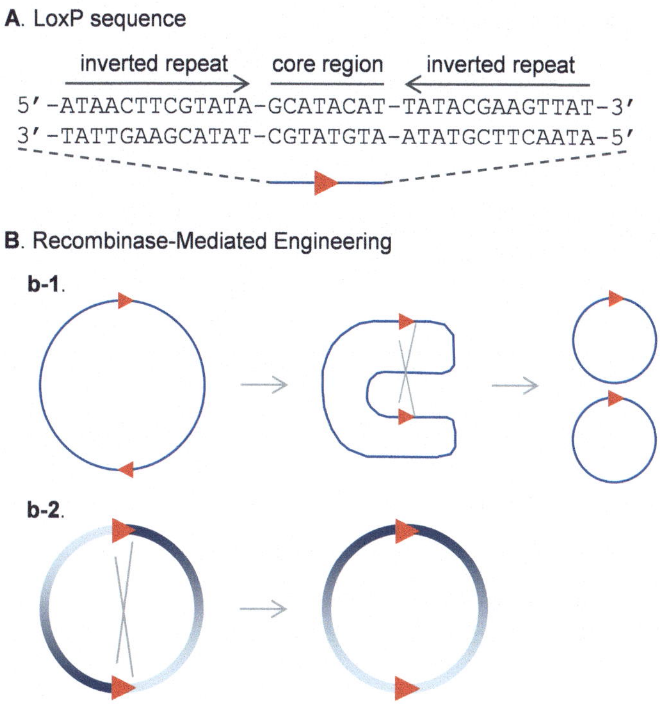

Fig. 1 The principle of mouse chromosome engineering is based on Cre/LoxP-mediated recombinations, which were first established in plasmids in bacterial cells. (**a**) The loxP sequence consists of a core region flanked by two inverted repeats, which spans 34 base pairs. (**b**) Recombinase-mediated recombination in plasmids. The illustration depicts the outcomes of recombination when two loxP sites are oriented in the same direction (**b-1**) and the opposite direction (**b-2**). Arrowhead, loxP site

2 Modeling Down Syndrome (DS) by Using Cre/LoxP-Mediated Chromosome Engineering in Mice

The development of Cre/loxP-mediated chromosome engineering technology has created unprecedented opportunities for projects that were previously not feasible. Prior to the introduction of Cre/loxP-mediated chromosome engineering, mouse genomic reagents, such as chromosomal deletions and translocations, were generated primarily through irradiation (Johnson et al. 1989; Russell et al. 1958; Davisson et al. 1990). However, this method has limitations because the effects of radiation mutagenesis are random, i.e., no control over the mutations, other than via radiation dose. Therefore, it could not efficiently produce important genetic reagents like duplications and balancer chromosomes, and the endpoints of generated chromosomal rearrangements were unpredictable. The implementation of Cre/loxP-mediated chromosome engineering successfully addressed these shortcomings. Now, chromosomal deletions, duplications, and balancer chromosomes can be

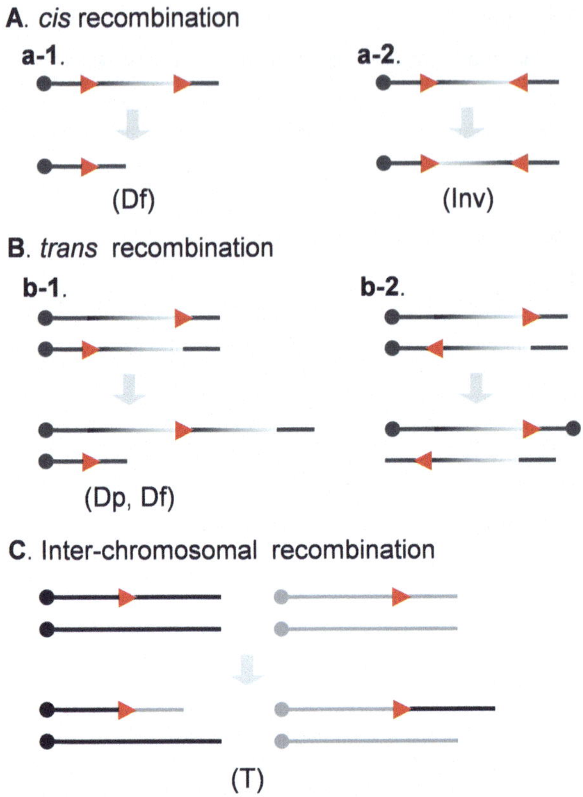

Fig. 2 Consequences of LoxP-mediated recombination associated with the mouse genome. (**a**) *Cis* recombination occurs when two loxP sites are targeted on the same chromosome homolog. Depending on the orientation of the loxP sequences, recombination results in a deletion (Df) when both loxP sites have the same orientation (**a-1**). When loxP sites are in opposite orientations (**a-2**), recombination leads to an inversion (Inv). (**b**) *Trans* recombination occurs when each loxP is targeted on each homolog of the same chromosome. When both loxP sites have the same orientation (**b-1**), recombination results in a duplication (Dp) and the reciprocal Df. When the two loxP sites have opposite orientations (**b-2**), recombination leads to dicentric and acentric chromosomes. (**c**) Inter-chromosomal recombination occurs when two loxP sites are targeted onto two different chromosomes, resulting in a translocation (T). Arrowhead, loxP site

precisely developed with predetermined endpoints (Figs. 2 and 3) (Zheng et al. 1999a; Yu and Bradley 2001; Brault et al. 2007), significantly facilitating genetic analysis in mice. Another significant application of this technology is the modeling and analysis of human chromosomal disorders based on syntenic conservation between the human and mouse genomes. An early achievement in this area was the development of a mouse model for DiGeorge syndrome by Antonio Baldini's laboratory and their collaborators using Cre/loxP-mediated chromosome engineering (Lindsay et al. 1999). Subsequent efforts led to the identification of *Tbx1* as a causative gene for certain major phenotypes associated with DiGeorge syndrome (Lindsay et al. 2001; Merscher et al. 2001).

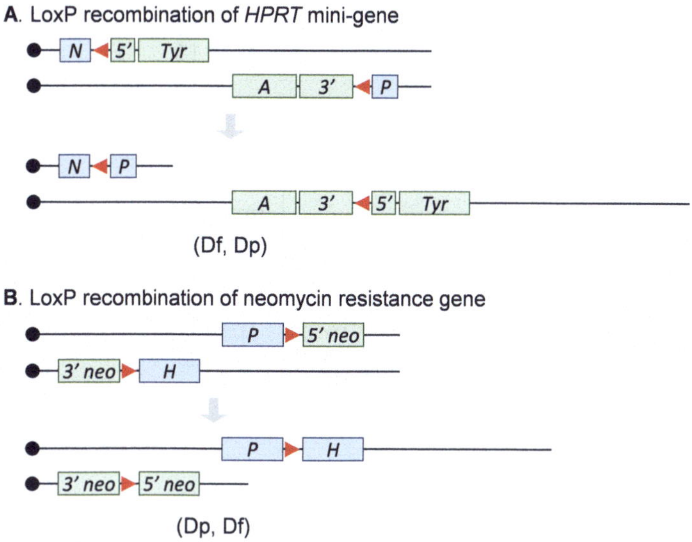

Fig. 3 LoxP-mediated recombinations using different types of targeting vector backbones. (**a**) The 5′-half and 3′-half of the *HPRT* mini-gene can be used as genetic cassettes. After recombination, hypoxanthine-aminopterin-thymidine (HAT) can be used to select ES cell clones carrying Dp and Df. (**b**) The 5′-half and 3′-half of the neomycin resistance gene can be used as genetic cassettes in complementary targeting vectors. After recombination, G418, an analogue of neomycin, can be used to select ES cell clones carrying Dp and Df. 5′ 5′-*HPRT* fragment, 3′ 3′-*HPRT* fragment, N neomycin-resistance gene, P puromycin-resistance gene, Tyr Tyrosinase transgene, A Agouti transgene, 5′ neo 5′-half of the neomycin resistance gene, 3′ neo 3′-half of the neomycin resistance gene, H hygromycin B-resistance gene, Arrowhead loxP site

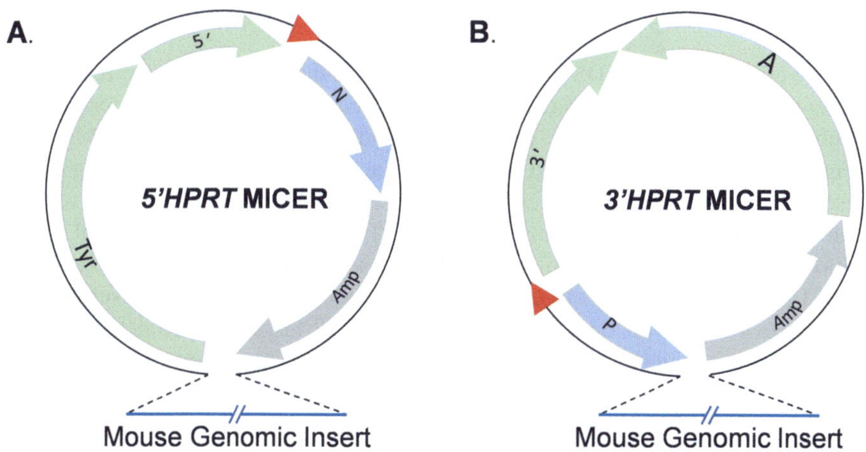

Fig. 4 Targeting vectors from mutagenic insertion and chromosome engineering resource (MICER). (**a**) A 5′-*HPRT* MICER vector. (**b**) A 3′-*HPRT* MICER vector. 5′ 5′-*HPRT* fragment, 3′ 3′-*HPRT* fragment, N neomycin-resistance gene, P puromycin-resistance gene, Tyr Tyrosinase transgene, A Agouti transgene, Amp ampicillin-resistance gene, Arrowhead loxP site

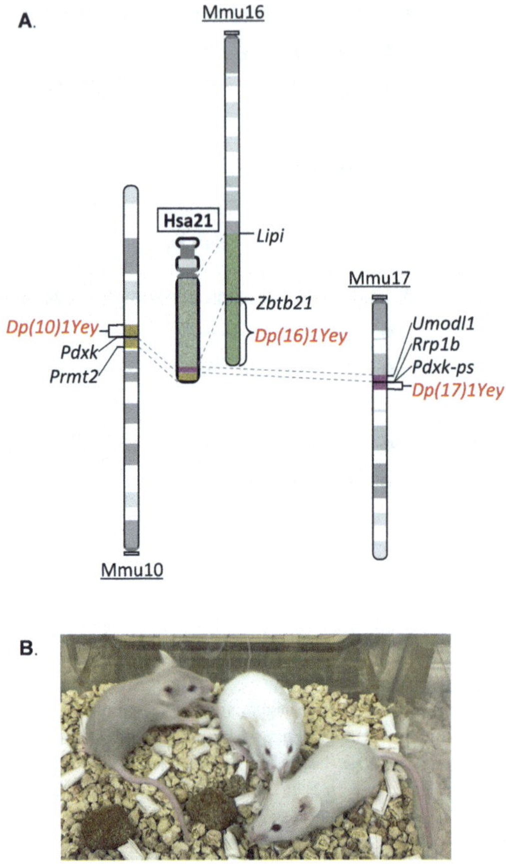

Fig. 5 Duplication mouse models of DS. (**a**) Syntenic genomic regions to Hsa21 and the duplicated regions in *Dp(10)1Yey*, *Dp(16)1Yey*, and *Dp(17)1Yey* are shown on Mmu10, 16, and 17, respectively. (**b**) Mice with different coat colors are shown. The tan-colored mouse on the left *is Dp(10)1Yey;Dp(16)1Yey;Dp(17)1Yey*, which is the consequence of the presence of two copies of the tyrosinase transgenes. The wild-type mouse in the middle has a white coat color, and the *Dp(16)1Yey* mouse on the right has a cream coat color, which is the consequence of the presence of one copy of the tyrosinase transgene

The efforts to model DS were concentrated on the mouse syntenic regions of human chromosome 21 (Hsa21), which are located on mouse chromosome 10 (Mmu10), mouse chromosome 16 (Mmu16), and mouse chromosome 17 (Mmu17) (Fig. 5). The first application of Cre/loxP-mediated chromosome engineering in DS research was reported by Reeves laboratory, which was carried out using vectors that reconstituted a functional neomycin gene (Fig. 3b) to create a reciprocal duplication/deletion that focused on the so-called Down syndrome critical region within Hsa21 syntenic region on Mmu16 (Olson et al. 2004). However, most subsequent DS mouse models were engineered using the MICER targeting vectors.

The effort to engineer *Dp(10)1Yey* was initiated before the availability of the mouse genome sequence, the extensive genetic characterizations of Hsa21 syntenic region on Mmu10 by Roger Reeves' laboratory and others substantially benefitted the efforts to engineer the duplication in this mouse genomic region (Cole et al. 1998; Cole et al. 1999; Pletcher et al. 2000; Wiltshire et al. 1999). Because repetitive sequences are inefficient targets for site-specific homologous recombination, careful selection of the targeting vectors was critical for engineering of *Dp(10)1Yey* and *Df(10)1Yey* (Fig. 5a) (Yu et al. 2010a; Yu et al. 2010b). The *Dp(16)1Yey* DS model carrying the duplication spanning the entire Hsa21 syntenic region on Mmu16 produces gene dosage imbalance for 115 Hsa21 gene orthologs, the largest duplication established in the mouse genome at the time (Fig. 5a) (Li et al. 2007). Later, *Dp(16Lipi-Zbtb21)1TybEmcf* (*Dp1Tyb*), another duplication mouse mutant spanning the same Hsa21 orthologous region on Mmu16, was reported by the team of Victor Tybulewicz and Elizabeth Fisher (Lana-Elola et al. 2016). The *Dp(17)1Yey* model is triplicated for the Hsa21 syntenic segment on Mmu17 (Yu et al. 2010b). This model completes the set of mice that together span all DS-syntenic genomic regions in the mouse germline. *Pdxk-ps* was used as one of the targeting sites, which is located immediately adjacent to the distal endpoint of the syntenic region. Interestingly, *Pdxk* is the proximal endpoint of the Hsa21 syntenic region on Mmu10 (Yu et al. 2010b), thus supporting the possibility that *Pdxk* is at the evolutionary breakpoint junction for the Hsa21 syntenic regions on Mmu10 and Mmu17 (Fig. 5a).

Diverse DS-related phenotypes have been analyzed using *Dp(16)1Yey* mice, including abnormalities in the cardiopulmonary system (Li et al. 2007; Colvin et al. 2023), motor functions (Aziz et al. 2018), reproduction (Dard et al. 2023), developmental cognitive deficits (Aziz et al. 2018; Patel et al. 2015; Pinto et al. 2020; Raveau et al. 2018; Yu et al. 2010c), interferonopathies (Sullivan et al. 2016; Waugh et al. 2023), myeloproliferative disorder (Liu et al. 2018), hearing impairment (Chen et al. 2022), craniofacial and skeleton anomalies (Sherman et al. 2022; Starbuck et al. 2014), alterations in metabolism (Peiris et al. 2016), changes in sleep pattern (Levenga et al. 2018) and vocal communication (Glass et al. 2023), and Alzheimer-associated neurodegeneration (Sawa et al. 2022; Chen et al. 2021a, 2023a, b). Many of these phenotypes have also been observed in *Dp1Tyb* mice (Table 1) (Lana-Elola et al. 2016, 2021; Redhead et al. 2023; Sloan et al. 2023; Thomas et al. 2020; Watson-Scales et al. 2018).

Experimental evidence suggests that Hsa21 syntenic regions on Mmu10 and Mmu17 contribute to DS-related phenotypes (Marechal et al. 2015; Pereira et al. 2009; Sahun et al. 2014; Zhang et al. 2014). The triple duplication mouse mutant, *Dp(10)1Yey;Dp(16)1Yey;Dp(17)1Yey*, contains a third copy of all mouse orthologs of Hsa21 genes, providing a complete mouse model of gene dosage effects in DS (Table 1) (Yu et al. 2010b; Belichenko et al. 2015), without a freely segregating extra chromosome (see below). This compound mutant can be efficiently generated with the aid of coat colors (Fig. 5a, b) (Li et al. 2021).

3 The Search for Genes of Major Effect in DS Models

Cre/loxP-mediated chromosome engineering played a crucial role in establishing major mouse models, such as *Dp(16)1Yey* and the compound model triplicated for all three Hsa21 syntenic regions (Yu et al. 2010b; Li et al. 2007). This technology was equally pivotal in the development of mouse mutants carrying smaller chromosomal duplications and deletions within the Hsa21 syntenic regions. These mutants

Table 1 Phenotypic comparison of the Dp models of Down syndrome

Dp(16)1Yey	*Dp1Tyb*	*Dp(10)1Yey;Dp(16)1Yey;Dp(17)1Yey*
Developmental cognitive deficits (Aziz et al. 2018; Pinto et al. 2020; Raveau et al. 2018; Yu et al. 2010c)	Developmental cognitive deficits (Lana-Elola et al. 2021; Canonica et al. 2024; Chang et al. 2020)	Developmental cognitive deficits (Yu et al. 2010b; Belichenko et al. 2015)
Motor function abnormalities (Aziz et al. 2018; Watson-Scales et al. 2018)	Motor function abnormalities (Lana-Elola et al. 2021; Watson-Scales et al. 2018)	Motor function abnormalities (Watson-Scales et al. 2018)
Cardiopulmonary system abnormalities (Li et al. 2007; Colvin et al. 2023)	Cardiopulmonary system abnormalities (Lana-Elola et al. 2016, 2021)	
Myeloproliferative disorders (Liu et al. 2018)	Myeloproliferative disorders (Lana-Elola et al. 2021)	
Craniofacial & skeletal anomalies (Sherman et al. 2022; Starbuck et al. 2014)	Craniofacial & skeletal anomalies (Lana-Elola et al. 2021; Redhead et al. 2023; Sloan et al. 2023; Thomas et al. 2020)	
Changes in sleep pattern (Levenga et al. 2018) and vocal communication (Glass et al. 2023)	Changes in sleep pattern & vocal communication (Lana-Elola et al. 2021)	
Hearing Impairment (Chen et al. 2022)	Hearing Impairment (Lana-Elola et al. 2021)	
Alterations in metabolism (Peiris et al. 2016)	Alterations in metabolism (Lana-Elola et al. 2021)	
Interferonopathies Pinto et al. 2020; Sullivan et al. 2016; Waugh et al. 2023)		
Reproduction (Dard et al. 2023)		
Alzheimer's associated neurodegeneration (Sawa et al. 2022; Chen et al. 2021a, 2023a, b)		

hold significance in complementing human genetic efforts aimed at establishing a more precise genotype–phenotype relationship.

The discovery of trisomy 21 in cells from individuals with DS (Lejeune et al. 1959a; Lejeune et al. 1959b) was followed quickly by the observation that some individuals with a clinical diagnosis of DS did not have full trisomy 21, but carried a third copy of only part of the chromosome due to deletion, duplication, translocation, or other chromosomal rearrangements (Niebuhr 1974). A number of groups began to study such patients (Delabar et al. 1993; Korbel et al. 2009; Korenberg 1990; Lyle et al. 2009; Sinet et al. 1993) with the aim of establishing genotype–phenotype relationships based on triplicated genomic regions and well-documented clinical features. Despite decades of effort, the goal of identifying causative genes through this human genetics-based approach for many DS phenotypes remains unachieved. The approach was limited by the number of informative segmental trisomy 21 cases and the high variability of occurrence and severity of human phenotypes. Further, it makes an implicit assumption that DS is a collection of single gene defects, i.e., that a variant in a single Hsa21 gene (or tightly linked genes) is sufficient to produce a pathologic outcome by itself (i.e., on a euploid background) when it is upregulated.

Mice offer a complementary approach to conventional human genetics for two main reasons: (Smith et al. 1995) regions on Hsa21 are orthologous to three regions in the mouse genome located on mouse chromosomes 10 (Mmu10), 16 (Mmu16), and 17 (Mmu17) (Fig. 5); and (Van Deursen et al. 1995) mouse models triplicated for Hsa21 orthologous regions exhibit phenotypes analogous to those that occur in DS. To pursue mouse-based genetic dissection, a large number of mouse mutants carrying different single duplications (Dp) or deletions (Df) in Hsa21 orthologous regions have been reported (Herault et al. 2017; Xing et al. 2016), including a genetic mapping panel (Lana-Elola et al. 2016).

Both additive and subtractive strategies can be employed for "genetic dissection" in mice. An additive strategy can be used to narrow down the region(s) associated with a phenotype and eventually identify the gene(s) contributing to a particular phenotype using overlapping duplication mutants (Lana-Elola et al. 2016; Redhead et al. 2023; Sloan et al. 2023; Watson-Scales et al. 2018). However, in DS, the triplicated genes not only affect phenotypes directly but also interact with other triplicated genes. Some of these interactions have been speculated to occur based on established knowledge (Siddiqui et al. 2008; Tsang et al. 2014), while others have been demonstrated experimentally (Marechal et al. 2015; Pereira et al. 2009; Zhang et al. 2014; Duchon et al. 2021). For example, non-linear (>1.5X) over-expression of the chromosome 21-linked interferon target gene *MX1* in aging fibroblasts with trisomy 21 can be attributed to the presence of both an extra copy of this gene and extra copies of two genes, *IFNAR1* and *IFNAR2*, which are on chromosome 21 and code for Type I interferon receptors (Li et al. 2006). For this reason, a mouse engineered to triplicate a single gene on an otherwise euploid background may not reflect its real contribution to DS because its impact manifested in such a mutant may be either diminished or increased by other triplicated genes in DS.

A subtractive strategy uses a deletion to normalize dosage of a part of the triplicated region, which allows for gradually narrowing down the critical region to reveal the identity of a gene or genes by compounding with a knockout allele(s) (Jiang et al. 2015; Liu et al. 2011, 2014). If phenotypic rescue occurs in the compound mutants, this will suggest that the genomic region or gene contributes to the phenotype in the context of triplications of the large number of Hsa21 gene orthologs, and phenotypic rescue implies that reducing the gene's impact can have consequences even though other Hsa21 gene orthologs are triplicated and interact among themselves (Sawa et al. 2022; Jiang et al. 2015; Garcia-Cerro et al. 2014; Salehi et al. 2006). Using "gene subtraction" in mouse models, researchers have confirmed that *Dyrk1a* (dual-specificity tyrosine phosphorylation-regulated kinase 1A gene) and *App* (amyloid beta precursor protein gene) are genes of major effect for developmental cognitive deficits and neurodegeneration in DS, respectively (Sawa et al. 2022; Jiang et al. 2015; Garcia-Cerro et al. 2014; Salehi et al. 2006; Brault et al. 2021; Cataldo et al. 2003). If normalizing a Hsa21 gene ortholog can alleviate a symptom in a mouse model of DS, the protein encoded by that gene can serve as a therapeutic target at the protein level. For these reasons, the protein products of *DYRK1A* and *APP* have served as the most important Hsa21 gene-based therapeutic targets for developmental cognitive deficits and neurodegeneration in DS, respectively (Arbones et al. 2019; Belichenko et al. 2016; Chen et al. 2021b; Feki and Hibaoui 2018; Stringer et al. 2017). Notably, inhibitors for tyrosine kinases, including DYRK1A encoded by the Hsa21 gene ortholog, have shown effectiveness in mouse DS models (Guedj et al. 2009; Nguyen et al. 2018; Souchet et al. 2019). Systemic genetic dissections have also been performed for DS-associated heart defects, motor dysfunctions, and skeleton anomalies in mice (Olson et al. 2004; Lana-Elola et al. 2016, 2024; Redhead et al. 2023; Sloan et al. 2023; Watson-Scales et al. 2018; Liu et al. 2011, 2014).

4 Effects of an Extra Chromosome Beyond Gene Dosage in a DS Model

The possibility that the presence of an extra chromosome may affect the phenotypes by exerting gene-dosage-independent impact has long been entertained by the research community (Starbuck et al. 2014; Duchon et al. 2011; Kazuki et al. 2020; Kemeny et al. 2018; Letourneau et al. 2014; O'Doherty et al. 2005; Omori et al. 2017; Puente-Bedia et al. 2021). This is partly because an extra chromosome requires the cell to replicate extra genetic material, organize the spindle at mitosis and/or affects three-dimensional interactions in interphase, which might interfere with the carefully orchestrated processes centered around cell replication and gene expression. For example, even a slight prolongation of mitosis could result in insufficient generation of anlage for many structures during the scores of cell divisions required for an "average" cell lineage between zygote and birth.

This question was explored experimentally in yeast by separating the impact of extra DNA independent of gene dosage effects (Torres et al. 2007; Tosh et al. 2022). The strategy entails transferring YAC (yeast artificial chromosome) clones carrying mammalian DNA into yeast. The YAC exists as an extra chromosome in the yeast cell; the human or mouse genes on YAC do not express in yeast. When compared to yeast without an extra chromosome (YAC), the genetically modified yeast exhibited abnormal phenotypes associated with the cell cycle. This raises the possibility that the extra chromosome contributes to phenotypes without an increased dosage of expressed genes. Of course, the yeast experiment in a single-cell organism does not capture the complexity of a mammalian system and its relative evolutionary proximity to people.

The early phenotypic analysis of *Dp(16)1Yey* mice suggests the possibility that this strain may have milder phenotypes than Ts65Dn (Gupta et al. 2016; Pritchard

Table 2 Possible reasons for the phenotypic differences between Ts65Dn and *Dp(16)1Yey*

No.	Explanation	Ts65Dn	*Dp(16)1Yey*
1	Strain background	B6EiC3Sn	B6129Sv
2	Number of triplicated Hsa21 gene orthologs	103	115
3	Triplication of Mmu17 genes that lack orthologs to Hsa21 genes	Yes	No
4	Presence of an extra chromosome	Yes	No

et al. 2008). Careful deliberations led to identification of four different possibilities that can explain the phenotypic differences between Ts65Dn and *Dp(16)1Yey* mice (Table 2). To address these possibilities, *Dp(16)1Yey* mice in the B6EiC3Sn background were generated by backcrossing. A comparison between Ts65Dn and *Dp(16)1Yey* on this same strain background showed more severe impairment of Ts65Dn in Morris water maze tests (Xing et al. 2023). This result suggests that the strain background may not be sufficient to explain all the phenotypic differences between Ts65Dn and *Dp(16)1Yey*. Sequencing of the breakpoints of the chromosomal translocation in Ts65Dn (Duchon et al. 2011; Reinholdt et al. 2011) made it feasible to examine other possibilities experimentally.

We considered two strategies to test the hypothesis that the extra chromosome itself may be a factor in developmental and functional effects of trisomy, independent of gene dosage effects (Fig. 6a, b). The strategy outlined in Fig. 6a was selected as being more similar to the situation in DS because the triplicated region near the centromere of Mmu17 is not syntenic to any genomic region on Hsa21 (Duchon et al. 2011; Reinholdt et al. 2011). Two deletion mutants, *Df(17)2Yey* and *Df(16)8Yey*, were generated using Cre/loxP-mediated chromosome engineering (Fig. 6a). Generation of the compound mutants Ts65Dn;*Df(17)2Yey* and *Dp(16)1Yey/Df(16)8Yey* rendered the identical Hsa21 gene orthologs on Mmu16 triplicated in both models. Using a specific breeding strategy, these two compound mutants achieved a comparable strain background (Xing et al. 2023). Possible outcomes for contributors of phenotypes expressed in the compound mice are listed in Table 3. With gene dosage normalized, the differences observed between Ts65Dn;*Df(17)2Yey* and *Dp(16)1Yey/Df(16)8Yey* can be assigned to the presence of the extra chromosome, manifesting as phenotypic and transcriptomic differences. Specifically, significant deficiencies in nesting were observed in both Ts65Dn;*Df(17)2Yey* and *Dp(16)1Yey/Df(16)8Yey*, which were indicative of the consequence of the gene dosage effect since they shared triplication of the identical 103 Hsa21 gene orthologs. On the other hand, Ts65Dn;*Df(17)2Yey* mice but not *Dp(16)1Yey/Df(16)8Yey* mice were impaired in T-maze tests, suggesting that the extra freely segregating chromosome in Ts65Dn;*Df(17)2Yey* plays a significant role in causing this phenotype (Table 3). The impact of the extra chromosome on the transcriptome were surprising and interesting (Fig. 6c). The observed gene dosage impact has been observed previously, and about 50% of the triplicated Hsa21 gene orthologs shared between the two compound models exhibited an elevated expression level. However, the number of disomic genes with altered expression (i.e., genes on other chromosomes that are not at dosage imbalance) was about

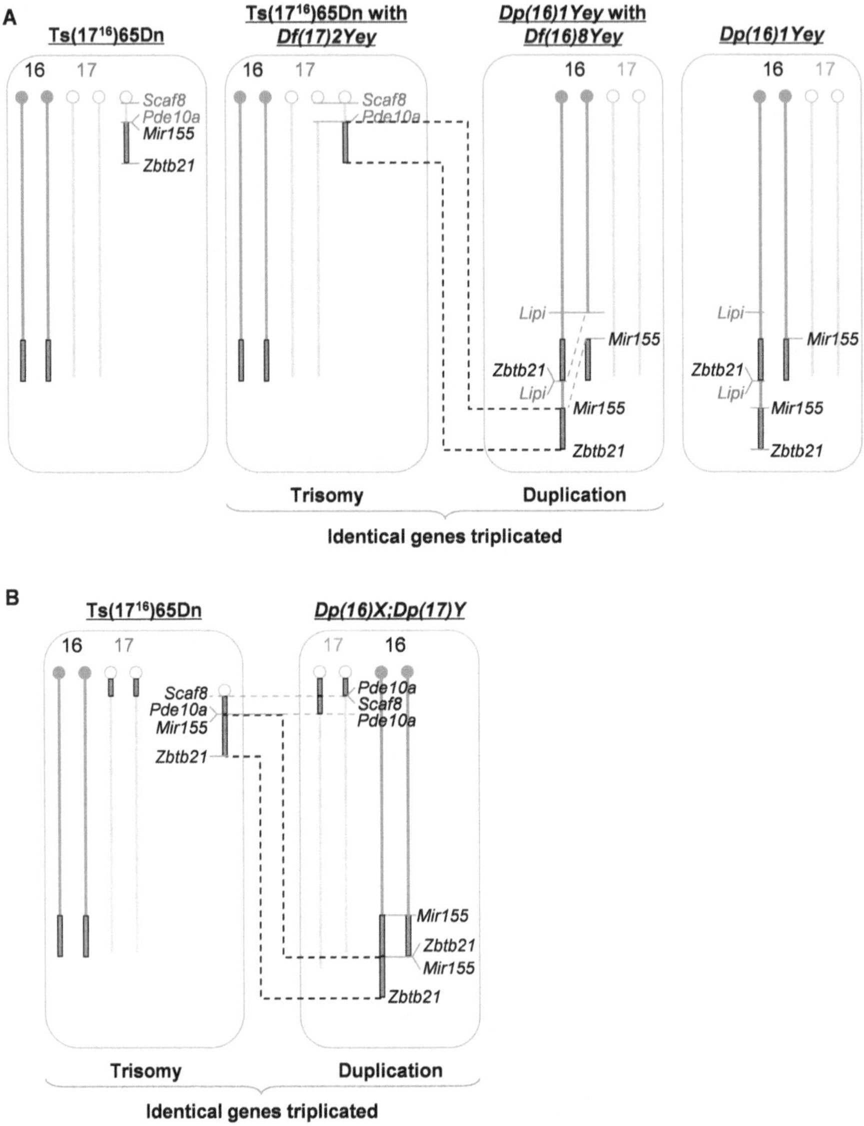

Fig. 6 Mouse model designs and transcriptomic analysis for dissecting the impacts of an extra chromosome. (**a**) A segmental trisomy model, Ts(17^{16})65Dn (Ts65Dn), compounded with *Df(17)2Yey*, and a duplication model, *Dp(16)1Yey*, compounded with *Df(16)8Yey*, have the same triplicated Hsa21 gene orthologs. (**b**) A segmental trisomy model, Ts(17^{16})65Dn, and a duplication model, *Dp(16)X;Dp(17)Y*, have the same triplicated Hsa21 gene orthologs. (**c**) RNA-seq analysis of the cerebral cortex revealed the impact of an extra chromosome at the transcriptomic level. The transcriptomic results shows about 50% of the same triplicated Hsa21 gene orthologs in both compound models exhibited elevated expression, Ts65Dn;*Df(17)2Yey* and *Dp(16)1Yey*/*Df(16)8Yey*, due to the gene dosage effect. On the other hand, a substantially larger number of the disomic genes exhibited alterations of expression in Ts65Dn;*Df(17)2Yey* mice than that in *Dp(16)1Yey*/*Df(16)8Yey* mice. The latter provides evidence for the impact of the extra chromosome beyond the gene dosage effect, because the triplicated Hsa21 gene orthologs are identical between Ts65Dn;*Df(17)2Yey* and *Dp(16)1Yey*/*Df(16)8Yey*

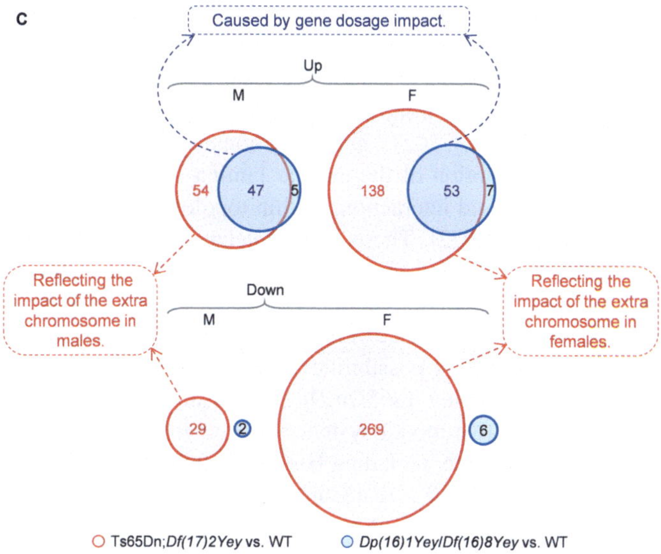

Fig. 6 (continued)

Table 3 Phenotypic impact of the gene dosage and the extra chromosome beyond gene dosage for the two genotypes versus wild-type (WT) mice. Sex-specific differences were observed

	Ts65Dn;*Df(17)2Yey* vs. WT			*Dp(16)1Yey/Df(16)8Yey* vs. WT		
Behavior	Both sexes	Males	Females	Both sexes	Males	Females
Nesting	+*	+*	–	++*	+*	–
T-maze	+**	+**	–	–	–	–

+, $P < 0.05$; ++, $P < 0.01$
*Evidence of the gene dosage effect
**Evidence of the gene dosage-independent effect

11.9-fold higher in Ts65Dn;*Df(17)2Yey* males that have an extra chromosome compared to *Dp(16)1Yey/Df(16)8Yey* males with the same gene dosage imbalance but a normal number of mouse chromosomes. The difference reached about 31.3-fold when the relevant female mice were compared (Xing et al. 2023), which suggests the possibility of a sex-specific difference, like many other features associated with DS (Fig. 6c).

Aforementioned altered expression of a large number of the disomic genes in Ts65Dn;*Df(17)2Yey*, but not in *Dp(16)1Yey/Df(16)8Yey* can be attributable to the presence of the extra chromosome, which in turn can be the cause for a more severe phenotypic presentation (Xing et al. 2023). According to the current state of research on nuclear architecture (Kempfer and Pombo 2020; Krumm and Duan 2019; Maass et al. 2019; Muller et al. 2019), a possible cause underlying changed expression of disomic genes in Ts65Dn;*Df(17)2Yey* may be the alteration of nuclear structure as a consequence of the presence of the extra chromosome. Recently, efforts have been made to examine the impact of an extra copy of Hsa21 or its mouse equivalent on intranuclear events and nuclear morphology at various levels (Kemeny et al. 2018; Omori et al. 2017; Puente-Bedia et al. 2021; Hwang et al. 2019), including impact on the spatial organization of chromosomal territories (Kemeny et al. 2018; Omori et al. 2017; Puente-Bedia et al. 2021), and 3D genome architecture (Meharena et al. 2022). Ts21 induces disruption of the nuclear lamina, reorganization of chromosomal looping, and increased interactions within topologically associated domains (TADs) (Meharena et al. 2022). These major nucleus-associated events could be caused by the triplication of some critical genes, such as *HLCS* and *HMGN1* or their orthologs (Letourneau et al. 2014; Narang et al. 2004; Postnikov and Bustin 2010; Singh et al. 2013) and/or the presence of the extra chromosome independent of gene dosage changes. Such different possibilities can now be dissected experimentally for the first time by employing Ts65Dn;*Df(17)2Yey* and *Dp(16)1Yey*/*Df(16)8Yey*. The combination of this two-model system and ever-improving technologies for analyzing genome architecture, including Hi-C and single-cell Hi-C (Kempfer and Pombo 2020; Ramani et al. 2020), creates an opportunity to garner new insights into important molecular mechanisms underlying DS.

Ts66Yah and Ts65Dn/*Df(17)2Yey* may exhibit some differences in cognitive phenotypes when compared to their respective wild-type controls (Xing et al. 2023; Duchon et al. 2022). These potential discrepancies may be attributed to the different strain backgrounds. Additionally, the potentially varying genomic configurations of Ts66Yah and Ts65Dn/Df(17)2Yey, resulting from the cis and trans normalization of the same triplicated Hsa21 gene orthologs on Mmu17, may also play a role.

In addition to the differences in cognitive behaviors between Ts65Dn and *Dp(16)1Yey* mice, phenotypic differences in deglutition and vocal communication have also been observed. Ts65Dn, but not *Dp(16)1Yey*, has shown small but significant differences at a juvenile age in some measures of mastication rates, and Ts65Dn shows significant differences at juvenile ages in some biological measures of muscles involved in deglutition (Glass and Connor 2016; Glass et al. 2019a). In adults, Ts65Dn shows significant reduction compared with the controls in the rates of swallowing of a pureed food consistency, leading to longer times between consecutive swallows during continuous eating (Fig. 7). The slower swallowing rate does not entail jaw movement cycles of longer duration, but rather is comprised of a greater number of jaw movement cycles preceding each swallow. While the causes of this phenotype are currently unknown, challenges that could manifest in this way might include inefficiencies in oral processing of the food bolus or deficits related to sensation of the food bolus. In contrast, concurrent assessment of the *Dp(16)1Yey* mouse model showed unaffected swallowing performance that was equivalent to the wild-type controls (Glass et al. 2019b).

Mice communicate through ultrasonic vocalization (USV), which, as a communication modality, has basic similarities to some mechanisms of vocal communication in humans. Multiple mouse models of DS have demonstrated USV phenotypes at early postnatal ages. While both Ts65Dn and *Dp(16)1Yey* have been found to demonstrate USV phenotypes at adult ages, the two models show phenotypes that are different from each other in some USV measures, and of the two models,

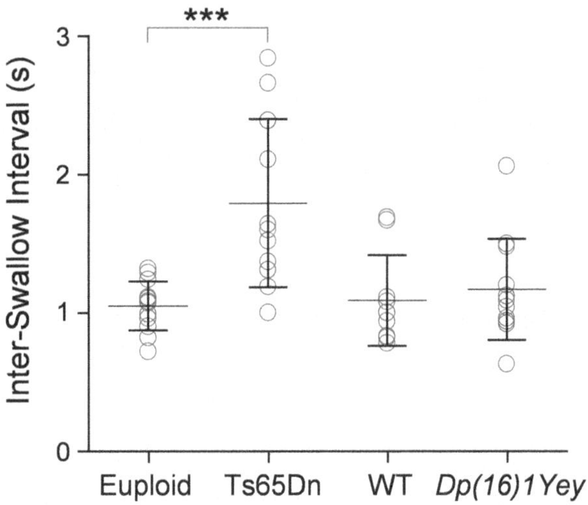

Fig. 7 Comparison of adult euploid/wild-type controls, *Dp(16)1Yey*, and Ts65Dn mice showed that Ts65Dn had a significantly greater time elapsing between consecutive swallows during continuous eating of a puree. Each data point indicates the mean value from one mouse. The y-axis measure(s) indicates seconds. Error bars indicate the group mean and standard deviation. ***$P < 0.001$. (Adapted from Glass et al. 2019b)

Ts65Dn is the more severely affected model across multiple USV measures (Glass et al. 2023; Zampieri et al. 2014).

Challenges related to vocal communication and deglutition (eating, drinking, and swallowing) are both prevalent in DS (Kent and Vorperian 2013; Lazenby 2008) and are important issues to engage with in translational biomedical research. Difficulties in speech and communication, in addition to affecting the quality of life, can compound barriers in accessing medical care for some individuals (Morris et al. 2013; Stransky et al. 2018). Difficulties related to eating, drinking, and swallowing have also been an area of care and treatment that infants with DS have historically been disproportionately and inappropriately excluded from (Fost 2020), which is a consideration that gives impetus for the inclusion of these phenotypes in current and future biomedical research seeking to benefit people with DS. Because differential phenotypes in vocal communication and deglutition were detected between Ts65Dn and *Dp(16)1Yey*, research could be pursued to determine if the gene-dosage-independent extra chromosome or the triplication of the centromeric region on Mmu17 plays roles in these phenotypes by using Ts65Dn, Ts65Dn;*Df(17)2Yey* and *Dp(16)1Yey*/*Df(16)8Yey*.

The development of Cre/loxP-mediated chromosome engineering was an important milestone in mouse genetics, facilitating DS research by allowing for the generation of duplications and deletions of the mouse genome with predetermined endpoints. Substantially more chromosomal rearrangements have been engineered for research on DS than for any other human disease. With the complementary strengths of trans-chromosomal strategies and CRISPR-mediated genome editing (Duchon et al. 2022; Birling et al. 2017; Schaeffer et al. 2023), rodent-based genetic modeling and analysis of DS have continued to expand, promising new discoveries for years to come.

Acknowledgments The authors would like to thank Roger H. Reeves and Benjamin Tycko for their reviews and input. Current work in the authors' laboratories was supported in part by the National Institutes of Health (R01HD090180, R01HD109750, R01DC019735, and R03TR003344) and The Children's Guild Foundation. In our research efforts, we utilized the services of the shared resources supported by P30CA016056 from the National Institutes of Health to Roswell Park Comprehensive Cancer Center, including the Genomics Shared Resource (GSR), Gene Targeting and Transgenic Shared Resource (GeTT), Comparative Oncology and Tumor Models Shared Resource (COTM), and Scientific Editing and Research Communications Core (SERCC).

References

Adams DJ, Biggs PJ, Cox T, Davies R, van der Weyden L, Jonkers J, Smith J, Plumb B, Taylor R, Nishijima I, Yu Y, Rogers J, Bradley A. Mutagenic insertion and chromosome engineering resource (MICER). *Nat Genet.* 2004;36:867–871. Epub 2004/07/06. https://doi.org/10.1038/ng1388.

Arbones ML, Thomazeau A, Nakano-Kobayashi A, Hagiwara M, Delabar JM. DYRK1A and cognition: A lifelong relationship. *Pharmacol Ther.* 2019;194:199–221. Epub 2018/10/01. https://doi.org/10.1016/j.pharmthera.2018.09.010.

Aziz NM, Guedj F, Pennings JLA, Olmos-Serrano JL, Siegel A, Haydar TF, Bianchi DW. Lifespan analysis of brain development, gene expression and behavioral phenotypes in the Ts1Cje, Ts65Dn and Dp(16)1/Yey mouse models of Down syndrome. *Dis Model Mech.* 2018;11. Epub 2018/05/03. https://doi.org/10.1242/dmm.031013. PMCID: PMC6031353.

Belichenko PV, Kleschevnikov AM, Becker A, Wagner GE, Lysenko LV, Yu YE, Mobley WC. Down Syndrome Cognitive Phenotypes Modeled in Mice Trisomic for All HSA 21 Homologues. *PLoS One.* 2015;10:e0134861. Epub 2015/08/01. https://doi.org/10.1371/journal.pone.0134861. PMCID: PMC4521889.

Belichenko PV, Madani R, Rey-Bellet L, Pihlgren M, Becker A, Plassard A, Vuillermot S, Giriens V, Nosheny RL, Kleschevnikov AM, Valletta JS, Bengtsson SK, Linke GR, Maloney MT, Hickman DT, Reis P, Granet A, Mlaki D, Lopez-Deber MP, Do L, Singhal N, Masliah E, Pearn ML, Pfeifer A, Muhs A, Mobley WC. An Anti-beta-Amyloid Vaccine for Treating Cognitive Deficits in a Mouse Model of Down Syndrome. *PLoS One.* 2016;11:e0152471. Epub 2016/03/31. https://doi.org/10.1371/journal.pone.0152471. PMCID: PMC4811554.

Birling MC, Schaeffer L, Andre P, Lindner L, Marechal D, Ayadi A, Sorg T, Pavlovic G, Herault Y. Efficient and rapid generation of large genomic variants in rats and mice using CRISMERE. *Sci Rep.* 2017;7:43331. Epub 2017/03/08. https://doi.org/10.1038/srep43331. PMCID: PMC5339700.

Brault V, Besson V, Magnol L, Duchon A, Herault Y. Cre/loxP-mediated chromosome engineering of the mouse genome. *Handb Exp Pharmacol.* 2007:29–48. Epub 2007/01/06. https://doi.org/10.1007/978-3-540-35109-2_2.

Brault V, Nguyen TL, Flores-Gutierrez J, Iacono G, Birling MC, Lalanne V, Meziane H, Manousopoulou A, Pavlovic G, Lindner L, Selloum M, Sorg T, Yu E, Garbis SD, Herault Y. Dyrk1a gene dosage in glutamatergic neurons has key effects in cognitive deficits observed in mouse models of MRD7 and Down syndrome. *PLoS Genet.* 2021;17:e1009777. Epub 2021/09/30. https://doi.org/10.1371/journal.pgen.1009777. PMCID: PMC8480849.

Canonica T, Kidd EJ, Gibbins D, Lana-Elola E, Fisher EMC, Tybulewicz VLJ, Good M. Dissecting the contribution of human chromosome 21 syntenic regions to recognition memory processes in adult and aged mouse models of Down syndrome. *Front Behav Neurosci.* 2024;18:1428146. https://doi.org/10.3389/fnbeh.2024.1428146.

Cataldo AM, Petanceska S, Peterhoff CM, Terio NB, Epstein CJ, Villar A, Carlson EJ, Staufenbiel M, Nixon RA. App gene dosage modulates endosomal abnormalities of Alzheimer's disease in a

segmental trisomy 16 mouse model of down syndrome. *J Neurosci.* 2003;23:6788–6792. Epub 2003/08/02. https://doi.org/10.1523/JNEUROSCI.23-17-06788.2003. PMCID: PMC6740714.
Chang P, Bush D, Schorge S, Good M, Canonica T, Shing N, Noy S, Wiseman FK, Burgess N, Tybulewicz VLJ, Walker MC, Fisher EMC. Altered Hippocampal-Prefrontal Neural Dynamics in Mouse Models of Down Syndrome. *Cell Rep.* 2020;30:1152–1163 e1154. Epub 2020/01/30. https://doi.org/10.1016/j.celrep.2019.12.065. PMCID: PMC6996020.
Chen XQ, Das U, Park G, Mobley WC. Normal levels of KIF5 but reduced KLC1 levels in both Alzheimer disease and Alzheimer disease in Down syndrome: evidence suggesting defects in anterograde transport. *Alzheimers Res Ther.* 2021a;13:59. Epub 2021/03/12. https://doi.org/10.1186/s13195-021-00796-6. PMCID: PMC7945332.
Chen XQ, Salehi A, Pearn ML, Overk C, Nguyen PD, Kleschevnikov AM, Maccecchini M, Mobley WC. Targeting increased levels of APP in Down syndrome: Posiphen-mediated reductions in APP and its products reverse endosomal phenotypes in the Ts65Dn mouse model. *Alzheimers Dement.* 2021b;17:271–292. Epub 2020/09/26. https://doi.org/10.1002/alz.12185. PMCID: PMC7984396.
Chen GD, Li L, McCall A, Ding D, Xing Z, Yu YE, Salvi R. Hearing impairment in murine model of Down syndrome. *Front Genet.* 2022;13:936128. Epub 2022/08/23. https://doi.org/10.3389/fgene.2022.936128. PMCID: PMC9385999.
Chen XQ, Sawa M, Becker A, Karachentsev D, Zuo X, Rynearson KD, Tanzi RE, Mobley WC. Retromer Proteins Reduced in Down Syndrome and the Dp16 Model: Impact of APP Dose and Preclinical Studies of a gamma-Secretase Modulator. *Ann Neurol.* 2023a;94:245–258. Epub 2023/04/13. https://doi.org/10.1002/ana.26659.
Chen XQ, Zuo X, Becker A, Head E, Mobley WC. Reduced synaptic proteins and SNARE complexes in Down syndrome with Alzheimer's disease and the Dp16 mouse Down syndrome model: Impact of APP gene dose. *Alzheimers Dement.* 2023b;19:2095–2116. Epub 2022/11/13. https://doi.org/10.1002/alz.12835. PMCID: PMC10175517.
Cole SE, Wiltshire T, Reeves RH. Physical mapping of the evolutionary boundary between human chromosomes 21 and 22 on mouse chromosome 10. *Genomics.* 1998;50:109–111. Epub 1998/06/20. https://doi.org/10.1006/geno.1998.5312.
Cole SE, Wiltshire T, Rue EE, Morrow D, Hieter P, Brahe C, Fisher EM, Katsanis N, Reeves RH. High-resolution comparative physical mapping of mouse chromosome 10 in the region of homology with human chromosome 21. *Mamm Genome.* 1999;10:229–234. Epub 1999/03/02. https://doi.org/10.1007/s003359900978.
Colvin KL, Nguyen K, Boncella KL, Goodman DM, Elliott RJ, Harral JW, Bilodeaux J, Smith BJ, Yeager ME. Lung and Heart Biology of the Dp16 Mouse Model of down Syndrome: Implications for Studying Cardiopulmonary Disease. *Genes (Basel).* 2023;14. Epub 2023/09/28. https://doi.org/10.3390/genes14091819. PMCID: PMC10530394.
Dard R, Tutunaru A, Bouassida M, Balde Camara A, Parizot E, Kassis N, Fortemps J, Cierniewski C, Racine C, Clemente ND, Vialard F, Janel N. Spermatogonial depletion and a spermatogenesis defect in the Dp(16)1Yey mouse model of Down syndrome. *Reproduction.* 2023. Epub 2023/12/08. https://doi.org/10.1530/REP-23-0233.
Davisson MT, Schmidt C, Akeson EC. Segmental trisomy of murine chromosome 16: a new model system for studying Down syndrome. *Prog Clin Biol Res.* 1990;360:263–280. Epub 1990/01/01.
Delabar JM, Theophile D, Rahmani Z, Chettouh Z, Blouin JL, Prieur M, Noel B, Sinet PM. Molecular mapping of twenty-four features of Down syndrome on chromosome 21. *Eur J Hum Genet.* 1993;1:114–124. https://doi.org/10.1159/000472398.
Duchon A, Besson V, Pereira PL, Magnol L, Herault Y. Inducing segmental aneuploid mosaicism in the mouse through targeted asymmetric sister chromatid event of recombination. *Genetics.* 2008;180:51–59. Epub 2008/09/02. https://doi.org/10.1534/genetics.108.092312. PMCID: PMC2535701.
Duchon A, Raveau M, Chevalier C, Nalesso V, Sharp AJ, Herault Y. Identification of the translocation breakpoints in the Ts65Dn and Ts1Cje mouse lines: relevance for modeling Down

syndrome. *Mamm Genome.* 2011;22:674–684. Epub 2011/09/29. https://doi.org/10.1007/s00335-011-9356-0. PMCID: PMC3224224.

Duchon A, Del Mar Muniz Moreno M, Lorenzo SM, de Souza MPS, Chevalier C, Nalesso V, Meziane H, de Sousa PL, Noblet V, Armspach JP, Brault V, Herault Y. Multi-influential genetic interactions alter behaviour and cognition through six main biological cascades in Down syndrome mouse models. *Human molecular genetics.* 2021. Epub 2021/03/12. https://doi.org/10.1093/hmg/ddab012.

Duchon A, Del Mar Muniz Moreno M, Chevalier C, Nalesso V, Andre P, Fructuoso-Castellar M, Mondino M, Po C, Noblet V, Birling MC, Potier MC, Herault Y. Ts66Yah, a mouse model of Down syndrome with improved construct and face validity. *Dis Model Mech.* 2022;15. Epub 2022/11/15. https://doi.org/10.1242/dmm.049721. PMCID: PMC9789398.

Feki A, Hibaoui Y. DYRK1A Protein, A Promising Therapeutic Target to Improve Cognitive Deficits in Down Syndrome. *Brain Sci.* 2018;8. Epub 2018/10/20. https://doi.org/10.3390/brainsci8100187. PMCID: PMC6210095.

Fost N. "The Hopkins Mongol Case": The Dawn of the Bioethics Movement. *Pediatrics.* 2020;146:S3–S8. Epub 2020/08/02. https://doi.org/10.1542/peds.2020-0818C.

Garcia-Cerro S, Martinez P, Vidal V, Corrales A, Florez J, Vidal R, Rueda N, Arbones ML, Martinez-Cue C. Overexpression of Dyrk1A is implicated in several cognitive, electrophysiological and neuromorphological alterations found in a mouse model of Down syndrome. *PLoS One.* 2014;9:e106572. Epub 2014/09/05. https://doi.org/10.1371/journal.pone.0106572. PMCID: PMC4154723.

Glass TJ, Connor NP. Digastric Muscle Phenotypes of the Ts65Dn Mouse Model of Down Syndrome. *PLoS One.* 2016;11:e0158008. Epub 2016/06/24. https://doi.org/10.1371/journal.pone.0158008. PMCID: PMC4919106.

Glass TJ, Twadell SL, Valmadrid LC, Connor NP. Early impacts of modified food consistency on oromotor outcomes in mouse models of Down syndrome. *Physiol Behav.* 2019a;199:273–281. Epub 2018/11/30. 10.1016/j.physbeh.2018.11.031. PMCID: PMC6358162.

Glass TJ, Valmadrid LCV, Connor NP. The Adult Ts65Dn Mouse Model of Down Syndrome Shows Altered Swallow Function. *Front Neurosci.* 2019b;13:906. Epub 2019/09/27. https://doi.org/10.3389/fnins.2019.00906. PMCID: PMC6727863.

Glass TJ, Lenell C, Fisher EH, Yang Q, Connor NP. Ultrasonic vocalization phenotypes in the Ts65Dn and Dp(16)1Yey mouse models of Down syndrome. *Physiol Behav.* 2023;271:114323. Epub 2023/08/14. https://doi.org/10.1016/j.physbeh.2023.114323. PMCID: PMC10592033.

Guedj F, Sebrie C, Rivals I, Ledru A, Paly E, Bizot JC, Smith D, Rubin E, Gillet B, Arbones M, Delabar JM. Green tea polyphenols rescue of brain defects induced by overexpression of DYRK1A. *PLoS One.* 2009;4:e4606. Epub 2009/02/27. https://doi.org/10.1371/journal.pone.0004606. PMCID: PMC2645681.

Gupta M, Dhanasekaran AR, Gardiner KJ. Mouse models of Down syndrome: gene content and consequences. *Mamm Genome.* 2016;27:538–555. Epub 2016/08/20. https://doi.org/10.1007/s00335-016-9661-8. PMCID: PMC5471624.

Herault Y, Rassoulzadegan M, Cuzin F, Duboule D. Engineering chromosomes in mice through targeted meiotic recombination (TAMERE). *Nat Genet.* 1998;20:381–384. Epub 1998/12/08. https://doi.org/10.1038/3861.

Herault Y, Delabar JM, Fisher EMC, Tybulewicz VLJ, Yu E, Brault V. Rodent models in Down syndrome research: impact and future opportunities. *Dis Model Mech.* 2017;10:1165–1186. Epub 2017/10/11. https://doi.org/10.1242/dmm.029728. PMCID: PMC5665454.

Hwang S, Williams JF, Kneissig M, Lioudyno M, Rivera I, Helguera P, Busciglio J, Storchova Z, King MC, Torres EM. Suppressing Aneuploidy-Associated Phenotypes Improves the Fitness of Trisomy 21 Cells. *Cell Rep.* 2019;29:2473–2488 e2475. Epub 2019/11/21. https://doi.org/10.1016/j.celrep.2019.10.059. PMCID: PMC6886690.

Jiang X, Liu C, Yu T, Zhang L, Meng K, Xing Z, Belichenko PV, Kleschevnikov AM, Pao A, Peresie J, Wie S, Mobley WC, Yu YE. Genetic dissection of the Down syndrome critical region. *Human molecular genetics.* 2015;24:6540–6551. https://doi.org/10.1093/hmg/ddv364. PMCID: 4614710.

Johnson DK, Hand RE, Jr., Rinchik EM. Molecular mapping within the mouse albino-deletion complex. *Proc Natl Acad Sci U S A.* 1989;86:8862–8866. Epub 1989/11/01. https://doi.org/10.1073/pnas.86.22.8862. PMCID: PMC298390.

Kazuki Y, Gao FJ, Li Y, Moyer AJ, Devenney B, Hiramatsu K, Miyagawa-Tomita S, Abe S, Kazuki K, Kajitani N, Uno N, Takehara S, Takiguchi M, Yamakawa M, Hasegawa A, Shimizu R, Matsukura S, Noda N, Ogonuki N, Inoue K, Matoba S, Ogura A, Florea LD, Savonenko A, Xiao M, Wu D, Batista DA, Yang J, Qiu Z, Singh N, Richtsmeier JT, Takeuchi T, Oshimura M, Reeves RH. A non-mosaic transchromosomic mouse model of down syndrome carrying the long arm of human chromosome 21. *Elife.* 2020;9. Epub 2020/07/01. https://doi.org/10.7554/eLife.56223. PMCID: PMC7358007.

Kent RD, Vorperian HK. Speech impairment in Down syndrome: a review. *J Speech Lang Hear Res.* 2013;56:178–210. Epub 2013/01/01. https://doi.org/10.1044/1092-4388(2012/12-0148). PMCID: PMC3584188.

Kemeny S, Tatout C, Salaun G, Pebrel-Richard C, Goumy C, Ollier N, Maurin E, Pereira B, Vago P, Gouas L. Spatial organization of chromosome territories in the interphase nucleus of trisomy 21 cells. *Chromosoma.* 2018;127:247–259. Epub 2017/12/15. https://doi.org/10.1007/s00412-017-0653-6.

Kempfer R, Pombo A. Methods for mapping 3D chromosome architecture. *Nat Rev Genet.* 2020;21:207–226. Epub 2019/12/19. https://doi.org/10.1038/s41576-019-0195-2.

Korbel JO, Tirosh-Wagner T, Urban AE, Chen XN, Kasowski M, Dai L, Grubert F, Erdman C, Gao MC, Lange K, Sobel EM, Barlow GM, Aylsworth AS, Carpenter NJ, Clark RD, Cohen MY, Doran E, Falik-Zaccai T, Lewin SO, Lott IT, McGillivray BC, Moeschler JB, Pettenati MJ, Pueschel SM, Rao KW, Shaffer LG, Shohat M, Van Riper AJ, Warburton D, Weissman S, Gerstein MB, Snyder M, Korenberg JR. The genetic architecture of Down syndrome phenotypes revealed by high-resolution analysis of human segmental trisomies. *Proc Natl Acad Sci U S A.* 2009;106:12031–12036. Epub 2009/07/15. https://doi.org/10.1073/pnas.0813248106. PMCID: 2709665.

Korenberg JR. Molecular mapping of the Down syndrome phenotype. *Prog Clin Biol Res.* 1990;360:105–115. Epub 1990/01/01.

Krumm A, Duan Z. Understanding the 3D genome: Emerging impacts on human disease. *Semin Cell Dev Biol.* 2019;90:62–77. Epub 2018/07/11. https://doi.org/10.1016/j.semcdb.2018.07.004. PMCID: PMC6329682.

Lana-Elola E, Watson-Scales S, Slender A, Gibbins D, Martineau A, Douglas C, Mohun T, Fisher EM, Tybulewicz VL. Genetic dissection of Down syndrome-associated congenital heart defects using a new mouse mapping panel. *Elife.* 2016;5. Epub 2016/01/15. https://doi.org/10.7554/eLife.11614. PMCID: PMC4764572.

Lana-Elola E, Cater H, Watson-Scales S, Greenaway S, Muller-Winkler J, Gibbins D, Nemes M, Slender A, Hough T, Keskivali-Bond P, Scudamore CL, Herbert E, Banks GT, Mobbs H, Canonica T, Tosh J, Noy S, Llorian M, Nolan PM, Griffin JL, Good M, Simon M, Mallon AM, Wells S, Fisher EMC, Tybulewicz VLJ. Comprehensive phenotypic analysis of the Dp1Tyb mouse strain reveals a broad range of Down syndrome-related phenotypes. *Dis Model Mech.* 2021;14. Epub 2021/09/04. https://doi.org/10.1242/dmm.049157. PMCID: PMC8543064.

Lana-Elola E, Aoidi R, Llorian M, Gibbins D, Buechsenschuetz C, Bussi C, Flynn H, Gilmore T, Watson-Scales S, Haugsten Hansen M, Hayward D, Song OR, Brault V, Herault Y, Deau E, Meijer L, Snijders AP, Gutierrez MG, Fisher EMC, Tybulewicz VLJ. Increased dosage of DYRK1A leads to congenital heart defects in a mouse model of Down syndrome. *Sci Transl Med.* 2024;16:eadd6883. Epub 2024/01/24. https://doi.org/10.1126/scitranslmed.add6883. PMCID: PMC7615651.

Lazenby T. The impact of aging on eating, drinking, and swallowing function in people with Down's syndrome. *Dysphagia.* 2008;23:88–97. Epub 2007/08/19. https://doi.org/10.1007/s00455-007-9096-1.

Lejeune J, Gauthier M, Turpin R. [Human chromosomes in tissue cultures]. *C R Hebd Seances Acad Sci.* 1959a;248:602–603. Epub 1959/01/26.

Lejeune J, Turpin R, Gautier M. [Mongolism; a chromosomal disease (trisomy)]. *Bull Acad Natl Med.* 1959b;143:256–265. Epub 1959/04/07.

Letourneau A, Santoni FA, Bonilla X, Sailani MR, Gonzalez D, Kind J, Chevalier C, Thurman R, Sandstrom RS, Hibaoui Y, Garieri M, Popadin K, Falconnet E, Gagnebin M, Gehrig C, Vannier A, Guipponi M, Farinelli L, Robyr D, Migliavacca E, Borel C, Deutsch S, Feki A, Stamatoyannopoulos JA, Herault Y, van Steensel B, Guigo R, Antonarakis SE. Domains of genome-wide gene expression dysregulation in Down's syndrome. *Nature.* 2014;508:345–350. Epub 2014/04/18. https://doi.org/10.1038/nature13200.

Levenga J, Peterson DJ, Cain P, Hoeffer CA. Sleep Behavior and EEG Oscillations in Aged Dp(16)1Yey/+ Mice: A Down Syndrome Model. *Neuroscience.* 2018;376:117–126. Epub 2018/02/20. https://doi.org/10.1016/j.neuroscience.2018.02.009. PMCID: PMC7255843.

Li CM, Guo M, Salas M, Schupf N, Silverman W, Zigman WB, Husain S, Warburton D, Thaker H, Tycko B. Cell type-specific over-expression of chromosome 21 genes in fibroblasts and fetal hearts with trisomy 21. *BMC Med Genet.* 2006;7:24. Epub 2006/03/17. 10.1186/1471-2350-7-24. PMCID: PMC1435874.

Li Z, Yu T, Morishima M, Pao A, LaDuca J, Conroy J, Nowak N, Matsui S, Shiraishi I, Yu YE. Duplication of the entire 22.9 Mb human chromosome 21 syntenic region on mouse chromosome 16 causes cardiovascular and gastrointestinal abnormalities. *Human molecular genetics.* 2007;16:1359–1366. Epub 2007/04/07. https://doi.org/10.1093/hmg/ddm086.

Li Y, Xing Z, Yu T, Pao A, Daadi M, Yu YE. Coat Color-Facilitated Efficient Generation and Analysis of a Mouse Model of Down Syndrome Triplicated for All Human Chromosome 21 Orthologous Regions. *Genes (Basel).* 2021;12. Epub 2021/08/28. https://doi.org/10.3390/genes12081215. PMCID: PMC8393392.

Lindsay EA, Botta A, Jurecic V, Carattini-Rivera S, Cheah YC, Rosenblatt HM, Bradley A, Baldini A. Congenital heart disease in mice deficient for the DiGeorge syndrome region. *Nature.* 1999;401:379–383. Epub 1999/10/12. https://doi.org/10.1038/43900.

Lindsay EA, Vitelli F, Su H, Morishima M, Huynh T, Pramparo T, Jurecic V, Ogunrinu G, Sutherland HF, Scambler PJ, Bradley A, Baldini A. Tbx1 haploinsufficieny in the DiGeorge syndrome region causes aortic arch defects in mice. *Nature.* 2001;410:97–101. Epub 2001/03/10. https://doi.org/10.1038/35065105.

Liu C, Morishima M, Yu T, Matsui S, Zhang L, Fu D, Pao A, Costa AC, Gardiner KJ, Cowell JK, Nowak NJ, Parmacek MS, Liang P, Baldini A, Yu YE. Genetic analysis of Down syndrome-associated heart defects in mice. *Hum Genet.* 2011;130:623–632. Epub 2011/03/29. https://doi.org/10.1007/s00439-011-0980-2. PMCID: PMC3257027.

Liu C, Morishima M, Jiang X, Yu T, Meng K, Ray D, Pao A, Ye P, Parmacek MS, Yu YE. Engineered chromosome-based genetic mapping establishes a 3.7 Mb critical genomic region for Down syndrome-associated heart defects in mice. *Hum Genet.* 2014;133:743–753. Epub 2013/12/24. https://doi.org/10.1007/s00439-013-1407-z. PMCID: PMC4024075.

Liu C, Yu T, Xing Z, Jiang X, Li Y, Pao A, Mu J, Wallace PK, Stoica G, Bakin AV, Yu YE. Triplications of human chromosome 21 orthologous regions in mice result in expansion of megakaryocyte-erythroid progenitors and reduction of granulocyte-macrophage progenitors. *Oncotarget.* 2018;9:4773–4786. Epub 2018/02/13. https://doi.org/10.18632/oncotarget.23463. PMCID: PMC5797011.

Lyle R, Bena F, Gagos S, Gehrig C, Lopez G, Schinzel A, Lespinasse J, Bottani A, Dahoun S, Taine L, Doco-Fenzy M, Cornillet-Lefebvre P, Pelet A, Lyonnet S, Toutain A, Colleaux L, Horst J, Kennerknecht I, Wakamatsu N, Descartes M, Franklin JC, Florentin-Arar L, Kitsiou S, Ait Yahya-Graison E, Costantine M, Sinet PM, Delabar JM, Antonarakis SE. Genotype-phenotype correlations in Down syndrome identified by array CGH in 30 cases of partial trisomy and partial monosomy chromosome 21. *Eur J Hum Genet.* 2009;17:454–466. Epub 2008/11/13. https://doi.org/10.1038/ejhg.2008.214.

Maass PG, Barutcu AR, Rinn JL. Interchromosomal interactions: A genomic love story of kissing chromosomes. *J Cell Biol.* 2019;218:27–38. Epub 2018/09/06. https://doi.org/10.1083/jcb.201806052. PMCID: PMC6314556.

Marechal D, Lopes Pereira P, Duchon A, Herault Y. Dosage of the Abcg1-U2af1 region modifies locomotor and cognitive deficits observed in the Tc1 mouse model of Down syndrome. *PLoS One.* 2015;10:e0115302. Epub 2015/02/24. https://doi.org/10.1371/journal.pone.0115302. PMCID: PMC4338106.

Meharena HS, Marco A, Dileep V, Lockshin ER, Akatsu GY, Mullahoo J, Watson LA, Ko T, Guerin LN, Abdurrob F, Rengarajan S, Papanastasiou M, Jaffe JD, Tsai LH. Down-syndrome-induced senescence disrupts the nuclear architecture of neural progenitors. *Cell Stem Cell.* 2022;29:116–130 e117. Epub 2022/01/08. https://doi.org/10.1016/j.stem.2021.12.002. PMCID: PMC8805993.

Merscher S, Funke B, Epstein JA, Heyer J, Puech A, Lu MM, Xavier RJ, Demay MB, Russell RG, Factor S, Tokooya K, Jore BS, Lopez M, Pandita RK, Lia M, Carrion D, Xu H, Schorle H, Kobler JB, Scambler P, Wynshaw-Boris A, Skoultchi AI, Morrow BE, Kucherlapati R. TBX1 is responsible for cardiovascular defects in velo-cardio-facial/DiGeorge syndrome. *Cell.* 2001;104:619–629. Epub 2001/03/10.

Muller H, Gil J, Jr., Drinnenberg IA. The Impact of Centromeres on Spatial Genome Architecture. *Trends Genet.* 2019;35:565–578. Epub 2019/06/16. https://doi.org/10.1016/j.tig.2019.05.003.

Morris MA, Dudgeon BJ, Yorkston K. A qualitative study of adult AAC users' experiences communicating with medical providers. *Disabil Rehabil Assist Technol.* 2013;8:472–481. Epub 2013/01/29. https://doi.org/10.3109/17483107.2012.746398.

Narang MA, Dumas R, Ayer LM, Gravel RA. Reduced histone biotinylation in multiple carboxylase deficiency patients: a nuclear role for holocarboxylase synthetase. *Human molecular genetics.* 2004;13:15–23. Epub 2003/11/14. https://doi.org/10.1093/hmg/ddh006.

Nguyen TL, Duchon A, Manousopoulou A, Loaec N, Villiers B, Pani G, Karatas M, Mechling AE, Harsan LA, Limanton E, Bazureau JP, Carreaux F, Garbis SD, Meijer L, Herault Y. Correction of cognitive deficits in mouse models of Down syndrome by a pharmacological inhibitor of DYRK1A. *Dis Model Mech.* 2018;11. Epub 2018/08/18. https://doi.org/10.1242/dmm.035634. PMCID: PMC6176987.

Niebuhr E. Down Syndrome. The possibility of a pathogenetic segment on chromosome 21. *Humangenetik.* 1974;21:99–101.

O'Doherty A, Ruf S, Mulligan C, Hildreth V, Errington ML, Cooke S, Sesay A, Modino S, Vanes L, Hernandez D, Linehan JM, Sharpe PT, Brandner S, Bliss TV, Henderson DJ, Nizetic D, Tybulewicz VL, Fisher EM. An aneuploid mouse strain carrying human chromosome 21 with Down syndrome phenotypes. *Science.* 2005;309:2033–2037. Epub 2005/09/24. https://doi.org/10.1126/science.1114535. PMCID: PMC1378183.

Olson LE, Richtsmeier JT, Leszl J, Reeves RH. A chromosome 21 critical region does not cause specific Down syndrome phenotypes. *Science.* 2004;306:687–690. Epub 2004/10/23. https://doi.org/10.1126/science.1098992. PMCID: PMC4019810.

Omori S, Tanabe H, Banno K, Tsuji A, Nawa N, Hirata K, Kawatani K, Kokubu C, Takeda J, Taniguchi H, Arahori H, Wada K, Kitabatake Y, Ozono K. A Pair of Maternal Chromosomes Derived from Meiotic Nondisjunction in Trisomy 21 Affects Nuclear Architecture and Transcriptional Regulation. *Sci Rep.* 2017;7:764. Epub 2017/04/12. https://doi.org/10.1038/s41598-017-00714-7. PMCID: PMC5429678.

Patel A, Yamashita N, Ascano M, Bodmer D, Boehm E, Bodkin-Clarke C, Ryu YK, Kuruvilla R. RCAN1 links impaired neurotrophin trafficking to aberrant development of the sympathetic nervous system in Down syndrome. *Nat Commun.* 2015;6:10119. Epub 2015/12/15. https://doi.org/10.1038/ncomms10119. PMCID: PMC4682116.

Peiris H, Duffield MD, Fadista J, Jessup CF, Kashmir V, Genders AJ, McGee SL, Martin AM, Saiedi M, Morton N, Carter R, Cousin MA, Kokotos AC, Oskolkov N, Volkov P, Hough TA, Fisher EM, Tybulewicz VL, Busciglio J, Coskun PE, Becker A, Belichenko PV, Mobley WC, Ryan MT, Chan JY, Laybutt DR, Coates PT, Yang S, Ling C, Groop L, Pritchard MA, Keating

DJ. A Syntenic Cross Species Aneuploidy Genetic Screen Links RCAN1 Expression to beta-Cell Mitochondrial Dysfunction in Type 2 Diabetes. *PLoS Genet.* 2016;12:e1006033. Epub 2016/05/20. https://doi.org/10.1371/journal.pgen.1006033. PMCID: PMC4873152.

Pereira PL, Magnol L, Sahun I, Brault V, Duchon A, Prandini P, Gruart A, Bizot JC, Chadefaux-Vekemans B, Deutsch S, Trovero F, Delgado-Garcia JM, Antonarakis SE, Dierssen M, Herault Y. A new mouse model for the trisomy of the Abcg1-U2af1 region reveals the complexity of the combinatorial genetic code of down syndrome. *Human molecular genetics.* 2009;18:4756–4769. Epub 2009/09/29. https://doi.org/10.1093/hmg/ddp438. PMCID: PMC2778371.

Pinto B, Morelli G, Rastogi M, Savardi A, Fumagalli A, Petretto A, Bartolucci M, Varea E, Catelani T, Contestabile A, Perlini LE, Cancedda L. Rescuing Over-activated Microglia Restores Cognitive Performance in Juvenile Animals of the Dp(16) Mouse Model of Down Syndrome. *Neuron.* 2020;108:887–904 e812. Epub 2020/10/08. https://doi.org/10.1016/j.neuron.2020.09.010. PMCID: PMC7736620.

Pletcher MT, Roe BA, Chen F, Do T, Do A, Malaj E, Reeves RH. Chromosome evolution: the junction of mammalian chromosomes in the formation of mouse chromosome 10. *Genome Res.* 2000;10:1463–1467. Epub 2000/10/24. https://doi.org/10.1101/gr.146600. PMCID: PMC310982.

Postnikov Y, Bustin M. Regulation of chromatin structure and function by HMGN proteins. *Biochim Biophys Acta.* 2010;1799:62–68. Epub 2009/12/02. https://doi.org/10.1016/j.bbagrm.2009.11.016. PMCID: PMC2818575.

Pritchard M, Reeves RH, Dierssen M, Patterson D, Gardiner KJ. Down syndrome and the genes of human chromosome 21: current knowledge and future potentials. Report on the Expert workshop on the biology of chromosome 21 genes: towards gene-phenotype correlations in Down syndrome. Washington D.C., September 28–October 1, 2007. *Cytogenetic & Genome Research.* 2008;121:67–77. Epub 2008/06/12. https://doi.org/10.1159/000124384.

Puente-Bedia A, Berciano MT, Tapia O, Martinez-Cue C, Lafarga M, Rueda N. Nuclear Reorganization in Hippocampal Granule Cell Neurons from a Mouse Model of Down Syndrome: Changes in Chromatin Configuration, Nucleoli and Cajal Bodies. *Int J Mol Sci.* 2021;22. Epub 2021/01/31. https://doi.org/10.3390/ijms22031259. PMCID: PMC7865916.

Ramani V, Deng X, Qiu R, Lee C, Disteche CM, Noble WS, Shendure J, Duan Z. Sci-Hi-C: A single-cell Hi-C method for mapping 3D genome organization in large number of single cells. *Methods.* 2020;170:61–68. Epub 2019/09/20. https://doi.org/10.1016/j.ymeth.2019.09.012. PMCID: PMC6949367.

Ramirez-Solis R, Liu P, Bradley A. Chromosome engineering in mice. *Nature.* 1995;378:720–724. Epub 1995/12/14. https://doi.org/10.1038/378720a0.

Raveau M, Lignon JM, Nalesso V, Duchon A, Groner Y, Sharp AJ, Dembele D, Brault V, Herault Y. The App-Runx1 region is critical for birth defects and electrocardiographic dysfunctions observed in a Down syndrome mouse model. *PLoS Genet.* 2012;8:e1002724. Epub 2012/06/14. https://doi.org/10.1371/journal.pgen.1002724. PMCID: PMC3364940.

Raveau M, Polygalov D, Boehringer R, Amano K, Yamakawa K, McHugh TJ. Alterations of in vivo CA1 network activity in Dp(16)1Yey Down syndrome model mice. *Elife.* 2018;7. Epub 2018/02/28. https://doi.org/10.7554/eLife.31543. PMCID: PMC5841929.

Redhead Y, Gibbins D, Lana-Elola E, Watson-Scales S, Dobson L, Krause M, Liu KJ, Fisher EMC, Green JBA, Tybulewicz VLJ. Craniofacial dysmorphology in Down syndrome is caused by increased dosage of Dyrk1a and at least three other genes. *Development.* 2023;150. Epub 2023/04/27. https://doi.org/10.1242/dev.201077. PMCID: PMC10163349.

Reinholdt LG, Ding Y, Gilbert GJ, Czechanski A, Solzak JP, Roper RJ, Johnson MT, Donahue LR, Lutz C, Davisson MT. Molecular characterization of the translocation breakpoints in the Down syndrome mouse model Ts65Dn. *Mamm Genome.* 2011;22:685–691. Epub 2011/09/29. https://doi.org/10.1007/s00335-011-9357-z. PMCID: PMC3505986.

Russell WL, Russell LB, Kelly EM. Radiation dose rate and mutation frequency. *Science.* 1958;128:1546–1550. Epub 1958/12/19. https://doi.org/10.1126/science.128.3338.1546.

Sahun I, Marechal D, Pereira PL, Nalesso V, Gruart A, Garcia JM, Antonarakis SE, Dierssen M, Herault Y. Cognition and hippocampal plasticity in the mouse is altered by monosomy of a genomic region implicated in Down syndrome. *Genetics*. 2014;197:899–912. Epub 2014/04/23. https://doi.org/10.1534/genetics.114.165241. PMCID: PMC4096369.

Salehi A, Delcroix JD, Belichenko PV, Zhan K, Wu C, Valletta JS, Takimoto-Kimura R, Kleschevnikov AM, Sambamurti K, Chung PP, Xia W, Villar A, Campbell WA, Kulnane LS, Nixon RA, Lamb BT, Epstein CJ, Stokin GB, Goldstein LS, Mobley WC. Increased App expression in a mouse model of Down's syndrome disrupts NGF transport and causes cholinergic neuron degeneration. *Neuron*. 2006;51:29–42. https://doi.org/10.1016/j.neuron.2006.05.022.

Sawa M, Overk C, Becker A, Derse D, Albay R, Weldy K, Salehi A, Beach TG, Doran E, Head E, Yu YE, Mobley WC. Impact of increased APP gene dose in Down syndrome and the Dp16 mouse model. *Alzheimers Dement*. 2022;18:1203–1234. Epub 2021/11/11. https://doi.org/10.1002/alz.12463. PMCID: PMC9085977.

Schaeffer L, Lindner L, Pavlovic G, Herault Y, Birling MC. CRISMERE Chromosome Engineering in Mouse and Rat. *Methods Mol Biol*. 2023;2631:277–297. Epub 2023/03/31. https://doi.org/10.1007/978-1-0716-2990-1_12.

Sherman KM, Williams DK, Welsh CA, Cooper AM, Falck A, Huggins S, Bokhari RS, Gaddy D, McKelvey KD, Dawson LA, Suva LJ. Low bone mass and impaired fracture healing in mouse models of Trisomy21 (Down syndrome). *Bone*. 2022;162:116471. Epub 2022/06/19. https://doi.org/10.1016/j.bone.2022.116471. PMCID: PMC9356441.

Siddiqui A, Lacroix T, Stasko MR, Scott-McKean JJ, Costa AC, Gardiner KJ. Molecular responses of the Ts65Dn and Ts1Cje mouse models of Down syndrome to MK-801. *Genes Brain Behav*. 2008;7:810–820. Epub 2009/01/08. https://doi.org/10.1111/j.1601-183X.2008.00428.x. PMCID: PMC2677016.

Sinet PM, Theophile D, Rahmani Z, Chettouh Z, Blouin JL, Prieur M, Noel B, Delabar JM. Molecular mapping of the Down syndrome phenotype on chromosome 21. *Prog Clin Biol Res*. 1993;384:63–86. Epub 1993/01/01.

Singh MP, Wijeratne SS, Zempleni J. Biotinylation of lysine 16 in histone H4 contributes toward nucleosome condensation. *Arch Biochem Biophys*. 2013;529:105–111. Epub 2012/12/12. https://doi.org/10.1016/j.abb.2012.11.005. PMCID: PMC3563257.

Sloan K, Thomas J, Blackwell M, Voisard D, Lana-Elola E, Watson-Scales S, Roper DL, Wallace JM, Fisher EMC, Tybulewicz VLJ, Roper RJ. Genetic dissection of triplicated chromosome 21 orthologs yields varying skeletal traits in Down syndrome model mice. *Dis Model Mech*. 2023;16:dmm049927. Epub 2023/03/21. https://doi.org/10.1242/dmm.049927. PMCID: PMC10163323.

Smith AJ, De Sousa MA, Kwabi-Addo B, Heppell-Parton A, Impey H, Rabbitts P. A site-directed chromosomal translocation induced in embryonic stem cells by Cre-loxP recombination. *Nat Genet*. 1995;9:376–385. Epub 1995/04/01. https://doi.org/10.1038/ng0495-376.

Souchet B, Duchon A, Gu Y, Dairou J, Chevalier C, Daubigney F, Nalesso V, Creau N, Yu Y, Janel N, Herault Y, Delabar JM. Prenatal treatment with EGCG enriched green tea extract rescues GAD67 related developmental and cognitive defects in Down syndrome mouse models. *Sci Rep*. 2019;9:3914. Epub 2019/03/10. https://doi.org/10.1038/s41598-019-40328-9. PMCID: PMC6408590.

Starbuck JM, Dutka T, Ratliff TS, Reeves RH, Richtsmeier JT. Overlapping trisomies for human chromosome 21 orthologs produce similar effects on skull and brain morphology of Dp(16)1Yey and Ts65Dn mice. *Am J Med Genet A*. 2014;164A:1981–1990. Epub 2014/05/03. https://doi.org/10.1002/ajmg.a.36594. PMCID: PMC4107150.

Stransky ML, Jensen KM, Morris MA. Adults with Communication Disabilities Experience Poorer Health and Healthcare Outcomes Compared to Persons Without Communication Disabilities. *J Gen Intern Med*. 2018;33:2147–2155. Epub 2018/08/26. https://doi.org/10.1007/s11606-018-4625-1. PMCID: PMC6258615.

Stringer M, Goodlett CR, Roper RJ. Targeting trisomic treatments: optimizing Dyrk1a inhibition to improve Down syndrome deficits. *Mol Genet Genomic Med.* 2017;5:451–465. Epub 2017/09/26. https://doi.org/10.1002/mgg3.334. PMCID: PMC5606891.

Sullivan KD, Lewis HC, Hill AA, Pandey A, Jackson LP, Cabral JM, Smith KP, Liggett LA, Gomez EB, Galbraith MD, DeGregori J, Espinosa JM. Trisomy 21 consistently activates the interferon response. *Elife.* 2016;5. Epub 2016/07/30. https://doi.org/10.7554/eLife.16220. PMCID: PMC5012864.

Thomas JR, LaCombe J, Long R, Lana-Elola E, Watson-Scales S, Wallace JM, Fisher EMC, Tybulewicz VLJ, Roper RJ. Interaction of sexual dimorphism and gene dosage imbalance in skeletal deficits associated with Down syndrome. *Bone.* 2020;136:115367. Epub 2020/04/20. https://doi.org/10.1016/j.bone.2020.115367. PMCID: PMC7262595.

Torres EM, Sokolsky T, Tucker CM, Chan LY, Boselli M, Dunham MJ, Amon A. Effects of aneuploidy on cellular physiology and cell division in haploid yeast. *Science.* 2007;317:916–924. Epub 2007/08/19. https://doi.org/10.1126/science.1142210.

Tosh J, Tybulewicz V, Fisher EMC. Mouse models of aneuploidy to understand chromosome disorders. *Mamm Genome.* 2022;33:157–168. Epub 2021/11/02. https://doi.org/10.1007/s00335-021-09930-z. PMCID: PMC8913467.

Tsang CK, Liu Y, Thomas J, Zhang Y, Zheng XF. Superoxide dismutase 1 acts as a nuclear transcription factor to regulate oxidative stress resistance. *Nat Commun.* 2014;5:3446. Epub 2014/03/22. https://doi.org/10.1038/ncomms4446. PMCID: PMC4678626.

Van Deursen J, Fornerod M, Van Rees B, Grosveld G. Cre-mediated site-specific translocation between nonhomologous mouse chromosomes. *Proc Natl Acad Sci U S A.* 1995;92:7376–7380. Epub 1995/08/01. https://doi.org/10.1073/pnas.92.16.7376. PMCID: PMC41342.

Watson-Scales S, Kalmar B, Lana-Elola E, Gibbins D, La Russa F, Wiseman F, Williamson M, Saccon R, Slender A, Olerinyova A, Mahmood R, Nye E, Cater H, Wells S, Yu YE, Bennett DLH, Greensmith L, Fisher EMC, Tybulewicz VLJ. Analysis of motor dysfunction in Down Syndrome reveals motor neuron degeneration. *PLoS Genet.* 2018;14:e1007383. Epub 2018/05/11. https://doi.org/10.1371/journal.pgen.1007383. PMCID: PMC5963810.

Waugh KA, Minter R, Baxter J, Chi C, Galbraith MD, Tuttle KD, Eduthan NP, Kinning KT, Andrysik Z, Araya P, Dougherty H, Dunn LN, Ludwig M, Schade KA, Tracy D, Smith KP, Granrath RE, Busquet N, Khanal S, Anderson RD, Cox LL, Estrada BE, Rachubinski AL, Lyford HR, Britton EC, Fantauzzo KA, Orlicky DJ, Matsuda JL, Song K, Cox TC, Sullivan KD, Espinosa JM. Triplication of the interferon receptor locus contributes to hallmarks of Down syndrome in a mouse model. *Nat Genet.* 2023;55:1034–1047. Epub 2023/06/06. https://doi.org/10.1038/s41588-023-01399-7. PMCID: PMC10260402 Inc. and serves in the advisory board of Perha Pharmaceuticals. The remaining authors declare no competing interests.

Wiltshire T, Pletcher M, Cole SE, Villanueva M, Birren B, Lehoczky J, Dewar K, Reeves RH. Perfect conserved linkage across the entire mouse chromosome 10 region homologous to human chromosome 21. *Genome Res.* 1999;9:1214–1222. Epub 1999/12/30. https://doi.org/10.1101/gr.9.12.1214. PMCID: PMC311004.

Xing Z, Li Y, Pao A, Bennett AS, Tycko B, Mobley WC, Yu YE. Mouse-based genetic modeling and analysis of Down syndrome. *Br Med Bull.* 2016;120:111–122. Epub 2016/10/30. https://doi.org/10.1093/bmb/ldw040. PMCID: PMC5146682.

Xing Z, Li Y, Cortes-Gomez E, Jiang X, Gao S, Pao A, Shan J, Song Y, Perez A, Yu T, Highsmith MR, Boadu F, Conroy JM, Singh PK, Bakin AV, Cheng J, Duan Z, Wang J, Liu S, Tycko B, Yu YE. Dissection of a Down syndrome-associated trisomy to separate the gene dosage-dependent and -independent effects of an extra chromosome. *Human molecular genetics.* 2023;32:2205–2218. Epub 2023/04/05. https://doi.org/10.1093/hmg/ddad056. PMCID: PMC10281752.

Yu Y, Bradley A. Engineering chromosomal rearrangements in mice. *Nat Rev Genet.* 2001;2:780–790. Epub 2001/10/05. https://doi.org/10.1038/35093564.

Yu T, Clapcote SJ, Li Z, Liu C, Pao A, Bechard AR, Carattini-Rivera S, Matsui S, Roder JC, Baldini A, Mobley WC, Bradley A, Yu YE. Deficiencies in the region syntenic to human 21q22.3 cause cognitive deficits in mice. *Mamm Genome.* 2010a;21:258–267. Epub 2010/06/01. https://doi.org/10.1007/s00335-010-9262-x. PMCID: PMC3023176.

Yu T, Li Z, Jia Z, Clapcote SJ, Liu C, Li S, Asrar S, Pao A, Chen R, Fan N, Carattini-Rivera S, Bechard AR, Spring S, Henkelman RM, Stoica G, Matsui S, Nowak NJ, Roder JC, Chen C, Bradley A, Yu YE. A mouse model of Down syndrome trisomic for all human chromosome 21 syntenic regions. *Human molecular genetics.* 2010b;19:2780–2791. Epub 2010/05/06. https://doi.org/10.1093/hmg/ddq179. PMCID: PMC2893810.

Yu T, Liu C, Belichenko P, Clapcote SJ, Li S, Pao A, Kleschevnikov A, Bechard AR, Asrar S, Chen R, Fan N, Zhou Z, Jia Z, Chen C, Roder JC, Liu B, Baldini A, Mobley WC, Yu YE. Effects of individual segmental trisomies of human chromosome 21 syntenic regions on hippocampal long-term potentiation and cognitive behaviors in mice. *Brain Res.* 2010c;1366:162–171. Epub 2010/10/12. https://doi.org/10.1016/j.brainres.2010.09.107. PMCID: PMC3027718.

Zampieri BL, Fernandez F, Pearson JN, Stasko MR, Costa AC. Ultrasonic vocalizations during male-female interaction in the mouse model of Down syndrome Ts65Dn. *Physiol Behav.* 2014;128:119–125. Epub 2014/02/19. https://doi.org/10.1016/j.physbeh.2014.02.020.

Zhang L, Meng K, Jiang X, Liu C, Pao A, Belichenko PV, Kleschevnikov AM, Josselyn S, Liang P, Ye P, Mobley WC, Yu YE. Human chromosome 21 orthologous region on mouse chromosome 17 is a major determinant of Down syndrome-related developmental cognitive deficits. *Human molecular genetics.* 2014;23:578–589. Epub 2013/09/18. https://doi.org/10.1093/hmg/ddt446. PMCID: 3888256.

Zheng B, Sage M, Cai WW, Thompson DM, Tavsanli BC, Cheah YC, Bradley A. Engineering a mouse balancer chromosome. *Nat Genet.* 1999a;22:375–378. Epub 1999/08/04. https://doi.org/10.1038/11949.

Zheng B, Mills AA, Bradley A. A system for rapid generation of coat color-tagged knockouts and defined chromosomal rearrangements in mice. *Nucleic Acids Res.* 1999b;27:2354–2360. Epub 1999/05/15. https://doi.org/10.1093/nar/27.11.2354. PMCID: PMC148802.

Humanized Rodent Models of Down Syndrome

Roger H. Reeves

Abstract Mouse and rat models have been created that carry most of human chromosome 21 (HSA21). These models have a nearly complete set of HSA21 genes and show little or no mosaicism. Transcript levels of HSA21 genes are regulated similarly to their mouse counterparts (orthologs) reflecting levels in Down syndrome-derived cells. Surveys of phenotypes reflect Down syndrome -related features that also occur in other mouse models. Rat tissues are generally more accessible than those in mice and rats historically have provided a rich and sensitive platform for behavioral studies involving learning and memory and for preclinical testing of pharmaceuticals.

Keywords Rat · Mouse · Mammal · Trisomy · Down syndrome · Animal model · Next generation

The most complete genetic representation of HSA21 is, of course, that chromosome itself. When introduced into a mouse (or rat), the resulting animal will have two copies of the mouse HSA21 homologous gene and one human copy of each conserved gene, creating dosage imbalance for that gene. While nearly complete in terms of gene content, the expression of human transcripts and proteins in a mouse carries its own caveats, which we consider here. Within this discussion we use the formal definition of trisomy which is the presence of an extra, freely segregating chromosome, as distinct from a model of gene dosage imbalance produced by duplication of a chromosome region.

R. H. Reeves (✉)
Department of Physiology, Johns Hopkins University School of Medicine,
Baltimore, MD, USA
e-mail: rreeves@jhmi.edu

B. Ye, R. H. Reeves (eds.), *Genetic Models of Down Syndrome*,
https://doi.org/10.1007/978-3-031-78611-2_3

1 Humanized Mouse Models for Down Syndrome

The advantages of mammalian models for the study of trisomy 21 have been discussed at length for several decades. Animal models give access to every tissue at all stages of life. Down syndrome has a profound effect on development, and several major hallmarks of the syndrome can be traced directly to the impact of trisomy on pre-, peri-, and postnatal development. These processes, especially those occurring prenatally, cannot be studied in people. Invasive, controlled experimental interventions provide data essential to preclinical development of therapeutic molecules and approaches. Stem cells and organoids can model some aspects of cell and tissue differentiation but do not create functional organs. These approaches cannot produce a functioning heart, demonstrate susceptibility or resistance to disease, or respond to stimuli eliciting specific behaviors. Complex higher order responses only occur in a living organism.

Two true aneuploid mouse models (with independently segregating extra chromosomes) have been studied extensively (see preceding chapters by Roper and Yu/ Chaps. 1 and 2). The first is Ts65Dn, identified in a screen for radiation-induced translocations by Muriel Davisson (1993). In Ts65Dn, most of the HSA21-orthologous region of mouse chromosome 16 (MMU16) is translocated near the centromere of MMU17. This model shows a number of parallels in development and function to people with trisomy 21. However, the presence of trisomic genes from proximal MMU17 complicates interpretation of results from this model. The second previous aneuploid mouse model, Tc1, contains a partial HSA21 creating mosaic Down syndrome and is discussed below (O'Doherty et al. 2005). Mosaicism for HSA21 occurs in about 3–5% of people with a clinical presentation of Down syndrome (Rubenstein et al. 2024). Due to the individual variation arising from mosaicism, Tc1 is no longer a leading choice for analysis of the etiology of Down syndrome phenotypes. Two new rodent models discussed here provide additional genetic representation of trisomy 21 in aneuploid mouse and rat models with low or no mosaicism.

Transfer of an intact HSA21 into embryonic stem cells (ESCs) between species was first demonstrated by Oshimura and colleagues using microcell-mediated chromosome transfer (MMCT) (Tomizuka et al. 1997). In this approach, cultured cells are arrested at metaphase. With an extended metaphase block, nuclear membranes eventually reform resulting in micronuclei, each containing few (even single) chromosomes. These can be recovered as microcells that have both a cell and a nuclear membrane. Using somatic cell hybrid fusion techniques, the microcells are fused to mouse ESCs. A detailed description of the MMCT process can be found in (Suzuki et al. 2020).

Tissue culture methods to maintain ESCs with the ability to form and populate the germ line of chimeras are exacting. The Oshimura group obtained several ESC lines that contained HSA21 and formed chimeras (Shinohara et al. 2001). Chimeras with a high percentage of HSA21-containing cells showed some effects of dosage imbalance seen in other Down syndrome models, but no chimeras were recovered in which the manipulated ESCs populated the germ line, i.e., there was no germ line transmission of cells that contained HSA21.

E.M.C. Fisher and colleagues observed that the Ts65Dn mouse model, which has trisomy, (defined as a freely segregating chromosome adding an extra centromere to the normal complement, as opposed to direct duplications that produce dosage imbalance but not additional "mechanical" effects; *see* Chaps. 1 and 2) transmitted the extra chromosome efficiently through the female but not the male germ line[1]; in people, trisomy for most chromosomes is more commonly or exclusively inherited maternally, as well (Hassold and Hunt 2001). Fisher et al. used MMCT with female ESCs and succeeded in creating the Tc1 mouse strain (B6129S-Tc(HSA21)1TybEmcf/J), in which females are able to transmit HSA21 to their offspring.

The original vision for this experiment (and for Oshimura's) included irradiation of cells donating the HSA21 in order to produce a panel of mice with different deleted chromosomes 21 that would be useful for correlating Down syndrome-related phenotypes with subsets of HSA21 genes (O'Doherty et al. 2005; Shinohara et al. 2001). As a result of this treatment, the HSA21 chromosome in Tc1 mice was later shown by whole genome sequencing to contain further small deletions, duplications, and inactivating mutations in a number of genes on the transferred HSA21 (Gribble et al. 2013). This analysis indicated that Tc1 mice retain roughly 70% of HSA21 genes as functional (non-mutated) single copies. About 5% of the human genes are present in more than one copy on the transferred HSA21 due to duplications, resulting in tetrasomy.

The most significant caveat with the Tc1 model is a substantial level of mosaicism (O'Doherty et al. 2005). That is, the oocyte contributes a copy of HSA21 to the embryo, but the human chromosome is lost randomly during development. Thus, all progeny of that cell are euploid with a normal mouse genome. The single time it was measured in Tc1, about half of adult cells in several tissues retained the partial HSA21.The time of loss and the cells affected will vary in every individual, creating a unique pattern of trisomic and eusomic cells with attendant differences in cell–cell communication during development and through life. While the Tc1 mouse made a significant contribution to understanding the impact of HSA21 gene dosage for many years, it is no longer recommended for studies of Down syndrome. As discussed below, a subset (20–30%) of TcRAT21 rats also show mosaicism, although the majority of trisomic rats retain HSA21 in most or all cells. TcMAC21 mice contain HSA21q as a mouse artificial chromosome and are not mosaic.

Those using somatic cell hybrids for various types of genetic mapping and complementation studies long noted the instability of human chromosomes in a mouse genetic background. The Oshimura group spent several years optimizing production of mouse artificial chromosomes (MAC) containing a mouse centromere and telomeres. Using this system, they inserted a nearly intact long arm of HSA21 into

[1] A subline of Ts65Dn in which males efficiently transmit the extra chromosome was later obtained by selective breeding (Moore et al. (2010).

mice. The MAC contains HSA21q, (*POTED* to *PRMT2*) where most of the protein coding genes are located. Again employing MMCT, Kazuki, Oshimura and colleagues successfully transferred the MAC to female ESCs, and this time the recovered chimeras transmitted the chromosome, establishing the TcMAC21 mouse line (Tc(HSA21,CAG-EGFP)1Yakaz/J (Kazuki et al. 2020). These mice contain 93% of HSA21 protein coding genes (PCG). A GFP gene inserted adjacent to the centromere provided by the MAC vector provides a useful marker for trisomy. When used to assess cells from ten different organs including nucleated peripheral blood cells, little or no mosaicism was detected. A number of phenotypic manifestations that are analogous to those in people with trisomy 21 and that have been seen in other mouse models are displayed in TcMAC21 mice, validating the occurrence of gene dosage effects.

These mice can be ordered from The Jackson Laboratory, https://www.jax.org/strain/035561 and from Riken, Strain RBRC05796. Important differences in phenotypes have been noted previously in "the same" mice raised in different commercial colonies. Like other true aneuploid strains (i.e., with an extra freely segregating chromosome) TcMAC21 cannot be inbred. The mice are available from both sites on a B6.D2 advanced intercross background. The Jackson Laboratory strain is currently being bred onto a B6.C3H background similar to that used previously for many trisomic and direct duplication mice.

2 Rat Models for Down Syndrome

As is the case with mouse models, there are two varieties of rat models relevant to Down syndrome. The first is a dosage imbalance model created by direct duplication of chromosome segments conserved between rat and human. As discussed, this type of model does not contain an extra chromosome and thus is not an actual trisomy. The second type introduces a functional, freely segregating chromosome, thereby introducing an extra centromere, telomeres and the attendant requirements for replication, pairing, and segregation at meiosis and mitosis, chromatin interactions, etc.

2.1 RNO11 and RNO20

HSA21 orthologous genes are found in two segments of the rat genome, on rat chromosomse (RNO) 11 and RNO20. Direct duplication of both of these segments has been accomplished by the CRISPR-based CRISMERE procedure as well as "classical" duplication/deletion to produce Rno11_DupLipi-Zfp295/YahIcsOrl and Rno20_DupUmodl1-Prmt2/YahIcsOrl rat strains (Birling et al. 2017). In addition, several genes of interest that are believed to contribute to the risk of Down syndrome-like phenotypes have been modified in substrains of these rats. Limited phenotyping of RNO20 duplication rats has been reported (Panagaki et al. 2022). The RNO11 and RNO20 models are both available from EMMA-INFRAFRONTIER, https://www.infrafrontier.eu/emma/strain-search/.

2.2 *TcRAT21*

The Kazuki/Oshimura group recently described another unique rodent model for Down syndrome, TcRAT21 (Kazuki et al. 2022). As opposed to the mouse artificial chromosome in TcMAC21 mouse, a complete HSA21 was utilized in making this rat model. The HSA21 came from the same somatic cell hybrid cell line as that used in TcMAC21 mice, and was marked by a constitutively expressed GFP introduced near the centromere. Again using MMCT, the human chromosome was introduced into male rat rBLK2i-1 ESCs. Chimeras were produced from these cells, including one male. Rather than attempt to transmit through breeding, round spermatids were recovered from GFP+ seminiferous tubules and injected into rat oocytes. Those that cleaved were then reintroduced to pseudopregnant females. No fertile male TcRAT21 animals have been observed; however, the human chromosome is transmitted through trisomic females. Of the first 3000 pups born, 31% were trisomic based on the presence of the GFP marker on the HSA21, and the number of trisomic males and females was roughly equal (48% and 52%).

Rodents appear to be much more tolerant of trisomy than are people. Trisomy 21 is not very compatible with human embryonic development; estimates of spontaneous abortion (miscarriage) are as high as 80% in human beings (Hassold and Hunt 2001). Because TcMAC21 females have small litters with a lower-than-expected number of trisomic pups, all trisomic females are typically retained for mating, and studies of the number of resorbed implantation sites have not yet been reported, which would indicate whether the sub-Mendelian occurrence of trisomic pups is due to embryonic loss or reduced number or efficacy of sperm carrying HSA21. It is interesting to note that in vitro fertilization with sperm from a male-fertile Ts65Dn substrain (Moore et al. 2010) results in a sub-Mendelian frequency of trisomic offspring suggesting the sperm formation and/or motility may be impaired by the presence of an extra chromosome.

Rats appear to tolerate the human centromere better than do mice, but 20–30% of GFP+ TcRAT21 offspring show significant mosaicism in peripheral blood lymphocytes. (Note: in some trisomic models, this tissue—blood—is especially prone to a loss of trisomic cells/ increase in the percentage of euploid cells with age (Cox et al. 1984; Gearhart et al. 1986). FISH analysis shows a very high degree of HSA21 retention across nine different solid tissues of trisomic rats (90–100%). Assessment of peripheral blood lymphocytes by flow cytometry at 1 month of age shows that about 65% of GFP+ animals retain HSA21 in >90% of lymphocytes, however, the remaining 35% retain HSA21 in fewer than 90%, i.e., they are mosaic (Kazuki et al. 2022). This mosaicism must be accounted for in any assessments of these animals, at least in target tissues and preferably overall since frequencies can vary considerably in different cell types. At the time of this publication, TcRAT21 is available from the NRBP Kyoto Rat Resource, search Strain name: STOCK Tc(HSA21)1Yakaz, http://www.anim.med.kyoto-u.ac.jp/nbr/strainsx/Strains_d.aspx?StrainID=1360&StrainsDir=Asc&StrainsPageSize=20&s_Strainname=Tc(HSA21)1Yakaz. Distribution in the West is planned from University of Missouri Mutant Mouse Resource & Research Center.

3 Genes, Orthologs, and Expression

Genes that are contiguous on HSA21 have mouse orthologs that occur in three blocks of conserved synteny found on mouse chromosomes MMU16, 17 and 10 (see Chap. 1). Comparison of human (build GRCh38.p12), rat (Rnor_6.0), and mouse (GRCm38.p6) PCG content excluding KRTAP sequences identifies 158 PCG in common between the three species (Kazuki et al. 2022). We may consider three classes of genes from HSA21, 164 genes that are recognized by Ensembl as protein coding (PCG); repeated pseudogenes, specifically the ca. 50 genes and pseudogenes of the keratin-related protein clusters in each species (KRTAP genes) which do not align precisely; and ~300+ non-PCG for which transcripts are observed. Five human PCG have no orthologs in mouse or rat and between them, mouse and rat have 48 PCG that are not present in human. Thus, *no combination of rodent direct duplication models can create dosage imbalance precisely equivalent to trisomy 21, and models containing a human chromosome will include and express (from a single allele) several genes not present in rodents*. A table with active links of each human, rat or mouse gene to its Ensembl record is available in Supplemental Table S4B of reference (Kazuki et al. 2022) at: https://www.cell.com/ajhg/fulltext/S0002-9297(21)00470-5?_returnURL=https%3A%2F%2Flinkinghub.elsevier.com%2Fretrieve%2Fpii%2FS0002929721004705%3Fshowall%3Dtrue#supplementaryMaterial.

3.1 Effective Gene Content in Mouse and Rat Models

For purposes of modeling Down syndrome, functional genes (i.e., not deleted, disabled by mutation, etc.) and their expression are immediately relevant. Given the primitive state of our understanding of the great majority of non-PCG and the difficulty in studying them, current focus is almost exclusively on PCG and indeed, is highly biased toward consideration of a handful of HSA21 PCG. There is a single published description of PCG in the TcMAC21 mouse and one in TcRAT21. The reader is referred to the original descriptions of these models from which the following summary is derived (Kazuki et al. 2020, 2022). Note that there is not an exact consensus on PCG number in mouse, rat or human. This is due in part to a large cluster of KRTAP genes (~50 sequences) whose number varies between rodents and people and of which the majority are pseudogenes.

3.2 *Number and Identity of Intact Orthologous PCG*

The mouse artificial chromosome (MAC) in TcMAC21 mice includes the long arm of HSA21, HSA21q. That chromosome was last in a person in the 1990s when an unidentified Japanese male donated cells to produce somatic cell hybrids (Kugoh et al. 1999). WGS of TcMAC21 reveals four deletions that presumably occurred during its 20+ year sojourn in cell culture and the vicissitudes of serial MMCT events during construction of the model. Whole genome sequence (WGS analysis) and assembly based on GRCh38.p12 (BioMart-Ensembl, May 2019) identified 199 of 213 annotated protein coding genes from HSA21q (93%) as present and intact in the TcMAC21 mouse. A list of encoded HSA proteins based on the DNA sequence is presented in (Kazuki et al. 2020).

TcRAT21 carries an HSA21 chromosome derived from the same somatic cell hybrid as that in TcMAC21 mice and its HSA21 shares one deletion. A second deletion in the TcRAT21 copy of HSA21 is unique to the rat model, while three of the deletions in the mouse-borne HSA21 are absent from rat. In all, 214 annotated HSA21 PCGs were identified as present and intact in the rat (Kazuki et al. 2022).

4 Allelic Variation in Human and Mouse Chromosomes

WGS analysis of HSA21 in TcMAC21 revealed 91 single nucleotide variants (SNVs). All but one of these were known human variants designated as native alternative alleles while the unique SNV was considered "likely to be tolerated" (Kazuki et al. 2020). This assessment applies to the human chromosome 21 in TcRAT21, as well.

Previous mouse models of Down syndrome have primarily been maintained on an inbred C57BL/6 J (B6) background or as an advanced intercross between B6 and one of various C3H substrains. The advanced intercross provides a more robust background to support trisomies which are generally difficult to breed and less robust than littermates and male sterile; human males carrying an extra chromosome are usually non-fertile, as well. Fertile male Ts65Dn mice were derived by selective breeding and provided significant practical advantages as well a better representation of the impact of trisomy on a somewhat diverse genetic background, more similar to the case in people (Moore et al. 2010). Efforts to identify male fertile TcMAC21 mice show promise. Like other trisomic models, TcMAC21 cannot be inbred for more than one or two generations. The strain was established on a different genetic background than previous models, an advanced intercross between B6/J and DBA/J. Efforts are underway to move the MAC onto a B6.C3 background to support comparison with previous decades of research in other trisomic mouse models. This might have a beneficial impact on husbandry, as well.

The RNO11 and 20 direct duplication models and the TcRAT21 are maintained on outbred Sprague-Dawley strain background. Males display reduced fertility in all three rat models.

5 Assessing Transcript Levels in Humanized Trisomic Rodents

Natural DNA sequence variation within species coupled with interspecific differences—human and mouse—may affect gene expression due to variation in the efficiency with which transcription factors, enhancers, etc. recognize and promote transcription and message processing. Species-specific amino acid differences could affect the stoichiometry of formation of protein complexes and small changes in the rates of enzymatic reactions due to such species differences could have a significant cumulative effect on development and function in transchromosomic mouse models. These changes are not necessarily dosage sensitive, but are more broadly analogous to within-species coding sequence variation that likely contributes to the variable presentation of Down syndrome in different individuals.

5.1 *Transcripts*

Several questions arise when a human chromosome is present in a rodent. Is transcription appropriately controlled? This can be considered at the tissue level as a function of total steady state transcripts of the human gene and its mouse orthologs. Bulk RNAseq was performed on forebrains from TcMAC21 mice and results compared with the same tissue from euploid littermates (n = 2 and 3, respectively) (Kazuki et al. 2020). In forebrain RNA of control euploid mice, 95% of the mouse orthologs of HSA21q genes from MMU 16, 17, and 10 (106/117) were detected at ≥1 fragment per kilobase of transcript per million mapped reads (FPKM). FPKM is a measure that normalizes the number of times a portion of a transcript is detected to its size. The same MMU16, 17 and 10 genes were expressed at similar levels in TcMAC21 forebrain (80–120% of the transcripts levels seen in euploid). In other words, the presence of the MAC did not greatly affect expression levels of most of the orthologous mouse genes. The remaining 5% of the mouse orthologs were present but reduced (2 genes) or over-expressed (9 genes) by more than 20% of euploid. Whether these greater deflections are noise or represent a true impact of trisomy remains to be established.

Forebrain steady state transcript levels of human PCG from the MAC were expressed at about 50% of the level of mouse genes on average (Kazuki et al. 2020). This is consistent with gene dosage in transchromosomic mice which have a single copy of the human genes and two copies of their mouse orthologs.[2] As a first

[2] While the average *collective* relative expression of HSA21 genes or their orthologs in trisomic mouse models is consistently around 50% of each euploid ortholog, roughly a third of overrepresented genes are expressed at higher than 50% of expected euploid levels and about a third are expressed at less than 50%. In other words, gene dosage, the number of templates available for transcription, does not provide a precise universal explanation of altered steady state transcript levels in DOWN SYNDROME.

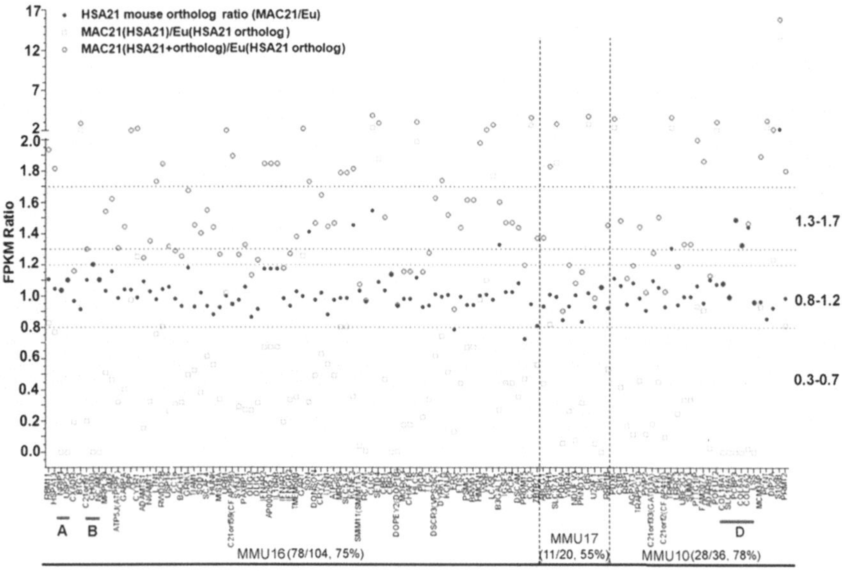

Fig. 1 HSA21 dosage imbalance analysis of TcMAC21 among 117 HSA21 mouse orthologs whose FPKM ≥1 in Eu. Three expression values are shown: (1) the FPKM ratio of HSA21 PCG of TcMAC21 to its ortholog in Eu mice (gray open squares); (2) the FPKM ratio of HSA21 mouse ortholog of TcMAC21 to that of Eu (blue dot); (3) the FPKM ratio of total expression (HSA21 PCG + its mouse ortholog) of TcMAC21 to the HSA21 mouse ortholog of Eu (red circles). The positions of HSA21q deleted regions are indicated in red. Eu and TcMAC21 littermates, $n = 3$ and 2, respectively. (Figure reproduced from Kazuki et al. 2020, used with permission from eLife)

approximation of trisomic transcript levels in TcMAC21, the FPKM of human and mouse transcripts were simply added (Fig. 1). On average across the entire gene set, the transcript levels were about 150% of euploid (130–170%). Measured individually, roughly 1/3 of the genes were expressed at less than 130%, 1/3 in the 130–170 range, and 1/3 greater than 170%. This is comparable to expression levels of trisomic mouse genes in other models of Down syndrome.

An analogous assessment was performed in TcRAT21 forebrain with comparable results (Kazuki et al. 2022). The availability of datasets from the same experiment in two different rodent models allowed a check on the utility of the "added FPKM" assessment method. The mouse + human transcript levels of fifty-five orthologous trisomic genes that were moderately to highly expressed (FPKM ≥10) in euploid forebrain and intact in both models were compared to euploid transcript levels in each model, and those ratios were examined. The overall expression pattern for most trisomic genes was not significantly different between mouse and rat (Wilcoxon matched pairs signed rank test, $p = 0.30$). Thus transcriptional regulation of HSA21 genes is largely consistent across two rodent species. As the subset of HSA21 genes expressed (or not) in trisomic mouse forebrain largely mirrored the subset of mouse orthologs expressed in that tissue, the simplest interpretation is that

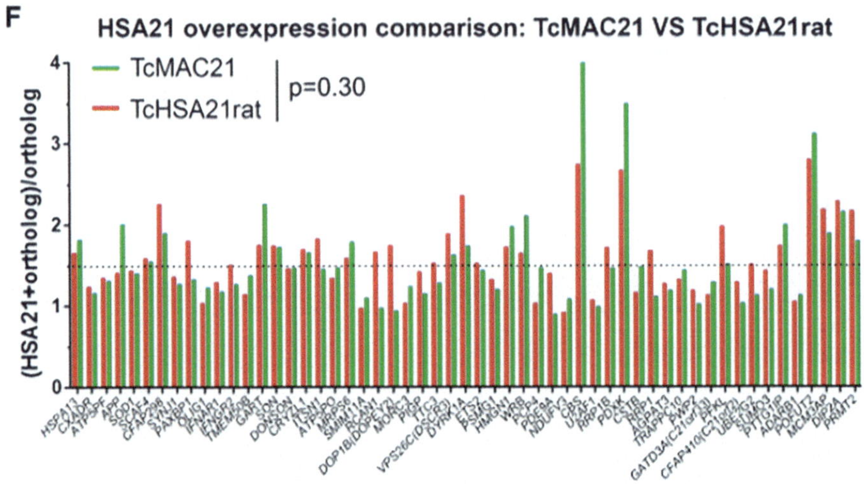

Fig. 2 RNAseq detects similar gene dosage effects on HSA21 transcript levels in rat and mouse models. Overexpression of HSA21 forebrain transcripts from TcHSA21RAT (red) and TcMAC21 mouse (green) for each ortholog is the ratio of the sum of FPKM for human plus rodent orthologs compared to rodent alone. The level of expression in the two models is not significantly different (Wilcoxin matched pairs signed-rank test, $p = 0.30$). (Figure reproduced from Kazuki et al. 2022 with permission)

regulation of HSA21 PCG transcription is broadly retained in ectopic rodent background.

Trisomy 21 and mouse models disrupt transcription of thousands of genes throughout the human and mouse genomes, respectively (Do et al. 2017; Letourneau et al. 2014; Saran et al. 2003). This is not surprising; for example, seven transcription factors are encoded on HSA21 along with many DNA-binding proteins that directly regulate transcription in trans. It would be surprising if this increase in levels of regulatory proteins did not affect gene expression outside of HSA21. Altered expression of disomic genes not orthologous to HSA21 is seen in both TcMAC21 mouse and TcRAT21 (Kazuki et al. 2020, 2022) (Fig. 2).

5.2 *Human Proteins in a Rodent*

While steady state transcript levels have a general correlation with protein levels in the handful of cases that have been examined in some detail, overall understanding of protein dynamics lags far behind that of nucleic acids. Changes due to relative over- or under-expression of proteins (either in molarity or in impact on function) must be operative in the expression of features diagnostic of Down syndrome. A survey using mass spectroscopy identified changes in ts21+ fibroblasts vs. isogenic controls including excessive degradation of proteins from various complexes in trisomic cells which affect their overall composition in ways not predicted from assessment of transcripts (Liu et al. 2017). Only a tiny fraction of proteins have been considered in depth. Fundamental research in this area will be informative across all of biology and medicine and is critical to understanding the effects of gene dosage imbalance in Down syndrome. Cross-species modeling may prove to be an especially useful tool in these investigations.

Variation in phenotypic outcome may be expected from evolved differences in protein sequence between human and rodents. In some cases, these variations may be effectively analogous to natural variation (single nucleotide variant, SNV) within species which can affect function, but is not demonstrated outside of a few specific diseases. Active sites of enzymes are often highly conserved across species, but amino acid residue changes that have subtle effects on protein structure may result in small changes in kinetics. Transcription factors are another class of proteins that frequently show highly conserved DNA binding sites, but may vary subtly in ability to bind cofactors that affect rates of transcription. Even synonymous differences in coding sequences between mouse and human that do not change the protein sequence but might alter secondary structures of mRNA and will require different tRNAs for translation might have an impact on the rate of translation, with downstream effects on kinetics of protein availability for further interactions. Many such changes, occurring below most current methods, even to detect them, could be consequential across tens of millions of differentiating cells.

6 Phenotypes

A similar survey of phenotypes expressed in TcMAC21 mouse and TcRAT21 was utilized to provide a baseline to inform further investigations.

A fundamental challenge of human disease modeling of any sort is to define how comparisons can be made between the model and the condition under investigation. We can now assess gene expression at the transcript level down to the level of single cells and apply quantitative measures of allele preference. Important first steps have been taken to understand expression at the protein level. However, the most complete understanding in a model is limited if the outcomes of that dosage cannot be measured with precision in phenotypes that accurately reflect the mechanisms of gene dosage that produced them.

For animal models of Down syndrome, an initial baseline typically begins with assessment of gene content, gene expression, sequence variation, and mosaicism (Besson et al. 2005; Kazuki et al. 2020; O'Doherty et al. 2005; Reeves et al. 1995; Yu et al. 2010). Careful husbandry records noting fertility of both sexes and longevity are important measures. A catalog of relevant baseline phenotypes for Down syndrome models would include physical features such as overall body size and growth rates, anomalies of craniofacial skeleton and skull formation, axial skeleton, occurrence and specific types of congenital heart defects, brain morphology and histology, changes that might suggest a link to Alzheimer pathology, a survey of blood cell types, and perhaps organ size and function. Measures of immune system function are highly relevant. Metabolic measurements for mouse have become quite specific over a number of areas of metabolism (Sarver et al. 2022, 2023). Numerous focused studies from various areas of expertise might then be suggested (or not) based on presence or absence and magnitude of effects in a given model (Jewett et al. 2023, n.d.; Waugh et al. 2023). Many assessments of these types have been assessed for TcMAC21 and TcRAT21 and phenotypes correspond at least qualitatively to those observed in other Down syndrome mouse models. The reader is referred to the primary reports for specifics regarding these models (Kazuki et al. 2020, 2022).

Of special interest in Down syndrome, because it is a prime concern of caregivers and a target for a substantial fraction of fundamental research, is the impact of trisomy on cognition. A number of assays of behavior, learning and memory have been employed in mouse models, some with correlated neurochemical, anatomical changes and/or electrophysiological changes. Many of these tests may be useful indicators of brain function but their relationship to the effects of trisomy 21 on brain development and function in Down syndrome are not necessarily clear. The tests available, their administration, interpretation and the reporting of results in inaccurate titles and abstracted summaries all lag far beyond the ability to understand how molecular level changes contribute to divergence from euploid.

To anthropomorphize just one example, imagine a human behavioral test in which a subject is dragged out of bed in the middle of the night by an enormous predator many times their size, has blinding lights shined in their face, is thrown into a threatening environment, and given repeated foot shocks. At a later time, the subject is assessed for memory of what could hardly be considered a normal experience. This is effectively what happens when diurnal investigators test nocturnal rodents during the day in the contextual and cued fear conditioning test. A difference between control and experimental groups in this procedure is typically reported as a "cognitive disability," a conclusion then equated with other reports in which the frequency of choice in receiving a treat in a Y maze or the ability to learn and remember a maze after being dropped in a pool of water differs between groups. Failure to reproduce results of carefully planned and conducted rodent behavioral tests has been demonstrated repeatedly (e.g. (Crabbe et al. 1999), but little progress has been made in applying effective standards across laboratories. Poor application and lack of critical interpretation of behavioral analyses in rodents is an Achilles heel of fundamental research into the cognitive impacts of trisomy and gene dosage. The impressively high resolution of molecular events is circumscribed without an understanding of the phenotypes to which they are alleged to contribute.

7 Concluding Remarks

In a few decades, rodent models that overexpress HSA21 genes or their rodent orthologs in mice (and now rats) have had a major effect in transforming Down syndrome research from a field in which research was considered a waste of time because trisomy 21 was too difficult to ever be understood, to one where understanding has been taken to a deep level, even supporting clinical trials to ameliorate cognitive outcomes of trisomy (Hart et al. 2017). While transchromosomic models provide the most complete genetic representation of HSA21, they have caveats that must be considered in interpreting results. It will be important to compare trisomic models to those with direct duplications producing gene dosage effects of rodent orthologs without the additional physical impact of an extra chromosome that must be duplicated and segregated at meiosis and mitosis (see Chap. 2, Eugene Yu). Transchromosomic models can also give a first level of understanding of the impact of the expression of HSA21 genes that do not have a rodent ortholog. As phenotyping assays become sufficiently sensitive, assessment of human genetic variation in the transchromosome and in the genome at large will become possible. The advent of Down syndrome modeling in rats with their rich history of pharmacological testing and very sophisticated (compared to mouse) assessment of learning and memory and its underlying molecular, electrophysiological, and cellular basis promises an even more robust future for advancing efforts to ameliorate the effects of trisomy 21 in people.

Bibliography

Besson, V., Duchon, A., Brault, V., Magnol, L., Bizot, J.-C., Dauphinot, L., Potier, M.-C., and Herault, Y. (2005). Studying dosage effects for human chromosome 21 homologous genes using chromosomal engineering in the mouse. Paper presented at: 19th International Mouse Genome Conference (Strausberg, FR).

Birling, M.C., Schaeffer, L., Andre, P., Lindner, L., Marechal, D., Ayadi, A., Sorg, T., Pavlovic, G., and Herault, Y. (2017). Efficient and rapid generation of large genomic variants in rats and mice using CRISMERE. Sci Rep *7*, 43331.

Cox, D.R., Smith, S.A., Epstein, L.B., and Epstein, C.J. Mouse trisomy 16 as an animal model of human trisomy 21 (Down syndrome): production of viable trisomy 16 diploid mouse chimeras.

Cox, D.R., Smith, S.A., Epstein, L.B., and Epstein, C.J. (1984). Mouse trisomy 16 as an animal model of human trisomy 21 (Down syndrome): production of viable trisomy 16 diploid mouse chimeras. Developmental biology *101*, 416–424.

Crabbe, J.C., Wahlsten, D., and Dudek, B.C. (1999). Genetics of mouse behavior: interactions with laboratory environment. Science (New York, NY) *284*, 1670–1672.

Davisson, M., Schmidt, C., Reeves, N., Irving, E., Akeson, E., Harris, B., and Bronson, R. (1993). Segmental trisomy as a mouse model for Down Syndrome. Prog Clin Biol Res *384*, 117–133.

Do, C., Xing, Z., Yu, Y.E., and Tycko, B. (2017). Trans-acting epigenetic effects of chromosomal aneuploidies: lessons from Down syndrome and mouse models.

Gearhart, J., Singer, H., Moran, T., Tiemeyer, M., Oster-Granite, M., and Coyle, J. (1986). Mouse chimeras composed of trisomy 16 and normal (2N) cells: Preliminary studies. Brain Research Bulletin *16*, 815–824.

Gribble, S.M., Wiseman, F.K., Clayton, S., Prigmore, E., Langley, E., Yang, F., Maguire, S., Fu, B., Rajan, D., Sheppard, O., *et al.* (2013). Massively parallel sequencing reveals the complex structure of an irradiated human chromosome on a mouse background in the tc1 model of down syndrome. PLoS One *8*, e60482.

Hart, S.J., Visootsak, J., Tamburri, P., Phuong, P., Baumer, N., Hernandez, M.C., Skotko, B.G., Ochoa-Lubinoff, C., Liogier D'Ardhuy, X., Kishnani, P.S., *et al.* (2017). Pharmacological inter-

ventions to improve cognition and adaptive functioning in Down syndrome: Strides to date. American journal of medical genetics Part A *173*, 3029–3041.
Hassold, T., and Hunt, P. (2001). To err (meiotically) is human: the genesis of human aneuploidy. Nat Rev Genet *2*, 280–291.
Jewett, C.E., McCurdy, B.L., O'Toole, E.T., Stemm-Wolf, A.J., Given, K.S., Lin, C.H., Olsen, V., Martin, W., Reinholdt, L., Espinosa, J.M., *et al.* (2023). Trisomy 21 induces pericentrosomal crowding delaying primary ciliogenesis and mouse cerebellar development. eLife *12*.
Kazuki, Y., Gao, F.J., Li, Y., Moyer, A.J., Devenney, B., Hiramatsu, K., Miyagawa-Tomita, S., Abe, S., Kazuki, K., Kajitani, N., *et al.* (2020). A non-mosaic transchromosomic mouse model of down syndrome carrying the long arm of human chromosome 21. eLife *9*.
Kazuki, Y., Gao, F.J., Yamakawa, M., Hirabayashi, M., Kazuki, K., Kajitani, N., Miyagawa-Tomita, S., Abe, S., Sanbo, M., Hara, H., *et al.* (2022). A transchromosomic rat model with human chromosome 21 shows robust Down syndrome features. American journal of human genetics *109*, 328–344.
Kugoh, H., K, M., Meguro, M., Shigenami, K., Schulz, T.C., and Oshimura, M. (1999). Mouse A9 cells containing single human chromosomes for analysis of genomic imprinting. DNA Res *6*, 165–172.
Letourneau, A., Santoni, F.A., Bonilla, X., Sailani, M.R., Gonzalez, D., Kind, J., Chevalier, C., Thurman, R., SanDown syndrometrom, R.S., Hibaoui, Y., *et al.* (2014). Domains of genome-wide gene expression dysregulation in Down's syndrome. Nature *508*, 345–350.
Liu, Y., Borel, C., Li, L., Muller, T., Williams, E.G., Germain, P.L., Buljan, M., Sajic, T., Boersema, P.J., Shao, W., *et al.* (2017). Systematic proteome and proteostasis profiling in human Trisomy 21 fibroblast cells. Nat Commun *8*, 1212.
Moore, C.S., Hawkins, C., Franca, A., Lawler, A., Devenney, B., Das, I., and Reeves, R.H. (2010). Increased male reproductive success in Ts65Dn "Down syndrome" mice. Mammalian genome : official journal of the International Mammalian Genome Society *21*, 543–549.
O'Doherty, A., Ruf, S., Mulligan, C., Hildreth, V., Errington, M.L., Cooke, S., Sesay, A., Modino, S., Vanes, L., Hernandez, D., *et al.* (2005). An aneuploid mouse strain carrying human chromosome 21 with down syndrome phenotypes. Science (New York, NY) *309*, 2033–2037.
Panagaki, T., Lozano-Montes, L., Janickova, L., Zuhra, K., Szabo, M.P., Majtan, T., Rainer, G., Maréchal, D., Herault, Y., and Szabo, C. (2022). Overproduction of hydrogen sulfide, generated by cystathionine β-synthase, disrupts brain wave patterns and contributes to neurobehavioral dysfunction in a rat model of Down syndrome. Redox Biol *51*, 2213–2217.
Reeves, R.H., Irving, N., Moran, T., Wohn, A., Kitt, C., Sisodia, S., Schmidt, C., Bronson, R., and Davisson, M.T. (1995). A mouse model for Down Syndrome exhibits learning and behaviour deficits. *Nature Genetics 11*, 177–183.
Rubenstein, E., Tewolde, S., Skotko, B.G., Michals, A., and Fortea, J. (2024). Occurrence of mosaic Down syndrome and prevalence of co-occurring conditions in Medicaid enrolled adults, 2016–2019. American journal of medical genetics Part C, Seminars in medical genetics, e32097.
Saran, N.G., Pletcher, M.T., Natale, J.E., Cheng, Y., and Reeves, R.H. (2003). Global disruption of the cerebellar transcriptome in a Down syndrome mouse model. Hum Mol Genet *12*, 2013–2019.
Sarver, D.C., Xu, C., Rodriguez, S., Aja, S., Jaffe, A.E., Gao, F.J., Delannoy, M., Periasamy, M., Kazuki, Y.A.-O., Oshimura, M., *et al.* (2023). Hypermetabolism in mice carrying a near-complete human chromosome 21. LID - https://doi.org/10.7554/eLife.86023 [doi] LID - e86023
Dysregulated systemic metabolism in a Down syndrome mouse model.
Sarver, D.C., Xu, C., Velez, L.M., Aja, S., Jaffe, A.E., Seldin, M.M., Reeves, R.H., and Wong, G.W. (2022). Dysregulated systemic metabolism in a Down syndrome mouse model.
Shinohara, T., Tomizuka, K., Miyabara, S., Takehara, S., Kazuki, Y., Inoue, J., Katoh, M., Nakane, H., Iino, A., Ohguma, A., *et al.* (2001). Mice containing a human chromosome 21 model behavioral impairment and cardiac anomalies of Down's syndrome. Hum Mol Genet *10*, 1163–1175.

Suzuki, T., Kazuki, Y., Hara, T., and Oshimura, M. (2020). Current advances in microcell-mediated chromosome transfer technology and its applications. Experimental cell research *390*, 111915.

Tomizuka, K., Yoshida, H., Uejima, H., Kugoh, H., Sato, K., Ohguma, A., Hayasaka, M., Hanaoka, K., Oshimura, M., and Ishida, I. (1997). Functional expression and germline transmission of a human chromosome fragment in chimaeric mice [see comments]. Nat Genet *16*, 133–143.

Waugh, K.A., Minter, R.A.-O., Baxter, J., Chi, C., Galbraith, M.D., Tuttle, K.D., Eduthan, N.P., Kinning, K.A.-O.X., Andrysik, Z.A.-O., Araya, P., *et al.* (2023). Triplication of the interferon receptor locus contributes to hallmarks of Down syndrome in a mouse model.

Jewett, C., McCurdy, B. L., O'Toole, E. T., Stemm-Wolf, A. J., Given, K. S., Lin, C. H., Olsen, V., Martin, W., Reinholdt, L., et al. (n.d.). Trisomy 21 induces pericentrosomal crowding delaying primary ciliogenesis and mouse cerebellar development. LID - https://doi.org/10.7554/eLife.78202 [doi] LID - e78202.

Yu, T., Li, Z., Jia, Z., Clapcote, S.J., Liu, C., Li, S., Asrar, S., Pao, A., Chen, R., Fan, N., *et al.* (2010). A mouse model of Down syndrome trisomic for all human chromosome 21 syntenic regions. Hum Mol Genet *19*, 2780–2791.

Simple Worm, Complex Disorder: Studying Down Syndrome Mechanisms in the Unassuming Model *C. elegans*

Sophia M. Sanchez and Jonathan T. Pierce

Abstract The nematode, *Caenorhabditis elegans*, may seem an unlikely choice for studying a condition as complex as Down syndrome. However, this tiny model has been studied extensively and exhibits a high degree of genetic orthology to humans and phenotypic conservation at the cellular and molecular levels – and even in some cases at the behavioral level. Its strengths include the ease of genetic engineering and availability of mutant strains, its transparency, fully-mapped nervous system, and the multitude of resources and experimental approaches available to *C. elegans* researchers. Here, we elaborate on strengths that researchers have leveraged to make discoveries in relation to Down syndrome. We also point out certain features of *C. elegans* that may represent limitations as well as ways to surmount some when modeling Down syndrome.

Keywords Nematode · Modelling · Overexpression · Genetic screening · *C. elegans* · Traits

In the 1860s, John Langdon Down characterized a group of people with common facial features and intellectual disabilities (Van Robays 2016; Down 1866). He meticulously documented their common traits and photographed his more than 200 subjects in a flattering way that was progressive for the time (Fig. 1a, b). Nearly a century later, it was discovered that individuals with Down syndrome (DS) carry a full extra or partial copy of the 21st chromosome, also known as *H*uman *s*omatic *a*utosome *21* or Hsa21 (Lejeune et al. 1959; Jacobs et al. 1959). And so began a quest to understand how trisomy 21 gives rise to the features enumerated by Dr. Down. The primary model organism used at the time was (and remains) the mouse. Over the past decades, dozens of DS mouse models have been generated that carry different combinations of Hsa21 genes (Herault et al.

S. M. Sanchez · J. T. Pierce (✉)
Department of Neuroscience, Center for Learning and Memory, Waggoner Center for Alcohol & Addiction Research, University of Texas at Austin, Austin, TX, USA
e-mail: jonps@utexas.edu

B. Ye, R. H. Reeves (eds.), *Genetic Models of Down Syndrome*,
https://doi.org/10.1007/978-3-031-78611-2_4

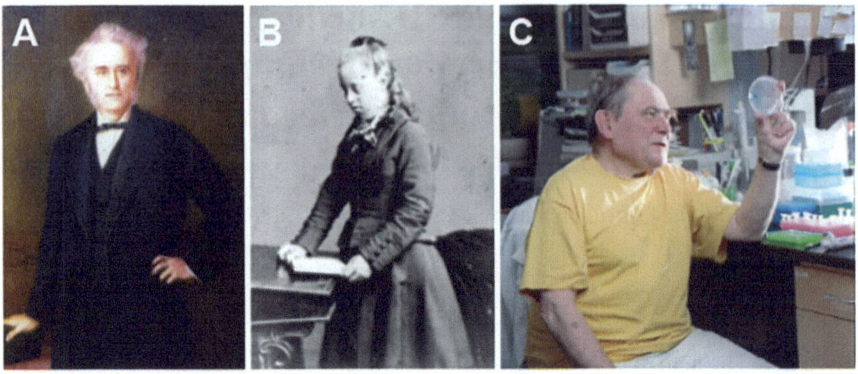

Fig. 1 Studying Down syndrome using *C. elegans*. The study of a complex human genetic condition involving the extra copy of a full chromosome benefits from an experimental model that enables the study of many genes. (**a**) Dr. John Langdon Down (Van Robays 2016). (**b**) One of his patients with DS, Mary Arnott (Van Robays 2016). (**c**) Sydney Brenner observing a plate of *C. elegans*, which he advanced as a model organism (Wade 2019)

2017). However, the challenge of understanding how the duplication of each of the over 200 genes on Hsa21 contributes to DS phenotypes remains daunting. To overcome this, additional systems with complementary strengths to the mouse have been recruited (Moyer et al. 2020).

The smallest animal to join this mission is a humble worm. Coincident with the discovery of trisomy 21, Sydney Brenner began to study how the tiny nematode, *Caenorhabditis elegans*, could be leveraged for biological discoveries (Fig. 1c; Brenner 1974; Corsi et al. 2015). At the time, it might have been unfathomable that "the worm," as it is affectionately called by researchers, could provide useful insights into a condition as complex as DS. However, over the past half century, many seminal discoveries have been made using *C. elegans*, proving its power for tackling biological questions that are often intractable with mouse models. (*n.b.*, as is common in the field, in this chapter we refer to *C. elegans* as a "worm" singularly, in reference to the species as a model organism).

Because this primer is aimed at researchers new to *C. elegans* with interest in DS, and because the etiology of many medical conditions in DS remains unclear, we chose to highlight the utility of the worm for studying broad issues that might one day be relevant to DS. We address several questions: What traits make *C. elegans* unique for making discoveries? What are some of those discoveries? What methods, tools, and reagents are conveniently at a worm researcher's disposal? How has *C. elegans* been useful for the study of DS so far? And lastly, how might discoveries from *C. elegans* be leveraged to therapeutically relieve deficits in DS?

1 Unique Traits for Studying Biological Questions

Brenner felt that the field of biology needed a new model. He had already taken advantage of the microscopic size and reproductive traits of bacteria and phage to make fundamental discoveries in molecular biology. He next proposed that analogous traits in *C. elegans* could be exploited to uncover cellular and molecular mechanisms underlying development and nervous system function.

In research, size matters. *C. elegans* is small—~1 mm as an adult. Thus, worms can be cultured by the thousands on plates with bacteria for food. This makes them much easier and cheaper to study en masse than mammalian models. For DS, where one might wish to study dozens to hundreds of genes, this is achievable because dozens of unique strains fit on a single lab bench. The small size also makes cryogenic archival easy, as one of their larval stages partly survives freezing (Fig. 2b). Additionally, the worm is not too small. Living, behaving worms can be visualized directly and picked up using a handheld spatula-shaped pick under a stereomicroscope (Fig. 2a).

Sex matters too. *C. elegans* occurs mostly as a self-fertilizing hermaphrodite with two XX chromosomes. Only 1 in 1000 of the population grown in a lab is male, carrying just one X chromosome. This population composition affords the best of both worlds for animal husbandry and genetics. Under basal conditions, one can easily study clonal populations by maintaining hermaphrodites. A single hermaphrodite gives rise to >200 progeny. Self-fertilization also enables the study of critical muscle or neuronal genes. Without needing to find a mate, even mutants that are severely impaired in movement or sensation can reproduce, unlike other animals (e.g. rodent, zebrafish, and fly). Alternatively, a genotype of interest can be introduced via sexual reproduction. This is simplified by the fact that hundreds of male sperm supplant the sperm of the hermaphrodite. These reproductive traits combined with small size make possible the parallel generation of dozens of unique strains in mere weeks, an impossibility in mice.

Unlike other animal models, *C. elegans* boasts an impressively complete set of anatomical and physiological "user manuals" that often surprise newcomers (Table 1). Brenner originally proposed that the worm could be studied at the level of each cell, thanks to its small size and transparency (Brenner 1974). In stark contrast to the 86 billion neurons in humans or 70 million neurons in mice, the nervous system in *C. elegans* hermaphrodites is composed of an invariant set of 302 neurons (Azevedo et al. 2009; Herculano-Houzel et al. 2006). As such, each neuron can be tractably studied one by one. Further, the nervous system was fully mapped, resulting in a complete wiring diagram of approximately 8000 chemical and electrical synapses (White et al. 1976). The *C. elegans* user manuals continue to be expanded. Recently, the transcriptional profile of each cell has been determined for each developmental stage across the worm lifespan (Hammarlund et al. 2018; Cao et al. 2017; Kaletsky et al. 2018). The thoroughness and accessibility of information on *C. elegans* via online compendia (Table 1) liberates scientists to pose and solve research questions that are unthinkable with other organisms. For example, one can ask: Which neurotransmitters are used in each neuron? What combination of gap

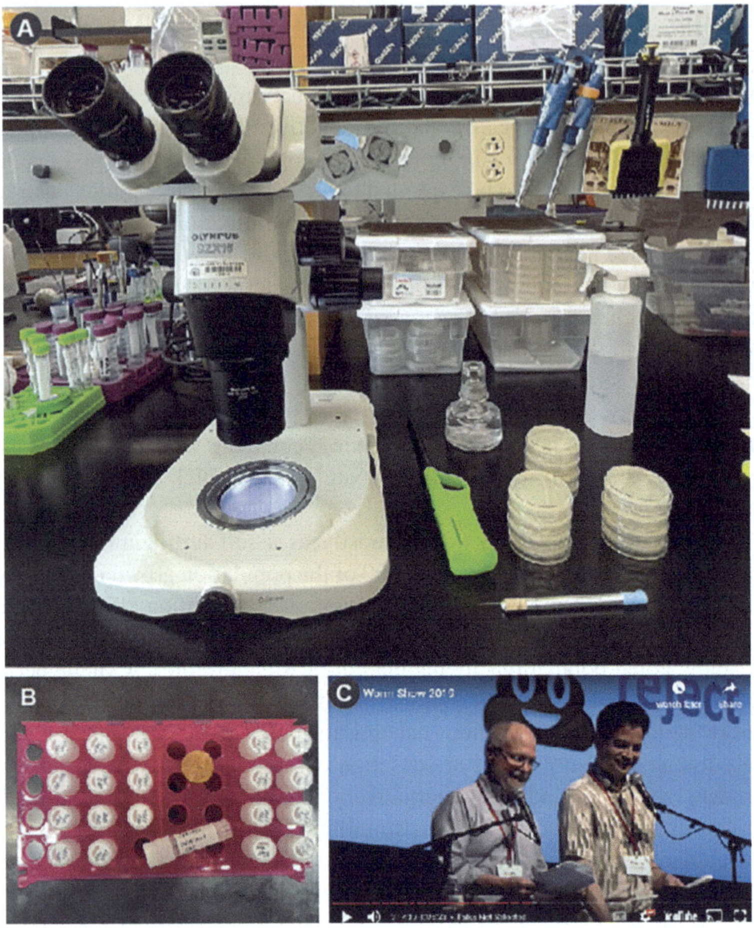

Fig. 2 Practical considerations in working with *C. elegans*. (**a**) Petri dishes (plates) containing agar seeded with a bacterial lawn of nonpathogenic *E. coli* conveniently contain up to ~2000 worms each. Plates housing worms are stored at room temperature stacked on a lab bench or within containers. A handheld platinum wire "pick" (middle) can be sterilized in an ethanol lamp flame before being used to pick up worms while viewed with a stereomicroscope. Reflected light visualizes internal features of the worm. Darkfield illumination shows the worm white on a dark background to obscure irregularities in the agar and lawn for clear video recording of worms. (**b**) Worm stocks can be frozen indefinitely in cryo-preservation vials (penny for size comparison). (**c**) The *C. elegans* community is famously collaborative and friendly, as demonstrated with a biannual international conference. The "Worm Show," a comedy routine involving members across the *C. elegans* community with hosts Curtis Loer and Morris Maduro, is often regarded as a highlight. (Archived on YouTube)

junction proteins connect each cell via electrical synapses? What is the combinatorial code of transcription factors that define the fate of each cell? These major questions have largely been answered in the worm.

Fundamental questions about most genes are easily addressed in *C. elegans*. The worm was the first multicellular organism to have its genome fully sequenced. It

Table 1 Selected resources and reagents at researchers' disposal when using *C. elegans*

Resource/reagent	Description and notes
WormAtlas (https://wormatlas.org/)	Anatomical atlas of hermaphroditic and male *C. elegans*, with wiring, organ, and cellular-level features, including across development, dauer stage, and aging.
WormBase (https://wormbase.org/#012-34-5)	Compendium of *C. elegans* tools, including a "genome browser" and detailed information about genes, gene variants, strains. Now available as part of the Alliance of Genome Resources (https://www.alliancegenome.org/), which combines *C. elegans* data with that from other model organisms. Includes a discussion forum (https://community.alliancegenome.org/c/model-organism-worms/7).
WormBook (http://www.wormbook.org/index.html)	An in-depth guide for reference in *C. elegans* research and experimental design, with insights into *C. elegans* development, evolution and ecology, the nervous system, methods (maintenance and phenotypic/behavioral tests), history, etc…
Caenorhabditis Genetics Center (https://cgc.umn.edu/)	Bio-repository of *C. elegans* researchers' mutant and transgenic strains available for order for your projects. *C. elegans* researchers submit strains to the CGC to be shared with others to promote further research.
OrthoList (http://www.greenwaldlab.org/ortholist/)	A resource of *C. elegans* genes with human orthologs gleaned from the convergence of 4 ortholog prediction programs (Shaye and Greenwald 2011; Kim et al. 2018, 2).
Textpresso (https://celegans.textpressocentral.org/tpc/home)	*C. elegans* publication search engine searches any text within publications, unpublished abstracts, and communications (Worm Breeder's Gazette). Great for searching obscure allele names. Ranks found text by occurrence and lists sentences for speed reading.
Micropublications (http://www.micropublication.org/)	Peer-reviewed articles on new methodology, many featuring *C. elegans*.
Fosmid Library (http://elegans.bcgsc.ca/perl/fosmid/CloneSearch)	Small ~20–40 Kb segments of the *C. elegans* genome harbored in a bacteriophage library.
Promoterome (http://worfdb.dfci.harvard.edu/html/promoteromedb/) Horizon Discovery Promoter Collection (https://horizondiscovery.com/en/non-mammalian-research-tools/products/c-elegans-promoter-collection)	Predicted promoters for *C. elegans* genes with forward and reverse primer sequences can be obtained to drive expression of your favorite gene in particular cell types or used as a database for identifying your favorite gene(s)' promoter (Dupuy et al. 2004).

(continued)

Table 1 (continued)

Resource/reagent	Description and notes
TransgeneOME	~15,000 C-terminally GFP-tagged genes (Sarov et al. 2012).
RNAi Library (https://www.dnaform.jp/en/products/clone/c_elegans_rnai_library/) *C. elegans* RNAi Collection (https://horizondiscovery.com/en/non-mammalian-research-tools/products/c-elegans-rnai-collection)	Bacteria that express the dsRNA can be grown and fed to *C. elegans* for convenient RNAi by feeding. One can also transform transgenes expressing dsRNA in cell-selective high-efficient manner (Timmons and Fire 1998; Esposito et al. 2007).
cDNA Library/ORFeome (http://worfdb.dfci.harvard.edu/html/index.php?page=searchwm) Horizon discovery ORF collection (https://horizondiscovery.com/en/non-mammalian-research-tools/products/c-elegans-orf-collection#:~:text=The%20C.%20elegans%20ORF%20collection%20%28also%20known%20as,interactions%2C%20RNAi%2C%20and%20protein%20expression%20in%20Caenorhabditis%20elegans)	A collection of *C. elegans* cDNAs/ORFs that one can obtain to express under your promoter of choice.
Transcription factor library (https://horizondiscovery.com/en/non-mammalian-research-tools/products/c-elegans-transcription-factors)	For studying *C elegans* TF/promoter interactions with yeast 1 hybrid assays.
CeNGEN (https://cengen.shinyapps.io/CengenApp/)	Single-cell neuronal gene expression database. Allows one to identify which identified neurons their favorite gene is expressed, or conversely, the expression profile of specific neurons (Taylor et al. 2021)
ModENCODE (https://wormbase.org/tools/rnaseq/fpkmmine.cgi)	Available within WormBase, this is a RNAseq database including transcriptional data from across *C. elegans* development, transcription factor–binding site information, and maps of chromatin organization (Gerstein et al. 2010).
NeuroPAL (https://pubmed.ncbi.nlm.nih.gov/33378642/)	A transgenic worm strain available through the CGC. These worms' neurons are all identifiable due to unique fluorophore combinations and their locations. These can be crossed with your favorite reporter strain to visualize expression and study the effect of various treatments or mutations on expression or neuronal properties.
Million Mutation Project (http://genome.sfu.ca/mmp/search.html)	A library of more than 2000 *C. elegans* strains that have been mutagenized (with EMS, ENU or a cocktail of EMS plus ENU) and full-genome sequenced such that the majority of *C. elegans* genes have a mutation in at least one of the strains, which can be obtained and outcrossed to isolate and study that mutant.

(continued)

Table 1 (continued)

Resource/reagent	Description and notes
C. elegans KO consortia	Knockout consortia from Oklahoma (https://www.zoology.ubc.ca/~alorch/kostrains.htm), Vancouver (https://www.zoology.ubc.ca/~alorch/ko/), and Japan (https://shigen.nig.ac.jp/c.elegans/mutants/index.xhtml)-generated hundreds of deletion alleles for most genes. The Japanese National Bioresource Project (NBRP) for the Experimental Animal *C. elegans* has begun to generate CRE lines and more for public use after filling out an MTA.
CaeNDR- *Caenorhabditis elegans* Natural Diversity Resource (https://caendr.org/)	Wild isolates of *C. elegans* from around the world with broad genetic diversity. One can leverage the variability of gene expression or the presence of particular variants to better understand ones' favorite gene(s) in ecological and evolutionary context. These can also be used for GWA and QTL studies with their online app. The collection also includes isolates of two closely related species; *Caenorhabditis briggsae* and *tropicalis*, more than 1000 unique isolates total (Cook et al. 2017; Crombie et al. 2024).

conveniently harbors about the same number of genes as humans (19,984 protein coding genes—46,927 including noncoding genes) ("*Caenorhabditis elegans* Genome Assembly WBcel235" n.d.). Up to 83% of protein-coding genes in worm are well conserved with human genes based on primary sequence (Lai et al. 2000). More recently, even human genes with low sequence conservation have been found to have functional homologs with conserved roles in the worm. For example, the BLAST algorithm will not report a worm ortholog for the cannabinoid, opioid, and melatonin receptors. Behavioral and pharmacological studies in the worm, however, provide evidence for functional equivalents of each receptor with roles in analogous behaviors. Further investigation in each case revealed that key residues important for ligand binding have conserved sequence and function (Oakes et al. 2017; Cheong et al. 2015; Woods et al. 2013; Niu et al. 2020). In addition, "humanized" chimeric worm models, in which the worm version of a gene is rescued with its human counterpart, provide strong complementary evidence for functional conservation (e.g. Davis et al. 2015; Mills et al. 2016; Scott et al. 2018).

There are, of course, differences between the worm and human genome. These include: significantly fewer redundancies of certain genes (e.g. mammals have three paralogs for *APP* while invertebrates only have one *APP*-like gene in worm called *apl-1*), fewer isoforms of most genes (e.g. at least 5 isoforms of *DYRK1A* in mammal, but only one isoform for the worm ortholog *mbk-1*), and shorter cis-regulatory regions in worm. Although these may be viewed as problematic, they can be helpful, as a model organism is intended to provide a simpler, more tractable system for rapid study and interpretation of scientific questions. For instance, worm genes are often smaller and thus easier to clone from genomic DNA, and worm genes with fewer paralogs make it easier to observe phenotypes in single mutants, avoiding the necessity for double or triple mutants.

To some, *C. elegans* may appear an overly reductive model, but these animals are far from simple. The worm exhibits a suite of sophisticated behaviors including learning and memory, as well as escape responses (Rankin et al. 1990; Becerra et al. 2023; Liu et al. 2018b). Other higher-order behaviors such as sleep, drug responses, motor pattern switching, and social interactions have also been observed (de Bono et al. 2002; de Bono and Bargmann 1998; Raizen et al. 2008; Chalfie et al. 2014). The worm also displays several behaviors resulting from the integration and processing of multiple sensory inputs, such as locomotor and egg laying decisions in the presence of noxious or pleasant stimuli (de Bono and Maricq 2005). Perturbations in any of these behaviors can serve as quantifiable readouts and, often, evidence of dysfunction in the nervous system or muscle. Because the function of most worm neurons is well characterized, a defect in a specific behavior can be readily traced to specific subsets of neurons. Importantly, when a gene is found to have a critical role for a behavior in worm, the equivalent mammalian gene often underlies a surprisingly analogous behavior in mammals. Three examples include: alcohol-induced activation of the BK potassium channel which elicits intoxication, melatonin signaling as a modifier of sleep, and mutations in autism risk genes that disrupt social behavior (Davis et al. 2015; Niu et al. 2020; Rawsthorne et al. 2021).

One primary advantage of *C. elegans* research is that worm researchers primarily use a single wild-type background for easy comparison in studies around the world. This strain, N2, was originally isolated in the UK. N2 is nearly isogenic (homozygous at every allele) and because it can be regularly restored from frozen stocks to minimize genetic drift due to spontaneous mutations, it serves as a stable reference for worm researchers all around the world.

C. elegans also offers a wealth of available strategies for genetic manipulation. To elucidate the genetic bases for phenotypes, specific genes can be easily targeted—often conditionally, with spatial and/or temporal control—to introduce desired mutations, including single-nucleotide polymorphisms, and small or large deletions and insertions. The worm genome can also be mutagenized at random with chemicals or UV light for forward genetic screens. Gene knockdown is also easy. After finding that double-stranded RNA worked better than antisense single-stranded RNA at knocking down expression, Andy Fire and Craig Mello discovered RNA interference, resulting in a Nobel prize (Fire et al. 1998; Timmons and Fire 1998). RNAi is routinely leveraged to knockdown genes by simply injecting or even feeding worms double-stranded RNA matching a target gene. The generation of transgenics is also very straightforward (Table 2). One popular technique simply entails injecting a cocktail of double-stranded DNA in linear or circular form directly into the gonad of the hermaphrodite. These concatenate to form a circular artificial chromosome, termed an "extrachromosomal array," which a portion of

Table 2 Example *C. elegans* genetic approaches

Approach & key steps	Example usage & notes	Strengths and weaknesses, potential applications/opportunities in DS
Forward genetic screen 1. Randomly mutagenize genes in worms 2. Select animals with a particular phenotype in F1, F2, or F3 progeny 3. Identify candidate causal mutation(s) 4. Test causality with transformation rescue of gene, or recapitulate with independent allele.	Example phenotypes screened: *Behavioral*: e.g. uncoordinated motion (Brenner 1974), yielded discovery of genes important for synaptic transmission; e.g. worm clumping/aggregation: yielded understanding of neuro- modulatory roles of IL-17 which is implicated in ASD and autoimmune disorders (Chen et al. 2017); e.g. locomotor hyperactivity (Bessou et al. 1998), yielded discovery of worm dystrophin & development of worm model of Duchenne muscular dystrophy *Morphological*: e.g. multivulva or vulvaless (Horvitz and Sulston 1980; Wang and Sternberg 2001); yielded discovery of genes involved in RAS/MAP Kinase and Notch signaling pathways *Cellular*: e.g. lack of programmed cell death during development; yielded discovery of apoptosis genes (Hedgecock et al. 1983), mislocalized tagged AChR (Pinan-Lucarré et al. 2014) yielded molecular mechanisms for receptor localization	Strengths: Unbiased; enables genome-wide gene discovery without a priori hypotheses for screenable phenotypes Ability to study many genes concurrently, can result in many hits that can fuel years of additional research If completed to saturation, frequency of mutations may reveal relative importance of different genes Unbiased mutagenesis may reveal unique alleles informing structure–function relationships in the gene or protein Weaknesses: Essential genes might be excluded Primarily identifies single genes of large effect, which may swamp out other genes Will miss genes with redundant roles Applications in DS: Could be used to study particular phenotypes especially relevant to DS (e.g. muscle tone) to shed light on potential genetic interactors beyond Hsa21 genes.
Suppressor or resistance screens 1. Randomly mutagenize genes in worm with mutation (or drug treatment) that produces phenotype of interest with room for suppression 2. Select worms with suppressed phenotype 3. Identify candidate suppressor mutation 4. Test causality with transformation rescue of gene, or recapitulate with independent allele. Reviewed in Fay and Johnson (2006)	For maternal effect lethal: used to identify genes involved in early embryonic division; yielded identification of *par* cell polarity genes (Kemphues et al. 1988) For suppressed vulvaless phenotype: yielded identification of worm ortholog of Ras-pathway proto-oncogene (Beitel et al. 1990) which has since been used to study several types of cancer (Kirienko et al. 2010) For resistance to intoxication by alcohol: led to discovery of a molecular target of ethanol (Davies et al. 2003)	Strengths: Can reveal genes in same or parallel pathways in unbiased manner Can reveal alleles without phenotypes on their own (i.e. synthetic phenotypes) Screens that leverage viability, growth rate, or egg laying phenotypes require little monitoring Weaknesses: Requires extra work to exclude nonspecific "hits" that suppress due to uninteresting reasons (e.g. mRNA or protein degradation mutants) Primarily identifies single-genes of large effect, which may swamp out other genes Applications in DS: Could screen for suppressors of phenotypes displayed by a strain that overexpresses an Hsa21 gene(s). This may reveal genetic interactors and new therapeutic targets.

(continued)

Table 2 (continued)

Approach & key steps	Example usage & notes	Strengths and weaknesses, potential applications/opportunities in DS
Enhancer or hypersensitivity screens 1. Randomly mutagenize genes in worm with mutation (or drug treatment) that produces a phenotype of interest that has room for enhancement 2. Select mutants with an enhanced phenotype 3. Identify candidate enhancer mutation 4. Test causality with transformation rescue of gene, or recapitulate with independent allele Reviewed in Fay (2006)	For enhanced *glp-1* germline deficits: yielded discovery of genes involved in mitosis and development (Qiao et al. 1995) For dominant enhancers of temperature-sensitive touch sensitivity defects: yielded discovery of other genes physically interacting with *mec-4* (Gu et al. 1996) For enhanced neurodegeneration with 6OHDA: revealed neuroprotective genes (Offenburger et al. 2018) For enhanced ventral-axon guidance defects of *unc-40*: revealed Hsa21 ortholog *eva-1* and their functional interaction (Fujisawa et al. 2007)	Strengths: Can reveal parallel pathways Can reveal alleles that have silent phenotypes on their own Weaknesses: Need to confirm whether the novel mutation causes the phenotype on its own or in synergy with the original mutation (or drug treatment) Applications in DS: Could reveal important genetic interactors beyond Hsa21 genes. One could screen for mutations that enhance a phenotype in Hsa21 OE strains, for example to reveal pathways that contribute to the same phenotype. Once identified, the pathway might contain a druggable target to treat the OE phenotype.
Reverse genetic loss-of-function (LF) or knock-down (KD) screen 1. Select list of genes to test for involvement in a phenotype of interest 2. Order LF mutants and/or RNAi reagents for those genes 3. Test which, if any, mutations or RNAi treatments cause phenotype 4. Confirm "hits" with transformation rescue and/or independent LF alleles	For interactors of TAT-5 in extracellular vesicle release: identified a novel role for Hsa21 gene *pad-1* (Beer et al. 2018) For genes involved in axon regrowth (Chen et al. 2011) For genes involved in longevity (Sutphin et al. 2017). Many labs have tested the role of genes throughout the genome in phenotypes by knocking them down one at a time using RNAi (Braukmann et al. 2017)	Strengths: Less biased when conducted in a systematic fashion (e.g. all Hsa21 orthologs, whole-genome RNAi catalog) Efficient: Readily available LF mutants and RNAi library, no strain generation step Reveals which of a subset of chosen, related genes are required or important for particular behaviors or phenotypes For RNAi screens, you can test KD of genes pre- and/or post-embryonically if you treat the parental generation or after eggs are laid, respectively For RNAi screens, one can choose to use WT background where gene KD is very efficient in nonneuronal tissue, use various strains that are hypersensitive to RNAi in nervous system, or use strains that are selectively sensitive to RNAi in subsets of cells Weaknesses: Biased to the extent of which genes are selected for KO or KD Takes much longer to KO or KD many genes in your list of interest versus mutating all genes randomly with forward genetic approach RNAi often shows partial penetrance KD Applications in DS: Led to identification of 10 Hsa21 genes required for neuronal and/or muscular function (Nordquist et al. 2018). Could be further leveraged to study less-conserved orthologs, human versions of Hsa21 genes, nongenetic elements, or non-Hsa21 genes.

Overexpression (OE) screen 1. Generate transgenic strains by transforming genes of interest (see Table 1) 2. Select independently derived strains as control for transgenesis 3. Phenotype OE strains with control strains 4. May choose to do broad phenotyping for a less-biased assessment, or conduct hypothesis-driven phenotyping.	To reveal mechanisms underlying cell polarization in embryonic development (Chan and Nance 2013) To reveal Hsa21 genes (with highly conserved orthologs) that may be involved in DS phenotypes (Sanchez et al. 2024)	Strengths: Less biased when conducted in a systematic fashion (e.g. all highly conserved Hsa21 orthologs) Reveals insights into chosen subsets of related genes and chosen phenotypes Weaknesses: Requires generation or attainment of many transgenic and control strains (easier now that the Sanchez et al. 2024 library of OE strains at the *Caenorhabditis* Genetic Center) Applications in DS: Can be further used to study particular phenotypes of interest in Hsa21 ortholog transgenic strains, with less-conserved orthologs, human versions of Hsa21 genes, nongenetic elements, or non-Hsa21 genes.
Reverse genetic mosaic analysis 1. Transform mutant strain with WT rescuing transgene linked to a fluorescent reporter on an extra chromosome. 2. Examine individual worms for the presence or lack of phenotype. Check pattern of reporter expression in each group to characterize which cells carry the extra chromosome. The extra chromosome will rescue function in the cells/lineages in which the gene acts, but can be lost in others 3. Converse experiment can be carried out with WT background and overexpressed WT or mutant transgene. Reviewed in Yochem and Herman (2003), Herman (1995)	*let-60* (RAS) mutant phenotype (lethality) is dependent on one cell (Yochem and Herman 2003) Innexin signaling occurs both cell autonomous and nonautonomously for left-/right differentiation of neurons (Chuang et al. 2007) Membrane fluidity is regulated non-cell-autonomously (Bodhicharla et al. 2018)	Strengths: Enables identification of cellular loci sufficient to rescue a phenotype or required to induce an abnormal phenotype May provide insight into whether a gene acts cell-autonomous or non-autonomously Weaknesses: Requires detailed inspection of cells that carry or lack the extra chromosome. Advancements in microfluidics, imaging, and automated detection systems are addressing these issues to render this approach higher-throughput. Perdurance (the presence of residual wild type protein product originating from a progenitor cell in the absence of the wild type gene) may conflate results Redundancy can mask results as well, can circumvent with double mutants if redundancy is known Difficult to interpret with incompletely penetrant phenotypes Applications in DS: Could be used to elucidate identified cells in which overexpression of Hsa21 genes is problematic or to better determine single Hsa21 genes' mechanisms and localization of their activity.

(continued)

Table 2 (continued)

Approach & key steps	Example usage & notes	Strengths and weaknesses, potential applications/opportunities in DS
Genome-wide association studies (GWA) or quantitative trait loci (QTL) studies 1. Measure a quantitative phenotype in genetically diverse, wild isolate *C. elegans.* 2. Identify quantitative trait loci associated with the presence or severity of that phenotype using CaeNDR online app. 3. Narrow down to particular loci through additional fine mapping. 4. Test for causation of the candidate allele by engineering the equivalent genotype in distinct backgrounds and checking for recapitulation of the phenotype. Reviewed in Widmayer et al. (2022)	Use to identify genes/QTLs responsible for natural variation in: Telomere length (Cook et al. 2016) Fecundity (Zhang et al. 2021) Identified autism risk gene ortholog, *gap-2*, as a modifier of the neuroendocrine TGF-beta ligand, DAF-7, by examining level of *daf-7::gfp* reporter expression in identified neuron (Lee et al. 2023)	Strengths: Can reveal genes underlying natural variation in phenotypes that are difficult to quantify across human populations and with high enough sample size Can identify genes in unbiased way that forward genetic approaches miss (redundant and essential genes) Relies on strains available in CaeNDR (no mutagenesis or transgenesis required) that are already whole-genome sequenced Study of traits in nature/ecological and evolutionary context can reveal genes that otherwise might not display obvious phenotypes or variation in a lab setting Weaknesses: May not reveal genes with strong effects that have pleiotropic effects that would impact fitness in nature. This may also represent a strength, because it can shed light on lesser-studied genes or ones that have not come out of conventional screens. Applications in DS: Could study how a phenotype caused by overexpression of a Hsa21 ortholog varies in genetically diverse strains. May identify genes and pathways that enhance or suppress the phenotype, leading to new ideas for therapeutics.
Epistasis analysis 1. Generate double mutants for genes hypothesized to be involved in the same pathway (implied by similar or opposite phenotypes). 2. Assess relative severity of mutant phenotypes to determine which mask the other and therefore act upstream, or if additive, are in parallel pathways. Reviewed in Avery and Wasserman (1992), Huang and Sternberg (2006)	Control of developmental timing (Ambros 1989) Identification of new components of G protein-coupled signaling pathways (Yu et al. 2013) Identification of a 3rd parallel pathway for handling apoptosed cell (Haley et al. 2018)	Strengths: Can reveal whether genes act in series or in parallel genetic pathway Can provide evidence for whether a gene functions up or downstream in a pathway Historically, genetic pathways often predict mechanistic molecular pathways Weaknesses: Requires the sometimes inconvenient generation of double- or multi-mutants and full/strong loss-of-function or null alleles, as well as gain-of-function or OE alleles to help order genes in a pathway Applications in DS: If OE and LF of a Hsa21 gene cause phenotypes, one can cross these to other mutants with similar phenotypes to gain insight into whether the genes act in same or parallel genetic pathways and in what order.

Screens and examples summarized from Jorgensen and Mango (2002), with minor modifications and additions. See Singh (2021) for examples and considerations for the design of forward genetic screens.

progeny inherit in a non-Mendelian manner, unlinked from the five autosomes and X chromosome (Woglar et al. 2020). Perhaps the most important aspect of worm genetics is the enormous library of freely available, genetically altered strains (Table 1). By navigating to the *C. elegans* genome browser in WormBase, one can select from hundreds of mutants available to order from the *Caenorhabditis* Genetic Center (Table 1). One may also toggle through a selection of viable, candidate hypomorphic alleles for essential genes supplied by the Million Mutant Project (Thompson et al. 2013). This approach enables in vivo structure–function analyses with mutants available through the mail in mere weeks.

2 Important Discoveries Made Using *C. elegans*

No better proof of the merits of *C. elegans* as a model exists than that of the seminal discoveries made using the worm. Each of the following discoveries, first made with worm, led to the identification of genes with highly conserved functions that have since been validated in human. Many, naturally, involve pathways and genes relevant to DS phenotypes.

In his first 1974 study, Brenner demonstrated the genetic utility of this model with a forward genetic screen. He chose to study a general behavior, locomotion, which is subserved by neurons and muscles, to discover genes required for many highly conserved processes in nervous system development and function. Brenner's foundational screen yielded many uncoordinated (Unc) mutants representing over 100 novel genes that include ion channels and receptors, as well as genes required for neurotransmitter synthesis and packaging, synaptic function, neuronal fate determination, cellular migration, and axon pathfinding. One such gene is *unc-26*, which was found to regulate synaptic function and later to have a human ortholog named *SYNJ1*/synaptojanin on Hsa21. Because *SYNJ1* loss-of-function mice are inviable, study of viable worm *unc-26* mutants led to a better understanding of its conserved role in synaptic vesicle recycling (Harris et al. 2000).

Forward genetic screens can be carried out in a variety of ways to identify novel or uncharacterized genes involved in biological processes. Since Brenner's seminal paper, hundreds of forward genetic screens have been conducted. These screens led to the discovery of genes necessary for fundamental biological phenomena including mechanosensation, chemosensation, dopamine signaling, and apoptosis (Chalfie and Sulston 1981; Bargmann et al. 1993; Sulston et al. 1975; Ellis and Horvitz 1986). Advancements in whole genome sequencing and cloning techniques have accelerated the identification of causal mutations. These and other genetic approaches are summarized in Table 3.

Completion of the wiring diagram of the adult worm nervous system laid the groundwork for further studies (White et al. 1976; Chen et al. 2006). It allowed researchers to study the development and maintenance of identified synapses on specific neurons. This led to the identification of numerous components of the Robo/Slit/Netrin axon guidance pathway, including an Hsa21 gene, *EVA1C*. Using

Table 3 Research tools in *C. elegans*

Goal	Method	Selected examples, subtypes, and worm-specific considerations
Measure cellular excitability/activity	Electrophysiology	Neurons in *C. elegans* are smaller (2–4 μm soma) than in other mammalian models, but methods have been developed to immobilize worms with glue and expose neurons and/or muscles. Because of the defined nervous system and connectome, one can record from identified cells and alter activity of their synaptic partners with optogenetics. Reviewed in Goodman et al. (2012) and Richmond (2006).
	Neural activity/ biological process reporters/biosensors, Fluorescence gain and loss techniques	Biosensor examples: HYlight for Hsa21-related phosphofructokinase/glycolytic activity (Wolfe et al. 2024), ERK-nKTR for extracellular signal-regulated kinase activity (de la Cova et al. 2017) Calcium imaging (Kerr et al. 2000) Fluorescence loss in photobleaching (FLIP), Fluorescence recovery after photobleaching (FRAP)- (Jang et al. 2021), Forster Resonance Energy Transfer (FRET) and Fluorescence Correlation Spectroscopy (FCS), reviewed in Veerapathiran and Wohland (2018)
Control cellular activity (Activate or inactivate neurons)	Optogenetics	Reviewed in Tsukada and Mori (2021). The transparent worm affords noninvasive optogenetics and/or illumination of multiple different locations simultaneously during live imaging. Photo-inducible Cre recombinase (Davis et al. 2021)
	Chemogenetics	DREADDs developed in worm for use through Gq signaling pathway: (Prömel et al. 2016) and inactivation of neurons with HisCl (Pokala et al. 2014)
	Laser ablation	One can perform an axotomy (Bourgeois and Ben-Yakar 2007) or fully destroy a neuron(s) to study its role in phenotype. Particularly tractable in the transparent worm with DIC optics and fully mapped connectome. Reviewed in Fang-Yen et al. (2012)
Measure cellular and neurite morphology	Image soma/neurite growth/development	Live time-lapse imaging (Li et al. 2022). New protocol for worm-sorting for live imaging during development: (Pan et al. 2023)
	Characterize synaptic connections	Split fluorophore trans-synaptic systems: GFP Reconstitution Across Synaptic Partners (GRASP) (Feinberg et al. 2008) and other variations like in vivo Biotin Labeling of INtercellular Contacts (iBlinc) (Desbois et al. 2015) WormPsyQi image analysis pipeline for quantification of fluorescence at synapses + new synaptic reporters (Majeed et al. 2024)
	Quantify cell death and apoptosis	Able to visualize cell corpses easily due to transparent worm, TUNEL staining and other procedures are used as well. Reviewed in Wu et al. (2012).

Phenotyping	Macroscopic phenotyping	Some defects are visible when viewing the whole animal, such as tumors, a blistering cuticle, rolling motion, uncoordinated motion, "variably abnormal" tail and head development, disrupted body plan/morphology leading to severe malformation, and others (Brenner 1974; Fay 2013).
	Development assays	Used to determine whether development is progressing normally according to expected stages and cell divisions (Porta-de-la-Riva et al. 2012)—a phenotype observed in several Hsa21 ortholog OE strains (Sanchez et al. 2024)
	Fitness assay	Embryonic and larval arrest are easily quantified, as well as brood size (Kwah and Jaramillo-Lambert 2023). Competitive fitness can be assessed by several methods compared here: Crombie et al. (2018).
	Discover natural products (hormones, pheromones, compounds, etc.) and assess metabolic/ biochemical function	Can take advantage of unique developmental pathways (larval arrest and entry into an alternative state called dauer that result from altered metabolism in response to environmental stimuli) in *C. elegans* to better understand highly conserved metabolic pathways like insulin signaling and cholesterol trafficking. Reviewed in Butcher (2019) Metabolic/biochemical pathways often studied with biosensors and other techniques highlighted in the imaging section.
	Direct assessment of synaptic function	The paralytic agents, aldicarb & levamisole, have been leveraged in many studies in *C. elegans* for perturbations at the synapse. By assessing the time to paralysis, one can identify presynaptic and postsynaptic dysfunction with these assays. Reviewed in Sammi et al. (2022)
	Drug screening	Very efficient in *C. elegans*, especially with multi-well plates and microfluidic devices that allow exposure to drugs through liquid or in agar on plates. Reviewed in Chen et al. (2015).
High-throughput behavioral phenotyping	Locomotion assays	For understanding E/I imbalance in neuropsychiatric disorders- Reviewed in Thapliyal and Babu (2018). Review on variety of automated trackers: Husson et al. (2018)
	Pharyngeal pumping assay	*C. elegans* pumps bacteria into its mouth with a pharynx and chews with a grinder. The pharynx has neuronal homologs and genetic homology to Central Pattern Generators (Avery and You 2012) and has been compared to the human heart, with similarities and differences discussed in Mango (2007). Pharyngeal pumping rate can indicate neuronal dysfunction and can be assessed with a handheld counter or via automated systems Avery and You (2012).
	Defecation assay	Like pumping, *C. elegans* defecate after eating through a 45-second enteric motor program that uses an intra-intestinal calcium wave that is modulated by a neuronal pattern generator (Dal Santo et al. 1999). Because GABA signaling controls the expulsion step, dysfunction in this GABAergic circuit can be interrogated by quantification of expulsion vs. muscle contractions (Mahoney et al. 2008).
	Egg laying/retention assay	*C. elegans* hermaphrodites lay eggs, a readily quantifiable behavior that requires integration of environmental/sensory cues and the activity of a single command egg laying neuron. Egg-laying and egg-retention assays are very common in *C. elegans* research and have been relied upon extensively for discoveries related to GPCR signaling in particular (Waggoner et al. 1998; Olson et al. 2023).

(continued)

Table 3 (continued)

Goal	Method	Selected examples, subtypes, and worm-specific considerations
	Aging/lifespan assay	Worms' short lifespan of ~2 weeks at room temperature makes them highly amenable to convenient studies of aging-related phenotypes and lifespan (Amrit et al. 2014)
	Sleep assay	Evidence suggests that *C. elegans* exhibit a sleep-like state with conserved roles like consolidating memory (Trojanowski and Raizen 2016; Chandra et al. 2023). One can measure differences in sleep and wakefulness with automated systems (Spies and Bringmann 2018; Lawler and Albrecht 2022).
	Touch sensitivity assay	*C. elegans* escapes when touched on the side and on the nose using mechanosensitive ENaC and TRP/TMC channels, respectively, like mammals (Al-Sheikh and Kang 2020)
	Learning assay	*C. elegans* can form associative memories, tested by classical conditioning. Common conditioning paradigms include the association of salt or temperature with starvation (chemo- and thermo- taxis plasticity), with easy quantification of worms that display learned avoidance from the unpleasant conditioned stimulus (Sasakura and Mori 2013; Rahmani and Chew 2021).
	Habituation, sensitization, and sensory adaptation assays	Studied primarily by gently tapping the plate that worms reside on and observing their response (backwards locomotion) (Kepler et al. 2020). Sensory adaptation can be achieved through several sensory modalities, including, most commonly, olfaction (Ferkey et al. 2021).
	Chemotaxis and thermotaxis	Worms are capable of sensing and moving along chemical and thermal gradients in response to their preferences, and perturbations in these behaviors are easily quantifiable (Ferkey et al. 2021).
	Social clumping and bordering assays	Individual worms join others in clumps around the border of the bacterial lawn (a behavior called "bordering") in which they collectively feed and achieve favored lower oxygen concentrations by burrowing into the clump. This can be quantified to assess the degree of socialness of a given genotype and/or degree of oxygen avoidance (Moreno et al. 2016).
Clone and tag genes	Classical cloning with restriction enzymes	Andy Fire's lab and Addgene have generated vector backbones and kits for efficient cloning with various types of fluorescent and other tags (His, GST, etc.). Many cell-specific promoters have been identified and are available in vectors that can be used to drive expression of a transgene or a fluorescent molecule in your favorite cell, neuron, or class of cells (see Table 1).
	High-efficiency cloning	Many kits and reagents have been developed that are useful for high-efficiency cloning in worm and beyond, including but not limited to HiFi assembly, Gibson Assembly, and Gateway.

Quantify and observe localization of transcripts/proteins	qRT-PCR	Not as popular in *C. elegans* due to the amount of tissue required. Instead, evidence of when and where genes are transcribed is better provided by fluorescent transcriptional reporters. For transgenes, expression of linked co-injection marker is often deemed sufficient evidence for expression. New method for higher-efficiency qRT-PCR in worm: Chauve et al. (2020)
	RNAseq, single-cell RNAseq	Expression profiling is popular using RNAseq, as well as single-cell RNAseq (Cao et al. 2017) snRNAseq (Levenson et al. 2023)
	Microarrays	Have largely been replaced by RNAseq (Guthmueller et al. 2011)
	smFISH	Small molecule Fluorescent In Situ Hybridization: Parker et al. (2021), Ji and van Oudenaarden (2018)
	Transcriptional reporter fusions	Tag your gene transcriptionally, by inserting GFP or your tag of choice after the gene's promoter. This quick method in most cases recapitulates full expression popularly used as an approximation of in vivo cellular expression patterns (Boulin et al. 2006; Tursun et al. 2009).
	Translational reporter fusions and other protein synthesis readout systems	Insert GFP or your tag of choice in frame following the full gene or within one of its introns (Boulin et al. 2006). Split fluorophore tagging for increased efficiency (example: split wrmScarlet- Goudeau et al. 2021) To avoid interference of a linked fluorophore, bicistronic expression methods enable co-expression of nonphysically linked genes by co-opting *C. elegans* operons (one caveat is that the genes are typically not expressed at the same level with this method) (Li and Wang 2012), or by linking transgenes (can have more than 2) with viral 2A self-cleaving peptide sequences (Ahier and Jarriault 2014).
	Western blotting, Immunohistochemistry, Immunofluorescence	Not used very commonly in *C. elegans* research for reasons like those for not using qRT-PCR, also worm cuticle is tough and requires permeabilization, there haven't been many *C. elegans*-specific antibodies produced. Nevertheless examples include: Western blot- Jeong et al. (2018). Immunohistochemistry- Duerr (2006). Immunofluorescence- Parker et al. (2021)
	Quantification of protein turnover	SNAP tagging & CLIP tagging for pulse-chase experiments, fluorescein arsenical hairpin (FlAsH/tetracysteine), newer bioorthogonal labeling (Ullrich et al. 2014). Using Dendra2: Pigazzini and Kirstein (2020)

(continued)

Table 3 (continued)

Goal	Method	Selected examples, subtypes, and worm-specific considerations
Conditional gene and protein activation, inactivation/ repression, and degradation	RNAi- by feeding and by injection Bipartite systems for temporal and spatial control of expression	Somatic control of expression by Gal4 driver lines (Wang et al. 2017a, b) FRT/FLP mediated gene activation (Davis et al. 2008; Voutev and Hubbard 2008) CRISPR inhibition (Long et al. 2015), or CRISPR activation (Gilbert et al. 2014) Conditional gene deletion: Cre and FLP recombinases- Kage-Nakadai et al. (2014), Muñoz-Jiménez et al. (2017) Several genetic approaches to restrict gene KD, KO, or expression to your favorite subset of cells have been developed. RNAi: Current advancements: single-cell, piRNAi (Priyadarshini et al. 2022)
	Auxin-inducible degradation GFP-nanobody-mediated degradation Zinc-finger 1 degradation	If the endogenous or transgenic protein can handle being tagged, it can be selectively degraded by removal from culture plates containing auxin with auxin-mediated degradation (Zhang et al. 2015), newly developed version: Ashley et al. (2021)
Identify epigenetic markings, chromosome organization	ChIPseq, ATACseq	There have been some barriers to these methods in worm (including the impermeability of the cuticle and relatively small ratio of DNA to tissue), but protocols have been developed to circumvent them. ChIPseq- (Sen et al. 2021). ATACseq- Steiner et al. (2012)
Increase incidence of male worms	Cross in *him-5* mutant Heat shock	Because *C. elegans* stocks are mostly hermaphroditic, it's often helpful to increase the ratio of males for crosses or if studying male phenotypes. *him-5* mutant is commonly used because its behaviors are indistinguishable from the N2 lab WT strain. Heat shock also achieves a higher rate of males in almost any background (Kenyon 1988)
Age-synchronize worms	Conduct timed egg-laying Dissolve in bleach	By immersing adult worms in bleach, their tough outer cuticles dissolve, leaving eggs on the plate that will be age-matched. Alternatively, one can place worms on a plate for a period of time to lay eggs and then remove the adults (Porta-de-la-Riva et al. 2012)
Induce dauer stage	Starve worms Dauermone treatment	Strains that are inadvertently neglected often survive for months thanks to the ability of *C. elegans* to enter this alternate lifestage. Study of dauers has yielded important information regarding genetic control of lifespan (Fielenbach and Antebi 2008).
Preserve for future & avoid genetic drift	Freeze worms	*C. elegans* can be frozen to prevent genetic drift and for preservation of strains when not in use (Hayden and Fang-Yen 2021; Stiernagle 2006).

Selected methods at researchers' disposal, often used in combination with one another and with genetic approaches a genetic enhancer screen, Fujisawa et al. (2007) sought to identify novel genes functioning with the *netrin* receptor. Axon guidance defects are readily apparent in the *DCC/netrin* receptor mutant, *unc-40*, as mutants exhibit an uncoordinated coiling motion while crawling due to miswired motor neurons (another Unc mutant from Brenner's seminal screen) (Brenner 1974; Chan et al. 1996). Initially, the miswiring of axons in *netrin* mutants was characterized using EM reconstruction (Hedgecock et al. 1990), but later studies visualized axon guidance in vivo using Green Fluorescent Protein (GFP), which was first developed for labeling cells and proteins in worm (Chalfie et al. 1994). To search for novel molecules in axon guidance, Fujisawa et al. (2007) mutagenized a strain carrying a moderate *unc-40* mutant allele in search of a novel mutation that would enhance the *unc-40* defect. The goal was to find a mutation that worsened the axon guidance defect, and thus exacerbated the defective movement phenotype. Such enhanced mutants have historically revealed molecules that operate in the same and/or parallel pathways. They isolated one mutant that they named *E*nhancer of *unc-40* *V*entral *A*xon guidance defects, or *eva-1*. Later, researchers noted that *eva-1* was homologous to a gene on Hsa21, *C21ORF63*, which others renamed *EVA1C* (James et al. 2013).

In complementary efforts, pioneering worm researchers John Sulston, Robert Horvitz, Judith Kimble, and others mapped the entire developmental lineage of cells in *C. elegans* by taking advantage of the worm's transparent body. They identified which cells differentiated into which daughter cells from the very first cell division to the last in adult. In doing so, they discovered that specific cells, often neurons, develop only to be culled through a conserved programmed cell death pathway—apoptosis. Relying on the ease of visually identifying mutants with extra or missing apoptotic cells in the transparent worm, they performed a combination of forward, enhancer, and suppressor genetic screens. Through these they discovered several key molecules required for apoptosis (Horvitz 2003; Hedgecock et al. 1983; Kimble et al. 1980). In recognition of their discoveries related to development and apoptosis, Brenner, Sulston, and Horvitz were awarded a Nobel Prize. None of the canonical apoptotic genes are on Hsa21, but several Hsa21 genes directly interact with those genes. Alterations in apoptosis have been linked to DS phenotypes, especially later in life in combination with Alzheimer's disease pathology (Rueda et al. 2013; Tramutola et al. 2016).

C. elegans has also been used to make seminal discoveries related to development and aging. The novel concept of protein phase separation of biomolecular condensates was first introduced using worm. These phase-separated compartments were discovered for their role in establishing polarity during the first few cell divisions with implications for human biology and disease (Folkmann et al. 2021; Boeynaems et al. 2018). Reverse genetic screens that knocked down genes across the genome identified additional genes critical for early development. Hits included chaperonin complex components, of which an Hsa21 gene ortholog called *cct-8* is a constituent (Timmons and Fire 1998; Sönnichsen et al. 2005). Worm studies from the 1980s onward led to concepts important in late development including the discovery of homeobox transcription factors (and their combinations) which determine and maintain the terminal fate of each neuron (Chalfie and Sulston 1981; Way and

Chalfie 1988; Hobert 2021). Finally, due to its tractable lifespan of 2 weeks, *C. elegans* has proven extremely useful in uncovering conserved pathways that genetically determine lifespan. Conserved mechanisms include the regulation of metabolism and protein synthesis via Daf/insulin and mTORC1 (mechanistic target of rapamycin complex 1) (Murphy and Hu 2013; Blackwell et al. 2019; Onken et al. 2022.). Intriguingly, the latter is controlled via the ISR (Integrated Stress Response) in worm, which is known to be dysregulated in patients with DS and may affect lifespan and age-related disease progression in DS (Statzer et al. 2022; Zhu et al. 2019).

A major challenge in neuroscience is to identify each cell in the nervous system. The solution to this problem began with a limitation. Aiming to overcome the inefficiency of immunohistochemistry labeling in worm, Martin Chalfie turned to GFP, the aforementioned fluorescent protein which Osamu Shimomura first isolated from jellyfish and found to glow under ultraviolet light. Chalfie co-opted GFP to label the touch neurons in the living worm green (Chalfie 1995; Chalfie et al. 1994). The green fluorophore was subsequently genetically altered to different versions that collectively emit a crayon-box worth of colors (Tsien 2007). With this palette in hand, researchers later generated NeuroPAL, a *C. elegans* strain that has an identifiable reporter in every neuron (Yemini et al. 2021). Unlike the Brainbow mouse which randomly color codes distinct neurons, NeuroPAL codes each of the 302 neurons deterministically with a specific color combination (excluding the color green) (Yemini et al. 2021; Weissman and Albert Pan 2015). Coupled with positional information, this color code can be used to identify each of the 302 neurons in the nervous system and to identify cellular expression patterns when combined with GFP transcriptional reporters. For more uses of fluorescent proteins, see Table 3.

3 Leveraging *C. elegans* Today: Methods, Tools, and Reagents

C. elegans researchers continue to add to the rigorous and thorough "user manuals" and tools developed by the founders. Here, to add to the background above, we highlight practical information that makes highly insightful, mechanistic studies convenient (Tables 2 and 3). In addition to these advantages, the *C. elegans* research community is tight-knit and known for being welcoming and highly collaborative (Fig. 2c), with a culture of accelerating research by making unpublished data available. For instance, anatomical, genomic, and transcriptomic datasets are available online. All literature, consisting of every paper, meeting abstract, and many unpublished communications can be accessed through the online tool Textpresso.

3.1 Genetic Approaches

The awesome power of *C. elegans* genetics is perhaps best reflected in forward genetic screens. When one does not understand how a biological process works, this approach provides rapid, unbiased insights. By leveraging random genome-wide mutagenesis, one can discover uncharacterized or novel genes required or important for a phenotype. When completed to near saturation, the list of mutant alleles often represents the degree of importance of single genes of large effect in a process (accounting for gene size). Secondary screens for mutations that enhance or suppress existing mutant phenotypes provide insights into genetic interactions, pathway epistasis, and redundancy, as do sensitized screens which require prior knowledge of the existence of parallel pathways and synthetic interactions. With the availability of so many mutants and RNAi reagents, reverse genetic screens are commonly employed now, in part to overcome the limitation of forward genetic screens that may miss essential genes, shorter genes, and genes of lower impact. To genetically engineer worms for experiments, there are several methods available to researchers, including tools discussed further in Tables 1 and 2.

3.2 Research Tools

When it comes to imaging, the small size and transparency of the worm makes it possible to image even the whole nervous system in freely moving animals. Calcium imaging using genetically encodable indicators was pioneered in worm (Kerr et al. 2000; Tian et al. 2009). Numerous biosensors have since been developed including the HYlight glycolytic sensor to study glycolysis in the context of the worm phosphofructokinase, PFK-1.1, an ortholog of the Hsa21 PFKL (Wolfe et al. 2024). This tool could be used in a transgenic strain that carries a GFP-tagged *pfk-1.1* transgene to better understand how overexpression of *pfk-1.1* may perturb glycolytic dynamics (Garde et al. 2022). Numerous studies have also used Fluorescence Recovery After Photobleaching (FRAP) in the worm to study how molecules dynamically shift in expression or location subcellularly in identified cells in vivo. Protein interactions and dynamics can be deduced in vivo using Forster Resonance Energy Transfer (FRET). A recent study utilized these approaches to discover that diffuse PFK-1.1 shifts into aggregates with other glycolytic enzymes, presumably to produce ATP via faster mass action at key locations in the cell (e.g. presynapse of neurons) in response to energy stress (Jang et al. 2021). In addition to measuring excitability, one may also activate or suppress the activity of particular neurons using opto- or chemogenetics (Tsukada and Mori 2021; Prömel et al. 2016). To circumvent the innate repulsion of the worm to blue light, new tools using green light are being developed (Oda et al. 2023). Transparency of the worm also allows ablation of particular neurons or axons by laser under the microscope to study their role or to study axon repair.

One of the primary advantages of *C. elegans* as a model is the ease of its use for high-throughput phenotyping (Table 3). Researchers often achieve sample sizes of 30 to 200 for a given condition, treatment, or genotype to compare to respective controls. The ease of running so many samples in parallel allows for yoked controls that account for potential confounds at play during the assay period. All strains are conveniently tested and stored on a lab bench and/or in an incubator. This workflow permits higher signal to noise with more samples and experiments; however, care is required to account for multiple comparisons to avoid false positive results. Phenotyping of worms with chemicals or stimuli during live imaging can also be scaled up using microfluidic devices with various degrees of automation (Ben-Yakar et al. 2009). For unrestrained worms, advanced imaging and automated tracking technologies work well to track locomotion as well as shape and size; the latter two measures may serve as a correlate for age (Husson et al. 2018). These platforms are capable of handling thousands of worms simultaneously and can run for hours or days at a time (Churgin et al. 2017; Kerr et al. 2022). Studies of Hsa21 genes have not yet benefited from such automated high-throughput approaches in worm.

3.3 Resources and Reagents

In addition to these methods, many resources and reagents have been developed to support *C. elegans* research (Table 1). These enable convenient studies and foster collaboration amongst scientists. A wealth of conveniently organized information across *C. elegans* studies can be found with the meta-search engine, WormBase (Davis et al. 2022). Here, each worm gene has its own webpage where its structure, expression patterns, associated mutant or transgenic strains, and related literature are all highlighted. Additional online resources provide valuable information on neuronal connectivity (WormAltas), gene promoters (Promoterome), and single-cell transcriptional profiles for each worm neuron and other tissues (CeNGEN). Other databases and libraries list transgenic strains and resources for convenient molecular biology, including fosmid, cosmid, and YAC libraries or genomic sequences, and a separate cDNA/ORFeome.

Worm researchers who frequently pivot to new hypotheses are accustomed to ordering new strains several times a month affordably through the mail. The primary repository of genetically modified worms, the *Caenorhabditis* Genetics Center (CGC), houses more than 25,000 strains (Aric L. Daul, "personal communication"). These include classical mutant alleles, deletion alleles from knockout consortia, and 2008 strains from the Million Mutation Project, which yielded over one million unique mutations in genes. Transgenic and mutant strains generated by individual *C. elegans* labs are also submitted daily. In addition to the CGC, the Andersen lab catalogs and shares over 1000 wild *C. elegans* strains and other *Caenorhabditis* species isolated from all over the world as part of the *Caenorhabditis elegans* Natural Diversity Resource (CaeNDR). Because most of these CaeNDR strains are whole-genome sequenced, they allow for easy QTL or GWAS studies to correlate genotype with phenotype via a convenient online pipeline.

Knocking down genes is similarly trivial in *C. elegans*. All it requires is raising worms on a bacterial strain that expresses double-stranded RNA matching your favorite gene. Worms consume the transgenic bacteria, whose dsRNA spreads systemically throughout much of the worm and many of its progeny (Timmons and Fire 1998). Bacterial strains that express dsRNA targeting most genes in the worm genome are available for purchase individually or as the entire library ("Gene Editing and Gene Modulation for Life Sciences" n.d.). Alternatively, worms may be soaked in or injected with dsRNA. It is worth noting that even though RNAi was discovered in *C. elegans*, unlike in other organisms, it is not very effective in the worm nervous system and germline. Thus, working with a sensitized background (e.g. *rrf-3* single or *lin-15 eri-1* double-mutant backgrounds) may be warranted to increase efficiency or penetrance of individuals exhibiting gene knock-down, but this is often a relatively minor disadvantage, given the efficiency of the technique otherwise (Conte et al. 2015).

Finally, to aid the study of DS, the Pierce lab recently generated a library of 50 transgenic worm strains that each overexpress a single Hsa21 gene ortholog and will be made available for public distribution through the CGC.

4 Considerations in Studying Down Syndrome and Hsa21 Genes with *C. elegans*

There are limitations in modeling certain aspects of DS with the worm. However, sometimes a difference between worm and human can be leveraged as an unexpected strength.

4.1 Genes

C. elegans genes have fewer and shorter introns than humans (Blumenthal and Steward 1997; Hodgkin 2001). Most undergo trans-splicing, and the *C. elegans* genome contains some operons, reminiscent of those in prokaryotes (Corsi et al. 2015). Such operon organization sometimes suggests functional relationships, given that genes in operons may be functionally linked, interact, and/or are co-expressed. For example, enzymes that work together to modify collagen are paired on one operon, while the two acetylcholine receptor subunits are paired on another operon (Blumenthal and Gleason 2003; Dossena et al. 2013; Treinin et al. 1998). While up to 80% of *C. elegans* genes have a homolog in humans, there are some for which there is no clear counterpart. These include the immune-pathway Interferon (IFN) genes, of which there are four on Hsa21 (Galbraith et al. 2023). These have an increasingly recognized role in DS pathology and have recently been targeted with promising results for potential therapeutic interventions (Galbraith et al. 2023). Some Hsa21 genes have worm orthologs that might not represent fully functional orthologs, as they are missing important domains. For example, the worm ortholog of Amyloid Precursor Protein (APP), called APP-like molecule 1 (APL-1), does not contain the amyloidogenic sequence, Amyloid beta (Daigle and Li 1993). However, sometimes a lack of conservation can be leveraged as a strength. For instance, the intracellular portion of APP that is well-conserved in worm APL-1 can be studied in isolation of Amyloid beta (Zheng and Koo 2006; Wang et al. 2017a). Alternatively, because no Amyloid beta exists in the worm, different versions and concentrations of Amyloid beta can be introduced, using the worm as a "blank slate" (Alvarez et al. 2022; Teo et al. 2019; Jongsma et al. 2023). Similar approaches have been taken to study other genes with imperfect homology. Even for genes in pathways that are not conserved with one-to-one equivalent steps and molecules in worm and mammal, their general biological processes often are analogous, e.g. apoptosis and necrosis (see pathways and immunity for more) (Oberst et al. 2008; Nikoletopoulou and Tavernarakis 2014). Likewise, general genetic phenomena such as microRNA-based gene regulation and factors for sperm function discovered in worm may pertain to DS, despite divergence in microRNA genes and sperm shape (Soutschek and Schratt 2023; Chu et al. 2006; Vavouri et al. 2007).

Matching genes and genomic regions between *C. elegans* and humans is sometimes tricky. In some cases, a gene (or gene family) has undergone an expansion in either humans or worms leading to numerous paralogs. For example, humans have three superoxide dismutase enzymes and worms have five. The copper-zinc-dependent, cytosolic *SOD1* gene on Hsa21 is orthologous to the worm gene *sod-1,* as well as the less-well characterized worm gene, *sod-5* (Shaye and Greenwald 2011). Likewise, the human ribosomal RNA-processing protein genes *RRP1* and *RRP1B* are orthologous to a single worm gene, *rrp-1* (Nordquist et al. 2018). Other genes have relatively low sequence homology but are clear functional homologs, such as the Hsa21 gene, *SON*, which only shares 11% identity with its worm version, *D1037.1* (Nordquist et al. 2018). However, this ortholog contains both DNA- and RNA-binding domains that are present in the human gene.

Another issue with using worms to study DS is that synteny of genes is not conserved; Hsa21 orthologs are spread across all six chromosomes in worm. In contrast, most Hsa21 orthologs are arranged in the same order (syntenic) on only three chromosomes in mouse. Although Hsa21 ortholog overexpression studies in worms have so far only used a single-gene approach, future technology should enable the generation of artificial chromosome to study multiple overexpressed Hsa21 genes at a time and their syntenic arrangement in worm. For now, different combinations of Hsa21 orthologs can be transformed together using established methods, albeit with inconsistent gene dosage and no control of gene arrangement on the extrachromosomal array.

4.2 Pathways and Immunity

Some molecular pathways display high conservation in *C. elegans*. These include the TGFβ pathway, insulin signaling, canonical Wnt signaling, P38/MAP Kinase, apoptosis, G protein-coupled receptor signaling, Notch signaling, and the heterochronic pathway (Gumienny and Savage-Dunn 2013; Murphy and Hu 2013; Sawa and Korswagen 2013; Kurz and Ewbank 2003; Hedgecock et al. 1983; Fernandez et al. 2020; Greenwald and Kovall 2013; Faunes and Larraín 2016). Other pathways are less well conserved or are unique to humans. For example, the Sonic Hedgehog signaling pathway important for developmental patterning is only partially conserved in *C. elegans* (Bürglin and Kuwabara 2006). This pathway has been implicated in DS phenotypes including failure of cell proliferation in the cerebellum after birth (Dahmane and Ruiz i Altaba 1999). Other partially or less-conserved pathways include the JAK/STAT pathway and TLR pathways, which have also been implicated in DS phenotypes (Galbraith et al. 2023). *C. elegans* does not have the vertebrate adaptive immune system, but does have conserved innate immune and cell-protection systems, including the Unfolded Protein Response and the aforementioned MAP Kinase and TGFβ pathways (Kim and Ewbank 2018). As mentioned, *C. elegans* does not have IFN genes which are considered specific to vertebrates, but functional similarities have been noted between the worm pathogen response and the IFN-1 pathway (Lažetić et al. 2023).

4.3 Neurobiology

Unlike humans, worms do not have a defined brain; however, most of its 302 neurons are concentrated in the head, connected by a nerve ring of axons and dendrites. This renders the worm unsuitable for studying complex mammalian-specific circuits, but handy for studying neuronal circuit motifs that are conserved, including feed-forward circuits, reciprocal circuits, and hubs (Uzel et al. 2022). To uncover more complex signaling patterns than originally anticipated, newer studies are recording and probing the activity of the entire nervous system simultaneously in vivo (Randi et al. 2023; Atanas et al. 2023; Chandra et al. 2023; Susoy et al. 2021). In a recent tour de force, synaptic and extrasynaptic neural signaling in the head were studied, measuring the activity of more than 20,000 pairwise connections (Randi et al. 2023). This revealed an unanticipated amount of extrasynaptic neurotransmission, including neuropeptide signaling, and prognosticates the importance of such non-anatomical signaling in other models.

C. elegans neurons signal via both graded potentials and action potentials. Although the worm does not have a direct ortholog for voltage-gated sodium channels, some worm neurons have the capacity to generate calcium-mediated regenerative currents, including plateau potentials and calcium-mediated action potentials (Mellem et al. 2008; Liu et al. 2018a; Goodman et al. 1998; Lockery and Goodman

2009). Despite these differences, worm behaviors have been shown to be modifiable through experience or training that requires synaptic plasticity (Kano et al. 2008; Hoerndli et al. 2015). The neuronal substrates for plasticity parallel those of human neurons, including the modification of dendritic spines on some neurons, the pruning or growth of new synapses, and the role of AMPA and NMDA receptors in LTP-like processes (Cuentas-Condori et al. 2019; Chandra et al. 2023; Lau et al. 2013; Morrison and van der Kooy 2001; Baidya et al. 2014).

Like mammals, *C. elegans* also has ciliated sensory neurons which utilize dozens of components of intraflagellar transport, many of which were discovered in worm (Inglis et al. 2006). These are nonmotile, primary cilia, which are the more ubiquitous type in humans (Mitchison and Valente 2017). Cilia have important roles not only in sensory neurons, but transiently during development and long-term in central nervous system neurons in mammals (Veland et al. 2009; Lee and Gleeson 2011; Sheu et al. 2022). Thus, uncovering the function of cilia components in worm may generalize to previously assumed mammalian-specific roles. Many ciliated sensory neurons in worm are easy to assess because they project into the environment, permitting lipophilic dye uptake when intact (Tong and Bürglin 2010). This has enabled researchers to identify and study genes involved in human ciliopathies, and deficits in cilia have been observed in DS (Galati et al. 2018).

C. elegans uses most of the same canonical neuronal signaling molecules as humans. The neurotransmitter profile in worms includes glutamate, gamma-aminobutyric acid (GABA), acetylcholine (ACh), dopamine (DA), and serotonin (5-HT). Synthetic pathways and transporters of neurotransmitters are also highly conserved (Hobert 2013). For instance, the first mutants for the dopamine synthetic enzyme, tyrosine hydroxylase, and the GABA transporter were discovered in worm before they were identified in mammals (McIntire et al. 1997; Sulston et al. 1975). Both the "wired" connectome of canonical rapid neurotransmitters and "wireless" connectome of neuromodulators have recently been well described, which aids in deducing which neurons and synapses contribute to behaviors (Fernandez et al. 2020; Ripoll-Sánchez et al. 2023; White et al. 1986; Chen et al. 2006).

Some differences in neuronal signaling and wiring between worm and mammal are noteworthy. For instance, rather than motor neurons projecting axons to innervate muscles, the worm body wall muscles project thin "arms" to reach the motor neurons. In addition, rather than exclusively using cholinergic signaling as in mammals, worm muscles receive dual excitatory cholinergic and inhibitory GABAergic inputs to shape the sinusoidal body bends (Zhou and Bessereau 2019). Invertebrates also lack epinephrine and norepinephrine, but use functional analogs, tyramine and octopamine, for similar behavioral roles (Hobert 2013). Similarly, certain neuropeptides are slightly mismatched. For instance, worms have versions of oxytocin and thyroid hormone peptides with weak sequence homology but functionally analogous roles (Scott et al. 2017; Van Sinay et al. 2017). The lack of certain signaling molecules has enabled orthogonal methods to perturb worm physiology. For instance, the absence of histamine signaling in worm means that a *Drosophila* histamine-gated chloride channel can be used as a chemogenetic tool; the function of neurons and/or muscles in these transgenic worms may be probed conditionally by hyperpolarization with external histamine (Pokala et al. 2014).

4.4 Organs, Systems, and Behavior

Certain human phenotypes, such as the facial and skeletal differences in DS, lack direct analogs to study in *C. elegans*. Despite that, as elaborated above, some cellular signaling pathways are conserved, including Notch, RAS/MAP Kinase, and Robo/Slit/Netrin, which orchestrate development in both worm and mammal (Wang and Sherwood 2011). Similarly, organ systems exclusively present in mammals, such as the heart and circulatory system, liver, thyroid, and complex brain circuits, are absent in the worm. However, aspects of cell biology and physiology that parallel their functionality exist in worm. Two examples are the neuroendocrine secretion of hormones in worm, which mimics glandular hormone release in humans, and life-long pumping of the worm muscular feeding organ, which mimics aspects of cardiac muscle and function (Hartenstein 2006; Benian and Epstein 2011).

C. elegans exhibits a wide repertoire of behaviors that are not as sophisticated, but still analogous to some human behaviors. Worm sensory transduction for olfaction, mechanosensation, and taste operates largely in the same way as in humans (Iliff and Shawn Xu 2020), as do fundamental aspects of patterned motor activity and gait switching, habituation, and association (Olivares et al. 2021; Ji et al. 2021; Vidal-Gadea et al. 2011; Rankin et al. 1990; Becerra et al. 2023). Worms also display behaviors and phenotypes that resemble those induced by different external stimuli or genetic disorders. For example, when worms drink alcohol, they become sluggish and uncoordinated, which researchers have leveraged to study acute intoxication (Davies et al. 2003). Worms display primitive learning such as conditioning to associate a cue (such as concentration of salt in the environment) with an unconditioned response (such as starvation) (Saeki et al. 2001). Recently, researchers found that like people, worms use sleep for memory consolidation, and they uncovered the first evidence for plasticity of an identified synapse in response to learning (Chandra et al. 2023). Worms also communicate via pheromones with a complex chemical code (McGrath and Ruvinsky 2019).

Other worm behaviors and phenotypes might not parallel those in human directly, but these can still be used to identify perturbations in cellular and molecular function and development. For example, egg laying has been used to understand broadly conserved cellular and molecular phenomena (Olson et al. 2023; Trent et al. 1983; Wolinsky and Way 1990). When the HSN pair of egg-laying neurons dysfunctions, worms cannot excite muscles to lay their eggs. They subsequently accumulate eggs which hatch as larvae that writhe within the mother. This striking "bag-of-worms" phenotype has been used as a proxy for degeneration of this neuron pair and has led to the discovery of apoptotic and other neurodegenerative processes (Conradt and Horvitz 1998; Sae-Lee et al. 2020; Fig. 3b). As in humans, the largest organ in the worm is its skin. Mutations in certain collagen genes result in the outer cuticle layer failing to connect with other layers. This causes a "roller"/Rol phenotype in which the worm moves forward, as if twisting in a loose sock, forming loopy tracks. One Hsa21 gene, *COL18A1*, encodes a collagen with homology to the worm gene *cle-1*. Future worm research may uncover consequences for this collagen in tissue integrity and perhaps aging (Ewald et al. 2015). See Table 3 for some common behaviors and phenotypes studied in worm.

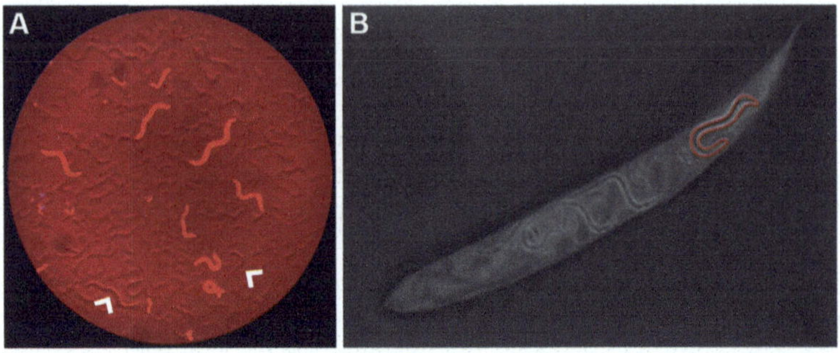

Fig. 3 Studying glowing worms: easy transgenesis and identification of defects using extrachromosomal arrays. (**a**) Transgenic worms that harbor an extrachromosomal array are identified by expression of mCherry in the body wall muscle. The array also contains linked copies of *Y74C10AL.2*, an ortholog of the Hsa21 gene *TMEM50B*, under its endogenous promoter. Worms raised on the same plate that do not carry the array are not red (white arrowheads) and therefore are excluded from the experiment or used as yoked controls. (**b**) The bag-of-worms phenotype. Loss of neuronal and/or muscle control for egg laying leads the hermaphrodite to accumulate eggs rather than depositing them on the plate, and they hatch within. Eventually, the parent dies and the larvae escape

4.5 Development and Life Cycle

Like many invertebrates, *C. elegans* grows through four larval stages before reaching the adult stage where its growth continues but slows. For the most part, all cells follow a deterministic developmental plan which follow specific checkpoints (Martin Chalfie 2019; Onken et al. 2022).

Although the worm body plan is different from that of mammals, underlying cellular and molecular processes are conserved. Horvitz and Victor Ambros also identified "heterochronic" genes that govern developmental transitions and checkpoints (Ambros and Robert Horvitz 1984; Abbott 2003). Perturbation of these genes results in mis-coordination of developmental timing for different tissues. One key member of this pathway includes the first described microRNAs (Lee et al. 1993; Wightman et al. 1993). Ruvkun and Ambros were awarded the Lasker Prize in 2008 for this discovery (Hofschneider n.d.). Since being discovered in worm, microRNAs have emerged as important players in coordinating developmental timing, and their disruption has consequences for developmental disabilities including in DS (Siew et al. 2013).

4.6 Sex

Although *C. elegans* is primarily maintained as hermaphrodites in the lab, there are also males, making the worm suitable for studying sex-dependent effects. Males and hermaphrodites exhibit behavioral and morphological differences, and the sexual fate of their cells is primarily determined cell-autonomously by the number of X chromosomes, rather than by hormones. This allows for the experimental exploration of sex differences at the cellular level by ectopically expressing sex-determinant transcription factors to create mosaic animals that are part hermaphrodite and part male (Lee and Portman 2007).

The hermaphroditic worms (XX genotype that generate sperm and self-fertilize) typically make up 99.9% of a laboratory stock, which can be convenient for studying clonal populations, but inconvenient for conducting crosses or studying males. The low frequency of males can be increased by crossing hermaphrodites with the rare males. Researchers often take advantage of the "high-incidence of males" *him-5* mutant strain that appears wild type but exhibits higher than 40% males. Crossing males with hermaphrodites via sexual reproduction leads to about 50% male F1 progeny. Alternatively, strains containing both males and hermaphrodites can be maintained on plates in a 15 °C incubator to slow growth and pull males from as needed for crossing.

5 Validity of the Model

In deciding how to best approach an experimental question relevant to DS in the worm, one should consider the following:

5.1 Construct Validity: Generation or Selection of the Model

1. Which genetic elements does one want to include? One should first consider orthology and functional conservation; ideally genes should have significant sequence homology or conservation of important functional domains.
2. Does one want to transform a single or multiple copies of the gene? One may also choose to study genes singly or in combination. Dozens of unique transgenes (>40) may be expressed in combination using extrachromosomal arrays (e.g. Yemini et al. 2021).
3. Does one want to insert a single copy of the gene, or for it to be carried in an extrachromosomal array? Traditionally, the latter approach is favored for its speed, but it must be balanced with several considerations. First, an extrachromosomal array contains many extra copies of a transgene—much more than the single extra copy in DS. Copy number may be estimated using RNAseq approaches. However, gross overexpression may be acceptable, or even desirable, if you want to "hit the biological system with a hammer." If many extra copies of a transgene do not result in overt phenotypes, one could (liberally) deduce that a single-extra copy would similarly have no effect. Second, an extrachromosomal array is only passed to a fraction (20–90%) of progeny, depending on the array. This necessitates a linked transgene reporter or a selection system.

Fluorescent reporters are handy for simultaneously labeling cells of interest, but due to their ectopic expression, may adversely affect phenotypes. Other selection systems involve transformation of a transgene(s) of interest along with a wild-type gene that rescues a severe phenotype in a mutant background (e.g. *unc-119* or *pha-1* mutants with uncoordinated motion or the temperature-sensitive absence of the feeding organ, respectively). However, the genetic background or rescue transgene may also inadvertently disrupt phenotypes. For instance, the once popular *lin-15b* selection system inadvertently interferes with RNAi efficiency. Therefore, controls expressing the transgenic reporter or rescue gene alone are essential.

4. Should one integrate the transgene(s) into the genome? An older but useful approach is to knock in the multi-copy extrachromosomal array randomly or precisely into the endogenous genome if more than one copy of the gene is desired (Mariol et al. 2013; Yoshina et al. 2015). Low-copy transgene integration at random locations may be accomplished with gene guns with minimal training, but precise targeted single-gene insertion is now common using CRISPR (Schweinsberg and Grant 2013; Dickinson and Goldstein 2016). Because random insertion may occasionally disrupt endogenous genes, characterized "safe harbor" insertion sites are utilized that do not cause overt side effects. See Table 4 for additional considerations.
5. Does one prefer to study genetically diverse wild isolates from across the world rather than the lab wild-type strain, N2? The N2 strain, isolated in 1956 from a mushroom compost in Bristol, England, exhibits some idiosyncrasies (Caswell-Chen et al. 2005). Famously, it carries a mutation in an ortholog of the neuropeptide Y receptor, *npr-1*, which results in pleiotropic phenotypes. More than any other artificially selected mutation, the *npr-1* gain-of-function variant makes the strain well-adapted for domestic life in the lab, but also eliminates hermaphrodite-hermaphrodite social attraction (Sterken et al. 2015). This anti-social behavior keeps worms apart which is convenient for picking individuals but may confound studies on some social behaviors. Studying *C. elegans* strains isolated from around the globe offers one the option to study how natural genetic variation affects phenotypes related to genes important to DS. These strains have been full genome sequenced and many have been subjected to RNAseq, meaning one can simply order strains that have predicted high-impact variants in, or specific expression levels for, combinations of genes of interest related to DS (Zhang et al. 2022).

5.2 Face Validity: Choosing Meaningful Phenotypes to Study

The worm as a whole does not resemble a human, with or without Down syndrome. However, there are similarities in behavior, some gross phenotypes, and, most obviously, cellular, molecular, and developmental events and processes that directly mirror those in human. As mentioned in previous sections, many aspects of neuronal signaling and intracellular signaling are highly conserved, leading to discoveries

Table 4 Transgenesis techniques in *C. elegans*: advantages and disadvantages in light of application in DS research

Method	Advantages & Disadvantages	Considerations for applications in DS research, proposed measures for circumventing disadvantages
Multi-copy insertion by injection of plasmids or PCR products which form an "extrachromosomal array" (Stinchcomb et al. 1985; Mello and Fire 1995)	(+) Most efficient, does not require cloning or special reagents (+/−) Mosaicism exists in some transgenics; can be leveraged for mosaic analyses (−) Requires selection of animals with a transgenic marker, often a fluorophore (included because the extrachromosomal array is not integrated into the genome and not fully transmissible). Therefore, have to select progeny carrying the array by presence of the marker for propagation/continued study. (−) Not easy to sequence due to highly repetitive nature of arrays, can include duplications and deletions (−) Might not perfectly mimic endogenous expression pattern and silenced in the germline (can be advantageous) (−) Can create RNAi-like suppression	To avoid studying mosaics, if studying behavior, use a fluorescent transgenic marker that is present in many cells (e.g. the body wall marker shown in Fig. 3) and do not study animals where that marker isn't in all of the expected cells. There are alternative transgenic markers that may enable more rapid identification of transgenic worms, but these may confound worm behavior. Depending on the phenotype being studied, one can opt for these other approaches (including gross phenotypic markers or antibiotic-selection). By using PCR products, one can include an endogenous promoter and UTRs, the concentration can be lower, and the number of integrants per injection is higher than with plasmids. One can also circumvent size-limitations with plasmids (although obtaining large PCR products themselves may be difficult). Perform control experiments to verify whether marker-only animals perform differently from wild type and run these controls yoked to your experimental strains (marker + gene) when possible. Some markers are known to be more disruptive than others.
Multi-copy integration by gamma/UV irradiation or trimethylpsoralen- Mello and Fire (1995), Evans (2006), and Kage-Nakadai et al. (2012)	(+) Efficient integration (+/−) Often silenced in the germline (−) First requires generation of animals harboring extrachromosomal arrays, followed by treatment to induce integration. (−) Requires UV irradiation or a mutagen (−) Introduces potential secondary mutations (−) Random, not site-directed integration	Best for use when high-copy integration is required and therefore might not be best option

(continued)

Table 4 (continued)

Method	Advantages & Disadvantages	Considerations for applications in DS research, proposed measures for circumventing disadvantages
Multi-copy integration by miniSOG (mini Singlet Oxygen Generator)- Noma and Jin (2018)	(+) One-step; allows integration by injection alone (+/−) Often silenced in the germline (−) Requires blue light (−) Introduces potential secondary mutations (−) Random, not site-directed integration	Best for use when high-copy integration is required and therefore might not be best option
Multi-copy integration by CRISPR/Cas9- Yoshina et al. (2015)	(+) Lower likelihood of secondary mutations (+) Site-directed integration (+/−) Often silenced in the germline (−) First requires generation of animals harboring extrachromosomal arrays (−) More labor-intensive than miniSOG or irradiation/mutagen-employing techniques	Best for use when high-copy integration is required and therefore might not be best option
Low-copy integration by microparticle bombardment/ biolistic transformation- Praitis et al. (2001), Schweinsberg and Grant (2013)	(+) Efficient relative to single-copy insertion methods (+) Doesn't require dexterity or training that microinjection does nor the microinjection set-up (+/−) Expresses in the germline (−) Need expensive equipment (a "gene gun") (−) Must be carried out on a mutant background for selection by complementation (−) Require many worms and a slow selection process (−) Random, not site-directed integration	Likely advantages of other approaches render them more useful than this one, except in a minority of cases. In instances in which one is interested in studying Hsa21 genes in the germline, this approach could be useful. It could also be useful for labs new to worm who wish to avoid costs of an injection rig or time spent learning to inject.

Single-copy integration by transposon-mediated transgenesis: Mos1-mediated single copy insertion (MosSCI)-Frøkjaer-Jensen et al. (2008)	(+) Integration into defined safe harbor loci (6 options) (+) Allows multiplexed integrations (−) Requires Mos1 transposable elements, and can lead to confounding effects from second-site Mos1 insertions (−) Relatively low insertion frequency relative to newer methods (−) Requires screening for positive and negative selection markers	Has largely been replaced by CRISPR-based methods
Single-copy or multi-copy integration by transposon-mediated transgenesis: Modular safe-harbor transgene insertion (MosTI)-El Mouridi et al. (2022)	(+) Integration into defined loci (+) Compatible with targeting vectors with popular promoters and fluorophores Split selection markers, so no negative selection (+) Selection markers can be excised with cre recombinase or exchanged for one another (+) Can be combined with in vivo recombineering to minimize cloning (+) Allows relatively efficient multiplexed integrations (−) Requires Mos1 transposable elements and can lead to confounding effects from second-site Mos1 insertions (−) Requires cloning unless combined with in vivo recombineering	Could enable the integration of many different transgenes, and while at random sites, would be efficient

(continued)

Table 4 (continued)

Method	Advantages & Disadvantages	Considerations for applications in DS research, proposed measures for circumventing disadvantages
Single-copy integration by transposon-mediated transgenesis by minimal Mos1 transposon (miniMOS)- Frøkjær-Jensen et al. (2014)	(+) Highly-efficient multiplexed integration (+) Integration into defined safe harbor loci (6 options) (+) Enables insertion of transgenes up to 45 Kb (−) Requires mapping of insertion sites	Could enable the integration of many different transgenes, and while at random sites, would be efficient
Single-copy integration with CRISPR alone- Dickinson and Goldstein (2016)	(+) Precise localization can be achieved/ highest flexibility in landing site determination (+) Avoids mutations from chemical mutagens or potentially introduced by transposable elements (+/−) Expresses in the germline (−) Requires specific vectors for targeting each location (−) Requires selection with substantial screening	Potentially inefficient with large inserts
Single-copy integration via SKI LODGE- single-copy knock-in loci for defined gene expression- Silva-García et al. (2019)	(+) Integration into four safe harbor landing sites, engineered for easily inserting your gene of interest in frame with one of four commonly used promoters (for ubiquitous, neuronal, muscle, or germline expression) (+) Enables tissue-specific epitope tagging (+) Allows for co-expression of gene of interest with nonphysically linked wrmScarlet (+) No cloning necessary (+) Can be used to insert two templates concurrently (−) Not yet optimized for insertions larger than 3 Kb (−) Requires outcrossing of selection marker and potential off-site mutations	Could enable study of Hsa21 genes in specific tissues, but to compare many would require significant time on preparation of reagents outcrossing. Other methods that allow multiplexing might be preferable, especially given insert size limitations of SKI LODGE at this time.

Single-copy integration by self-selection based on a split antibiotic selection marker-Stevenson et al. (2020)	(+) High-efficiency screening; makes use of in vivo recombineering so no cloning nor molecular-validation necessary for selection of transformants Selection marker can be excised with cre recombinase (−) Only one antibiotic (hygromycin) and landing site to work with thus far (−) Relatively low-efficiency transformation (−) Doesn't enable integration of extrachromosomal arrays	The slightly more recently developed MosTI method is able to achieve what this one does with additional capabilities (like alternative sites available) and therefore likely can be leaned upon instead.
Single-copy integration by Recombinase-Mediated Cassette Exchange- Nonet (2020)	(+) Integration into specific landing sites (+) Selection marker can be excised by Cre (+) Enables bipartite expression systems of genes nonphysically linked genes (e.g. drivers and reporters for efficient expression of single-copy transgenes) (+/−) Higher efficiency than most other methods but relatively low-efficiency of transformation in comparison to MosTI (−) Could be limited in size of transgenes it allows (−) Doesn't enable integration of extrachromosomal arrays (−) Enables multiplexing in theory but not advanced enough yet to be preferable to doing separate injections	The slightly more recently developed MosTI method is able to achieve what this one does with higher insertion efficiency but could be combined with the recombineering portion of this approach to reduce cloning.

Reviewed and discussed in further detail in Nance and Frøkjær-Jensen (2019) relevant to humans in axon guidance, apoptosis, mitochondrial function, and beyond. For further discussion of relevant phenotypes and behaviors, see *Organs, Systems, and Behavior*, Table 3, and *Modeling specific DS-associated phenotypes or focusing on particular genes in C. elegans*.

5.3 Predictive Validity: Ensuring Relevance for Human Biological and Pathological Phenomena

To test for predictive validity of findings in *C. elegans*, one should be prepared to test if they generalize to other organisms. By working with orthologs or genes with conserved functions or conservation of protein domains or motifs, one can increase the likelihood of findings that are valid across species. One may also choose to "humanize" the worm and study a human gene in lieu of the worm version.

C. elegans has also been used to screen for drugs that benefit human patients. For instance, a worm model was generated for Phosphomannomutase 2 deficiency (PMM2-CDG), which is the most common congenital disorder of glycosylation (Iyer et al. 2019). A high-throughput screen found that the diabetic neuropathy drug, Epalrestat, counters the slow growth and inviability of the PMM2-CDG mutant worm. Follow-up studies found that Epalrestat increased PMM2 enzymatic activity in PMM2-CDG patient fibroblast lines. Epalrestat showed promise in reducing symptoms in PMM2-CDG patients and advanced to phase III clinical trials (Maggie's Pearl, LLC 2024). This study highlights the promise of moving from a worm model of a human genetic disorder to novel drug treatment for patients in record time.

Predictions using *C. elegans* may go beyond specific genes to higher-order concepts. For instance, the Pierce lab recently used data mining to identify genetic factors that explain the pattern of which genes do or do not cause overexpression phenotypes in worm (Sanchez et al. 2024). These predictive factors could drive additional analyses in human or vertebrate studies. We discuss one such example in the following section.

6 Approaches Leveraged for the Study of Down Syndrome in *C. elegans*

6.1 Studies on Single Hsa21 Genes

Thus far, only a handful of Hsa21 genes have been the focus of individual *C. elegans* studies. These include *APP/apl-1*, *SOD1/sod-1/sod-5*, *DOPEY2*/pad-1, *DYRK1A/mbk-1*, *DSCR1/RCAN1/rcan-1*, *DONSON/dnsn-1*, *EVA1C/eva-1*, *PFKL/pfk-1.1*, and *SYNJ1/unc-26*. Several such examples are highlighted below.

APP has primarily been studied for its involvement in Alzheimer's disease (AD). The worm equivalent of *APP* is a gene called *apl-1*, or *AP*P-*l*ike protein *1*. Because APL-1 lacks the key amyloidogenic portion of APP that leads to toxic plaque formation, it is not a straightforward one-to-one ortholog (Alvarez et al. 2022). Therefore, some labs have opted instead to model aspects of AD by expressing human Amyloid beta peptide or full-length human *APP* in *C. elegans* (McColl et al. 2012). Using the latter approach, the Pierce lab observed overt neurodegeneration in a particular class of cholinergic neurons which was further exacerbated by overexpression of human *APOE* (Mondal et al. 2018; Sae-Lee et al. 2020; Yi et al. 2017). This occurred with the ε 4 but not ε 3 variant of *APOE*, matching the heightened risk of AD found in people carrying ε 4 over ε 3 variants. Homozygous *APOE* ε4 is now thought to represent a genetic cause of AD and is the major risk factor for hastening early-onset AD in individuals with DS (Bejanin et al. 2021; Xu et al. 2024). While neurodegeneration is well documented in cases of human AD, many mouse models of AD do not recapitulate this important phenotype, so its presence in worm offers an attractive model (Jankowsky and Zheng 2017; Shekari and Fahnestock 2021). The worm is poised to reveal more about *APP* and its genetic interactors in the context of DS in future studies.

The Hsa21 gene, *SOD1* (Superoxide Dismutase 1) has been studied in worm primarily due to its association with another neurodegenerative disorder, Amyotrophic Lateral Sclerosis (ALS). *SOD1* has two worm orthologs, *sod-1* and *sod-5*. Superoxide dismutase enzymes play a protective role against reactive oxygen species (ROS). Neomorphic variants of *SOD1* cause ALS (Ghasemi and Brown 2018). To model ALS, worm researchers opted to overexpress or CRISPR-replace the mutant, rather than the wild-type version of *sod-1*(which would better model DS) (Baskoylu et al. 2018). However, *SOD* genes have also been associated with longevity, so some labs have studied the overexpression of wild-type *sod-1* in worm to uncover longevity mechanisms. Beneficial effects of overexpressed *sod-1* on longevity were intriguingly traced to altered Daf/insulin signaling rather than anti-ROS mechanisms (Cabreiro et al. 2011). Sanchez et al. (2024) found that overexpressing wild-type *sod-1* caused worms to be significantly developmentally delayed. It remains to be seen whether the developmentally delayed worms are in fact longer lived. Future studies will extend study on the potential dual detrimental and beneficial effects of overexpressed *SOD1* in the context of DS.

The Hsa21 genes that were initially targeted for study in *C. elegans* because of their potential association with DS are *DOPEY2/pad-1*, *DYRK1A/mbk-1*, and *RCAN1/rcan-1* (previously named *DSCR1*) (Guipponi et al. 2000; Raich et al. 2003; Lee et al. 2003). *pad-1* is a large gene with many introns, so researchers initially opted to study effects of knocking it down with RNAi, and found that it is required for patterning of the embryo (hence its name, *pa*tterning-*d*efective *1*). Recently, through an RNAi screen, *pad-1* was found to modulate synapse vesicle release in worm (Beer et al. 2018), which appears to warrant further investigation in other DS models. In contrast, the kinase *DYRK1A/mbk-1* has been studied extensively in other model organisms, but not as much in worm. This is likely due to a lack of salient loss-of-function and overexpression phenotypes in worm (Raich et al. 2003).

More recently, *mbk-1* was found to have a role in longevity, a phenotype which has since been observed in *Drosophila* but not yet followed up in mammalian models (Mack et al. 2017; Lowe et al. 2019). Finally, RCAN-1 was found to inhibit phosphatase activity by calcineurin (Lee et al. 2003). Coincidentally, two studies found a role for RCAN-1 in longevity in worm potentially by activating, rather than inhibiting, calcineurin (Sutphin et al. 2017; Li et al. 2016). Further work is needed to unravel the pleiotropic relationships of these genes.

C. elegans studies of loss-of-function mutations of *DONSON/dnsn-1, EVA1C/eva-1, PFKL/pfk-1.1* and *SYNJ1/unc-26* have revealed their required roles in DNA replication, neurodevelopment, cellular metabolism, and synaptic transmission, respectively. Interest in *DONSON* was piqued after it was identified as causal for microcephalic dwarfism and intellectual disabilities (Reynolds et al. 2017). Recently, work in *C. elegans* led to the discovery of its requirement in replication fork assembly (Xia et al. 2023). There is still much to learn about *dnsn-1* and its potential involvement in DS. Similarly, *eva-1* has only recently begun to be studied in general, much less for its potential roles in DS. *eva-1* was first characterized in worm as a co-receptor for Slit in axon guidance (Fujisawa et al. 2007), and its corpus callosum expression pattern in mice corroborates an analogous function in mammal (James et al. 2013). PFKL was identified early for its potential role in DS, but it has only recently received attention in worm. As mentioned previously, the Colon-Ramos lab found that PFK-1.1 forms intracellular condensates upon metabolic stress (Jang et al. 2021). Given this finding, overexpression of *pfk-1.1* may conceivably cause mislocalization or inappropriate accumulation of PFK-1.1 in hypoxic or even normal conditions. Finally, synaptojanin, *SYNJ1/unc-26*, has been studied in worm for its role in neurotransmission and recycling synaptic vesicles, since its identification through Sydney Brenner's landmark mutant screen (Harris et al. 2000; Brenner 1974). The *unc-26* gene is sizable, with four isoforms that include 9 or 10 exons each. The Pierce lab recently found that overexpressing *unc-26* caused developmental delay and a deficit in egg-laying that can be leveraged in future studies to understand how this gene may play into DS phenotypes (Sanchez et al. 2024). Overall, each of these genes appears to function in a highly conserved pathway in worm, suggesting that future studies may yield useful insights regarding their roles in DS phenotypes.

6.2 Studying Multiple Hsa21 Genes

Historically, forward genetic screens carried out in *C. elegans*, including the original one by Brenner, perhaps surprisingly did not produce many mutants that represent Hsa21 genes. This may be explained, in part, by the fact that 15 of the 85 conserved Hsa21 orthologs appear essential in *C. elegans* (Fig. 4b) (Nordquist et al. 2018). Why, however, haven't many of the other Hsa21 orthologs surfaced with overt phenotypes? Because conventional forward genetic screens tend to catch loss-of-function mutations, Hsa21 orthologs with redundant functions may have dodged

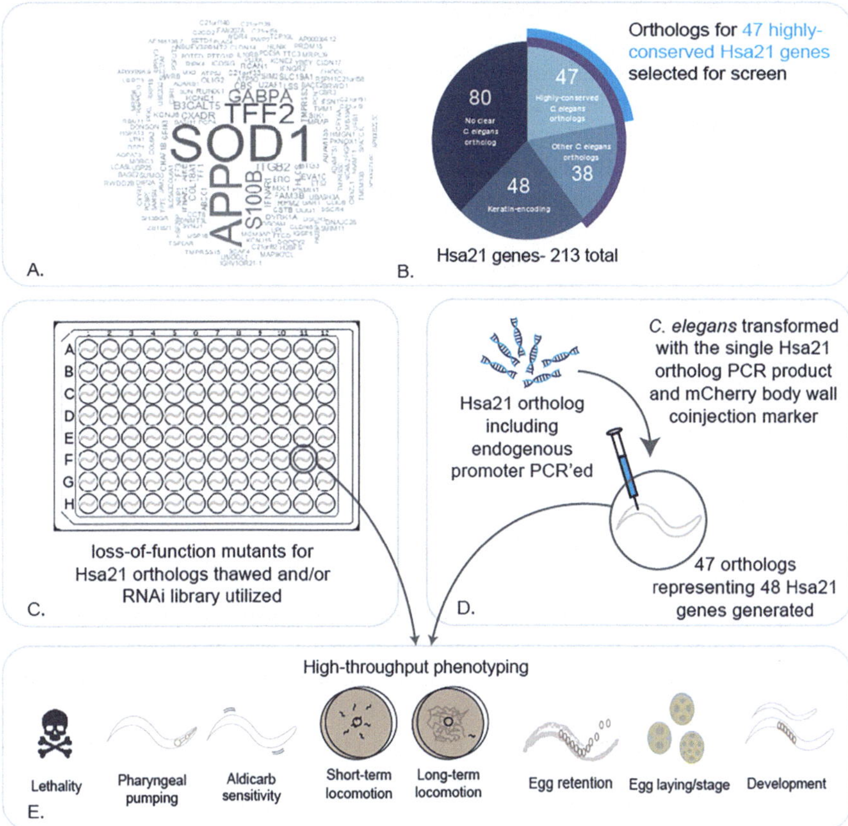

Fig. 4 *C. elegans* screens of Hsa21 genes have revealed in vivo functions, overexpression phenotypes, and dosage-sensitivity mechanisms. (**a**) Word cloud depicting the uneven study of Hsa21 genes across all publications on PubMed represented by font size on a log scale (as of 2023). (**b**) About a third of 213 protein-coding Hsa21 genes (or half, excluding those that encode keratin), have a *C. elegans* ortholog (purple arc). Of these, about half are highly conserved (blue arc). (**c**) One can easily thaw loss-of-function (LF) mutants *C. elegans* or utilize an RNAi library stored in 96-well plates from cryopreservation. (**d**) Setup for an overexpression (OE) screen in worm. (**e**) Example phenotypes used in recent OE and LF screens (Nordquist et al. 2018; Sanchez et al. 2024)

identification when mutated singly in screens. This especially may be the case for the 38 less-well conserved Hsa21 orthologs, each of which has numerous paralogs (Fig. 4b). It is also possible that loss of certain Hsa21 orthologs is not problematic, while overexpression of these genes would produce phenotypes.

To gain unbiased insight into numerous Hsa21 genes simultaneously, the Pierce lab systematically studied a collection of Hsa21 orthologs using *C. elegans* (Fig. 4). Using orthology prediction programs Ortholist2 and inParanoid, they determined that, excluding keratin-encoding genes, about one third of Hsa21 genes have a highly conserved worm ortholog (Fig. 4b) (Shaye and Greenwald 2011; Kim et al.

2018; Nordquist et al. 2018). Initially, because many of these ~50 genes had not been studied extensively before, they conducted a systematic loss-of-function (LF) screen to identify genes with roles in muscle, neurons, or synaptic transmission. To do this, they performed simple tests of locomotion and feeding, as well as a pharmacological assay using aldicarb to assess synaptic transmission (for more on these, see Table 3, Fig. 4c, e). They identified 10 Hsa21 genes required for normal muscle, neural, or synaptic function, and another 15 genes that appear to be essential for survival. This highlights the power of a systematic loss-of-function screen for identifying overlooked Hsa21 genes that may contribute to DS phenotypes and warrant further study across DS models.

More recently, the Pierce lab completed the converse screen by systematically overexpressing each Hsa21 ortholog one at a time in *C. elegans* (Sanchez et al. 2024, Fig. 4d). Compared to controls, 28 Hsa21 orthologs gave rise to overexpression (OE) phenotypes, while 20 did not. These complementary studies provide a foundation for future research to elucidate mechanistic roles of single Hsa21 orthologs when overexpressed.

In addition to providing information on single Hsa21 orthologs, the Pierce lab used the pattern of hits and misses from their OE screen to mine for clues on general properties of genes that may be intolerant to duplication (Sanchez et al. 2024). They used logistic regression to determine which of dozens of factors associated with each gene predict OE phenotypes. Interestingly, genes that displayed a ceiling in their expression level across wild *C. elegans* collected from around the world were more likely to give rise to OE phenotypes. This suggests that worms expressing genes beyond this ceiling may be selected against in nature. Intriguingly, this pattern extended beyond *C. elegans* to human. They examined the natural expression distribution of genes in healthy humans using the GTEx database. Remarkably, human genes that cause severe medical conditions when duplicated, such as Alzheimer's disease (*APP*), Parkinson's disease (alpha synuclein), and Familial hypercholesterolemia (*PCSK9*), each exhibited a ceiling in expression level in healthy subjects (Carithers et al. 2015). They next inspected over 1000 genes located within segmental copy-number variation (CNV) regions that cause severe phenotypes. About half of these CNV genes displayed a population expression ceiling. Conversely, they found that almost all of >1000 human genes that were documented to be duplication tolerant did not display population expression ceilings (Liao et al. 2023). This work highlights how the study of many genes at a time in worm may help reveal fundamental patterns that are challenging to notice in less genetically tractable systems.

7 Opportunities in Applying *C. elegans* to Answer Outstanding Questions in Down Syndrome

Many outstanding questions remain regarding the basic etiology of DS. The majority of Hsa21 genes have only been studied incompletely or for their involvement in conditions beyond DS. We found that when considering all protein-coding Hsa21 genes on Pubmed, 95% of the body of literature focused on only 20 of the 213 Hsa21 genes as depicted in Fig. 4a. This leaves much to be discovered. With the new addition of the library of overexpression *C. elegans* strains by the Pierce lab to the *Caenorhabditis* Genetic Center, researchers inside and out of the worm community may order strains to study LF or OE of their favorite Hsa21 ortholog.

Identifying genes that contribute to phenotypes in DS could direct the development of therapeutics to improve quality of life for individuals with DS. Attempts have been made using mouse models to discover drugs to treat DS phenotypes, with hints of promise from targeting specific Hsa21 genes. Most efforts have been directed at strategies to mitigate early-onset Alzheimer's pathology in DS caused by the extra copy of *APP*, including anti-amyloid drugs (Johannesson et al. 2021). The worm may play a role in this effort with its ease of high-throughput drug screening for neurodegenerative phenotypes and uncovering drug mechanisms (Mondal et al. 2018). The next-most targeted gene is *DYRK1A*, the aforementioned kinase whose ortholog is *mbk-1* in worm. Numerous studies point to the detrimental effects of *DYRK1A* overexpression, but some have identified positive effects, and drugs developed that target it have yet to prove effective in patients (Delabar et al. 2024; Neumann et al. 2018; Delabar et al. 2023; Moyer et al. 2023).

7.1 Future Study of DS-Related Phenotypes or Hsa21 Genes

DS is caused not just by one extra gene but many across Hsa21. For studying combinations of Hsa21 genes and their potential interactions, even an efficient genetic model like *C. elegans* will reach a practical limit on the number of unique combinations that can be tested. Even if one chose to focus on the ~50 highly conserved Hsa21 worm orthologs, testing all unique pairwise comparisons for overexpression phenotypes would require generating 1.12×10^{15} strains. Alternatively, one could perform an unbiased "transgenic screen" in which worms are transformed with a dilute cocktail of unique Hsa21 orthologs, such that transgenic progeny carry unique combinations of transgenes. The progeny may then be sorted into those with or without overexpression phenotypes and sequenced to reveal the combination of genes being overexpressed. For overexpression or knockdown screens, computational modeling may be helpful to hone in on gene interactions worth prioritizing.

One way to overcome the limitations of generating strains with multiple genes mutated or overexpressed is to take advantage of CaeNDR (Crombie et al. 2024). CaeNDR is a biobank of genetically diverse *C. elegans* (and closely related species) worms collected from across the globe (Table 1). This powerful resource for worm researchers provides valuable insights on natural variation of genes and can help save time spent on model generation. Currently, over 1000 unique strains have been full-genome sequenced and over 200 strains transcriptionally profiled (Zhang et al. 2022). One could select and phenotype strains that tend to express their favorite combination of Hsa21 orthologs higher than average. Positive results found with wild strains could be followed up in the wild backgrounds and/or in the lab wild-type strain, N2, in further studies. Potential relationships among the expression level of Hsa21 genes and all genes may be probed conveniently in silico from this collection and followed up with empirical experimentation using gene overexpression and knockdown.

7.2 Towards a Holistic Model of DS in **C. elegans**

In addition to generating models that focus on particular Hsa21 genes, a worm model of DS with arguably the highest construct validity will include just one extra copy of each Hsa21 gene or ortholog. Synthetic biologists have succeeded at reconstructing the yeast genome de novo with artificial, synthesized DNA (Zhao et al. 2023). Current approaches allow overexpression of dozens of Hsa21 genes, albeit in random order and relatively uncontrolled in dosage. However, in the future a worm strain may be generated that carries many Hsa21 transgenes that preserves the synteny of genes on human Hsa21.

There are merits to several, nonmutually exclusive approaches. A conservative approach could entail the co-transformation of just the 48 highly conserved Hsa21 orthologs. Another could include all known Hsa21 orthologs. Further, one could include all Hsa21 genes in a worm model of DS, using human cDNAs driven by worm promoters. Nevertheless, careful consideration of different modeling approaches reveals various pros and cons, mirroring the current situation with dozens of unique transgenic mouse models of DS (Herault et al. 2017). As with mouse, the more DS worm models, the higher the potential for discovery.

In conclusion, we encourage others to address fundamental questions in DS by turning to the convenient and powerful *C. elegans*:

- When and how do Hsa21 genes contribute to DS-related phenotypes throughout analogous stages of development, adulthood, and aging?
- How do Hsa21 genes interact to give rise to DS-related phenotypes? Does overexpression of Hsa21 genes and modifiers, such as *APOE*, suppress or potentiate *APP*-induced neurodegeneration, and how?
- What generalized principles and mechanisms underlie the propensity for genes to give rise to CNV phenotypes? How might these be predicted and countered?
- Does nontranscriptional dosage compensation occur in DS?
- Which factors contribute to incomplete penetrance of certain genes and phenotypes that relate to comorbidities in DS?
- What pathways might be most amenable to therapeutics in DS without side effects?
- What currently available and FDA-approved drugs could be used to counter DS-related phenotypes, or what novel compounds are good candidates?

Acknowledgements We are grateful for the insights and edits from Jesse Cohn, Benjamin Clites, Andy Cardona, and reviewers for feedback on drafts. We also thank members of the *C. elegans* community who have spent years curating and painstakingly developing the impressive tools and "user manuals" we highlight here. This work was funded in part by the NIH (R01GM122463, RF1AG057355, R21OD032463—J.P. and F31-HD105424 to S.S.).

Works Cited

Abbott, Allison L. 2003. "Heterochronic Genes." *Current Biology* 13 (21): R824–25. https://doi.org/10.1016/j.cub.2003.10.011.

Ahier, Arnaud, and Sophie Jarriault. 2014. "Simultaneous Expression of Multiple Proteins under a Single Promoter in *Caenorhabditis elegans* via a Versatile 2A-Based Toolkit." *Genetics* 196 (3): 605–13. https://doi.org/10.1534/genetics.113.160846.

Al-Sheikh, Umar, and Lijun Kang. 2020. "Mechano-Gated Channels in *C. elegans*." *Journal of Neurogenetics* 34 (3–4): 363–68. https://doi.org/10.1080/01677063.2020.1832091.

Alvarez, Javier, Pilar Alvarez-Illera, Jaime Santo-Domingo, Rosalba I. Fonteriz, and Mayte Montero. 2022. "Modeling Alzheimer's Disease in *Caenorhabditis elegans*." *Biomedicines* 10 (2): 288. https://doi.org/10.3390/biomedicines10020288.

Ambros, Victor. 1989. "A Hierarchy of Regulatory Genes Controls a Larva-to-Adult Developmental Switch in *C. elegans*." *Cell* 57 (1): 49–57. https://doi.org/10.1016/0092-8674(89)90171-2.

Ambros, Victor, and H. Robert Horvitz. 1984. "Heterochronic Mutants of the Nematode *Caenorhabditis elegans*." *Science* 226 (4673): 409–16. https://doi.org/10.1126/science.6494891.

Amrit, Francis Raj Gandhi, Ramesh Ratnappan, Scott Alexander Keith, and Arjumand Ghazi. 2014. "The *C. elegans* Lifespan Assay Toolkit." *Methods*, *C. elegans* Methods, 68 (3): 465–75. https://doi.org/10.1016/j.ymeth.2014.04.002.

Ashley, Guinevere E., Tam Duong, Max T. Levenson, Michael A. Q. Martinez, Londen C. Johnson, Jonathan D. Hibshman, Hannah N. Saeger, et al. 2021. "An Expanded Auxin-Inducible Degron Toolkit for *Caenorhabditis elegans*." *Genetics* 217 (3): iyab006. https://doi.org/10.1093/genetics/iyab006.

Atanas, Adam A., Jungsoo Kim, Ziyu Wang, Eric Bueno, McCoy Becker, Di Kang, Jungyeon Park, et al. 2023. "Brain-Wide Representations of Behavior Spanning Multiple Timescales and States in *C. elegans*." *Cell* 186 (19): 4134–4151.e31. https://doi.org/10.1016/j.cell.2023.07.035.

Avery, L., and S. Wasserman. 1992. "Ordering Gene Function: The Interpretation of Epistasis in Regulatory Hierarchies." *Trends in Genetics: TIG* 8 (9): 312–16. https://doi.org/10.1016/0168-9525(92)90263-4.

Avery, Leon, and Young-Jai You. 2012. "*C. elegans* Feeding." *WormBook: The Online Review of C. Elegans Biology*, May, 1–23. https://doi.org/10.1895/wormbook.1.150.1.

Azevedo, Frederico A. C., Ludmila R. B. Carvalho, Lea T. Grinberg, José Marcelo Farfel, Renata E. L. Ferretti, Renata E. P. Leite, Wilson Jacob Filho, Roberto Lent, and Suzana Herculano-Houzel. 2009. "Equal Numbers of Neuronal and Nonneuronal Cells Make the Human Brain an Isometrically Scaled-up Primate Brain." *The Journal of Comparative Neurology* 513 (5): 532–41. https://doi.org/10.1002/cne.21974.

Baidya, Melvin, Marx Genovez, Marissa Torres, and Michael Y. Chao. 2014. "Dopamine Modulation of Avoidance Behavior in *Caenorhabditis elegans* Requires the NMDA Receptor NMR-1." *PloS One* 9 (8): e102958. https://doi.org/10.1371/journal.pone.0102958.

Bargmann, C. I., E. Hartwieg, and H. R. Horvitz. 1993. "Odorant-Selective Genes and Neurons Mediate Olfaction in *C. elegans*." *Cell* 74 (3): 515–27. https://doi.org/10.1016/0092-8674(93)80053-h.

Baskoylu, Saba N., Jill Yersak, Patrick O'Hern, Sarah Grosser, Jonah Simon, Sarah Kim, Kelsey Schuch, et al. 2018. "Single Copy/Knock-in Models of ALS SOD1 in *C. elegans* Suggest Loss and Gain of Function Have Different Contributions to Cholinergic and Glutamatergic Neurodegeneration." *PLoS Genetics* 14 (10): e1007682. https://doi.org/10.1371/journal.pgen.1007682.

Becerra, Diego, Andrea Calixto, and Patricio Orio. 2023. "The Conscious Nematode: Exploring Hallmarks of Minimal Phenomenal Consciousness in *Caenorhabditis elegans*." *International Journal of Psychological Research* 16 (2): 87–104. https://doi.org/10.21500/20112084.6487.

Beer, Katharina B., Jennifer Rivas-Castillo, Kenneth Kuhn, Gholamreza Fazeli, Birgit Karmann, Jeremy F. Nance, Christian Stigloher, and Ann M. Wehman. 2018. "Extracellular Vesicle Budding Is Inhibited by Redundant Regulators of TAT-5 Flippase Localization and Phospholipid Asymmetry." *Proceedings of the National Academy of Sciences* 115 (6): E1127–36. https://doi.org/10.1073/pnas.1714085115.

Beitel, G. J., S. G. Clark, and H. R. Horvitz. 1990. "*Caenorhabditis elegans* Ras Gene Let-60 Acts as a Switch in the Pathway of Vulval Induction." *Nature* 348 (6301): 503–9. https://doi.org/10.1038/348503a0.

Bejanin, Alexandre, Maria Florencia Iulita, Eduard Vilaplana, Maria Carmona-Iragui, Bessy Benejam, Laura Videla, Isabel Barroeta, et al. 2021. "Association of Apolipoprotein E ε4 Allele With Clinical and Multimodal Biomarker Changes of Alzheimer Disease in Adults With Down Syndrome." *JAMA Neurology* 78 (8): 937–47. https://doi.org/10.1001/jamaneurol.2021.1893.

Benian, Guy M., and Henry F. Epstein. 2011. "*Caenorhabditis elegans* Muscle: A Genetic and Molecular Model for Protein Interactions in the Heart." *Circulation Research* 109 (9): 1082–95. https://doi.org/10.1161/CIRCRESAHA.110.237685.

Ben-Yakar, Adela, Nikos Chronis, and Hang Lu. 2009. "Microfluidics for the Analysis of Behavior, Nerve Regeneration, and Neural Cell Biology in *C. elegans*." *Current Opinion in Neurobiology* 19 (5): 561–67. https://doi.org/10.1016/j.conb.2009.10.010.

Bessou, Catherine, Jean-Bernard Giugia, Christopher J. Franks, Lindy Holden-Dye, and L. Ségalat. 1998. "Mutations in the *Caenorhabditis elegans* Dystrophin-like Gene Dys-1 Lead to Hyperactivity and Suggest a Link with Cholinergic Transmission." *Neurogenetics* 2 (1): 61–72. https://doi.org/10.1007/s100480050053.

Blackwell, T. Keith, Aileen K. Sewell, Ziyun Wu, and Min Han. 2019. "TOR Signaling in *Caenorhabditis elegans* Development, Metabolism, and Aging." *Genetics* 213 (2): 329–60. https://doi.org/10.1534/genetics.119.302504.

Blumenthal, Thomas, and Kathy Seggerson Gleason. 2003. "*Caenorhabditis elegans* Operons: Form and Function." *Nature Reviews Genetics* 4 (2): 110–18. https://doi.org/10.1038/nrg995.

Blumenthal, Thomas, and Keith Steward. 1997. "RNA Processing and Gene Structure." In *C. elegans II*, edited by Donald L. Riddle, Thomas Blumenthal, Barbara J. Meyer, and James R. Priess, 2nd ed. Cold Spring Harbor (NY): Cold Spring Harbor Laboratory Press. http://www.ncbi.nlm.nih.gov/books/NBK19975/.

Bodhicharla, Rakesh, Ranjan Devkota, Mario Ruiz, and Marc Pilon. 2018. "Membrane Fluidity Is Regulated Cell Nonautonomously by Caenorhabditiselegans PAQR-2 and Its Mammalian Homolog AdipoR2." *Genetics* 210 (1): 189–201. https://doi.org/10.1534/genetics.118.301272.

Boeynaems, Steven, Simon Alberti, Nicolas L. Fawzi, Tanja Mittag, Magdalini Polymenidou, Frederic Rousseau, Joost Schymkowitz, et al. 2018. "Protein Phase Separation: A New Phase in Cell Biology." *Trends in Cell Biology* 28 (6): 420–35. https://doi.org/10.1016/j.tcb.2018.02.004.

Bono, M. de, and C. I. Bargmann. 1998. "Natural Variation in a Neuropeptide Y Receptor Homolog Modifies Social Behavior and Food Response in *C. elegans*." *Cell* 94 (5): 679–89. https://doi.org/10.1016/s0092-8674(00)81609-8.

Bono, Mario de, and Andres Villu Maricq. 2005. "Neuronal Substrates of Complex Behaviors in *C. elegans*." *Annual Review of Neuroscience* 28:451–501. https://doi.org/10.1146/annurev.neuro.27.070203.144259.

Bono, Mario de, David M. Tobin, M. Wayne Davis, Leon Avery, and Cornelia I. Bargmann. 2002. "Social Feeding in *Caenorhabditis elegans* Is Induced by Neurons That Detect Aversive Stimuli." *Nature* 419 (6910): 899–903. https://doi.org/10.1038/nature01169.

Boulin, Thomas, John F. Etchberger, and Oliver Hobert. 2006. "Reporter Gene Fusions." In *WormBook: The Online Review of C. elegans Biology [Internet]*. WormBook. https://www.ncbi.nlm.nih.gov/books/NBK19738/.

Bourgeois, Frederic, and Adela Ben-Yakar. 2007. "Femtosecond Laser Nanoaxotomy Properties and Their Effect on Axonal Recovery in *C. elegans*." *Optics Express* 15 (14): 8521–31. https://doi.org/10.1364/oe.15.008521.

Braukmann, Fabian, David Jordan, and Eric Miska. 2017. "Artificial and Natural RNA Interactions between Bacteria and *C. elegans*." *RNA Biology* 14 (4): 415–20. https://doi.org/10.1080/15476286.2017.1297912.

Brenner, S. 1974. "The Genetics of *Caenorhabditis elegans*." *Genetics* 77 (1): 71–94. https://doi.org/10.1093/genetics/77.1.71.

Bürglin, Thomas R., Kuwabara, Patricia E. 2006. "Homologs of the Hh signalling network in *C. elegans*." WormBook 28 (Jan): 1–14. https://doi.org/10.1895/wormbook.1.76.1.

Butcher, Rebecca A. 2019. "Natural Products as Chemical Tools to Dissect Complex Biology in *C. elegans*." *Current Opinion in Chemical Biology* 50 (June):138–44. https://doi.org/10.1016/j.cbpa.2019.03.005.

Cabreiro, Filipe, Daniel Ackerman, Ryan Doonan, Caroline Araiz, Patricia Back, Diana Papp, Bart P. Braeckman, and David Gems. 2011. "Increased Life Span from Overexpression of Superoxide Dismutase in *Caenorhabditis elegans* Is Not Caused by Decreased Oxidative Damage." *Free Radical Biology & Medicine* 51 (8): 1575–82. https://doi.org/10.1016/j.freeradbiomed.2011.07.020.

"*Caenorhabditis elegans* Genome Assembly WBcel235." n.d. NCBI. Accessed May 15, 2024. https://www.ncbi.nlm.nih.gov/data-hub/assembly/GCF_000002985.6/.

Cao, Junyue, Jonathan S. Packer, Vijay Ramani, Darren A. Cusanovich, Chau Huynh, Riza Daza, Xiaojie Qiu, et al. 2017. "Comprehensive Single-Cell Transcriptional Profiling of a Multicellular Organism." *Science (New York, N.Y.)* 357 (6352): 661–67. https://doi.org/10.1126/science.aam8940.

Carithers, Latarsha J., Kristin Ardlie, Mary Barcus, Philip A. Branton, Angela Britton, Stephen A. Buia, Carolyn C. Compton, et al. 2015. "A Novel Approach to High-Quality Postmortem Tissue Procurement: The GTEx Project." *Biopreservation and Biobanking* 13 (5): 311–19. https://doi.org/10.1089/bio.2015.0032.

Caswell-Chen, E. P., J. Chen, E. E. Lewis, G. W. Douhan, S. A. Nadler, and J. R. Carey. 2005. "Revising the Standard Wisdom of *C. elegans* Natural History: Ecology of Longevity." *Science of Aging Knowledge Environment: SAGE KE* 2005 (40): pe30. https://doi.org/10.1126/sageke.2005.40.pe30.

Chalfie, M. 1995. "Green Fluorescent Protein." *Photochemistry and Photobiology* 62 (4): 651–56. https://doi.org/10.1111/j.1751-1097.1995.tb08712.x.

Chalfie, M., Y. Tu, G. Euskirchen, W. W. Ward, and D. C. Prasher. 1994. "Green Fluorescent Protein as a Marker for Gene Expression." *Science (New York, N.Y.)* 263 (5148): 802–5. https://doi.org/10.1126/science.8303295.

Chalfie, Martin. 2019. "Building a Worm: Complete Development One Cell at a Time." *Developmental Biology* 455 (2): 113–361. https://doi.org/10.1016/j.ydbio.2019.09.015.

Chalfie, Martin, Anne C. Hart, Catharine H. Rankin, and Miriam B. Goodman. 2014. "Assaying Mechanosensation." *WormBook: The Online Review of C. elegans Biology*, July. https://doi.org/10.1895/wormbook.1.172.1.

Chalfie, Martin, and John Sulston. 1981. "Developmental Genetics of the Mechanosensory Neurons of *Caenorhabditis elegans*." *Developmental Biology* 82 (2): 358–70. https://doi.org/10.1016/0012-1606(81)90459-0.

Chan, Emily, and Jeremy Nance. 2013. "Mechanisms of CDC-42 Activation during Contact-Induced Cell Polarization." *Journal of Cell Science* 126 (Pt 7): 1692–1702. https://doi.org/10.1242/jcs.124594.

Chan, S. S.-Y., H. Zheng, M.-W. Su, R. Wilk, M. T. Killeen, E. M. Hedgecock, and J. G. Culotti. 1996. "UNC-40, a *C. elegans* Homolog of DCC (Deleted in Colorectal Cancer), Is Required in

Motile Cells Responding to UNC-6 Netrin Cues." *Cell* 87 (2): 187–95. https://doi.org/10.1016/S0092-8674(00)81337-9.

Chandra, Rashmi, Fatima Farah, Fernando Muñoz-Lobato, Anirudh Bokka, Kelli L. Benedetti, Chantal Brueggemann, Mashel Fatema A. Saifuddin, et al. 2023. "Sleep Is Required to Consolidate Odor Memory and Remodel Olfactory Synapses." *Cell* 186 (13): 2911–2928.e20. https://doi.org/10.1016/j.cell.2023.05.006.

Chauve, Laetitia, Jérémie Le Pen, Francesca Hodge, Pia Todtenhaupt, Laura Biggins, Eric A. Miska, Simon Andrews, and Olivia Casanueva. 2020. "High-Throughput Quantitative RT-PCR in Single and Bulk *C. elegans* Samples Using Nanofluidic Technology." *Journal of Visualized Experiments: JoVE*, no. 159 (May), https://doi.org/10.3791/61132.

Chen, Beth L., David H. Hall, and Dmitri B. Chklovskii. 2006. "Wiring Optimization Can Relate Neuronal Structure and Function." *Proceedings of the National Academy of Sciences of the United States of America* 103 (12): 4723–28. https://doi.org/10.1073/pnas.0506806103.

Chen, Changchun, Eisuke Itakura, Geoffrey M. Nelson, Ming Sheng, Patrick Laurent, Lorenz A. Fenk, Rebecca A. Butcher, Ramanujan S. Hegde, and Mario de Bono. 2017. "IL-17 Is a Neuromodulator of *Caenorhabditis elegans* Sensory Responses." *Nature* 542 (7639): 43–48. https://doi.org/10.1038/nature20818.

Chen, Lizhen, Zhiping Wang, Anindya Ghosh-Roy, Thomas Hubert, Dong Yan, Sean O'Rourke, Bruce Bowerman, Zilu Wu, Yishi Jin, and Andrew D. Chisholm. 2011. "Axon Regeneration Pathways Identified by Systematic Genetic Screening in *C. elegans*." *Neuron* 71 (6): 1043–57. https://doi.org/10.1016/j.neuron.2011.07.009.

Chen, Xi, Jeff W. Barclay, Robert D. Burgoyne, and Alan Morgan. 2015. "Using *C. elegans* to Discover Therapeutic Compounds for Ageing-Associated Neurodegenerative Diseases." *Chemistry Central Journal* 9 (November):65. https://doi.org/10.1186/s13065-015-0143-y.

Cheong, Mi Cheong, Alexander B Artyukhin, Young-Jai You, and Leon Avery. 2015. "An Opioid-like System Regulating Feeding Behavior in *C. elegans*." Edited by Oliver Hobert. *eLife* 4 (April):e06683. https://doi.org/10.7554/eLife.06683.

Chu, Diana S., Hongbin Liu, Paola Nix, Tammy F. Wu, Edward J. Ralston, John R. Yates, and Barbara J. Meyer. 2006. "Sperm Chromatin Proteomics Identifies Evolutionarily Conserved Fertility Factors." *Nature* 443 (7107): 101–5. https://doi.org/10.1038/nature05050.

Chuang, Chiou-Fen, Miri K. VanHoven, Richard D. Fetter, Vytas K. Verselis, and Cornelia I. Bargmann. 2007. "An Innexin-Dependent Cell Network Establishes Left-Right Neuronal Asymmetry in *C. elegans*." *Cell* 129 (4): 787–99. https://doi.org/10.1016/j.cell.2007.02.052.

Churgin, Matthew A., Sang-Kyu Jung, Chih-Chieh Yu, Xiangmei Chen, David M. Raizen, and Christopher Fang-Yen. 2017. "Longitudinal Imaging of *Caenorhabditis elegans* in a Microfabricated Device Reveals Variation in Behavioral Decline during Aging." *eLife* 6 (May):e26652. https://doi.org/10.7554/eLife.26652.

Conradt, B., and H. R. Horvitz. 1998. "The *C. elegans* Protein EGL-1 Is Required for Programmed Cell Death and Interacts with the Bcl-2-like Protein CED-9." *Cell* 93 (4): 519–29. https://doi.org/10.1016/s0092-8674(00)81182-4.

Conte, Darryl, Lesley T. MacNeil, Albertha J. M. Walhout, and Craig C. Mello. 2015. "RNA Interference in *Caenorhabditis elegans*." *Current Protocols in Molecular Biology* 109 (January):26.3.1–26.3.30. https://doi.org/10.1002/0471142727.mb2603s109.

Cook, Daniel E., Stefan Zdraljevic, Joshua P. Roberts, and Erik C. Andersen. 2017. "CeNDR, the *Caenorhabditis elegans* Natural Diversity Resource." *Nucleic Acids Research* 45 (D1): D650–57. https://doi.org/10.1093/nar/gkw893.

Cook, Daniel E, Stefan Zdraljevic, Robyn E Tanny, Beomseok Seo, David D Riccardi, Luke M Noble, Matthew V Rockman, et al. 2016. "The Genetic Basis of Natural Variation in *Caenorhabditis elegans* Telomere Length." *Genetics* 204 (1): 371–83. https://doi.org/10.1534/genetics.116.191148.

Corsi, Ann K., Bruce Wightman, and Martin Chalfie. 2015. "A Transparent Window into Biology: A Primer on *Caenorhabditis elegans*." *Genetics* 200 (2): 387–407. https://doi.org/10.1534/genetics.115.176099.

Cova, Claire de la, Robert Townley, Sergi Regot, and Iva Greenwald. 2017. "A Real-Time Biosensor for ERK Activity Reveals Signaling Dynamics during *C. elegans* Cell Fate Specification." *Developmental Cell* 42 (5): 542–553.e4. https://doi.org/10.1016/j.devcel.2017.07.014.

Crombie, Timothy A., Ryan McKeown, Nicolas D. Moya, Kathryn S. Evans, Samuel J. Widmayer, Vincent LaGrassa, Natalie Roman, et al. 2024. "CaeNDR, the Caenorhabditis Natural Diversity Resource." *Nucleic Acids Research* 52 (D1): D850–58. https://doi.org/10.1093/nar/gkad887.

Crombie, Timothy A., Sayran Saber, Ayush Shekhar Saxena, Robyn Egan, and Charles F. Baer. 2018. "Head-to-Head Comparison of Three Experimental Methods of Quantifying Competitive Fitness in *C. elegans*." *PLoS ONE* 13 (10): e0201507. https://doi.org/10.1371/journal.pone.0201507.

Cuentas-Condori, Andrea, Ben Mulcahy, Siwei He, Sierra Palumbos, Mei Zhen, and David M. Miller. 2019. "*C. elegans* Neurons Have Functional Dendritic Spines." *eLife* 8 (October):e47918. https://doi.org/10.7554/eLife.47918.

Dahmane, N., and A. Ruiz i Altaba. 1999. "Sonic Hedgehog Regulates the Growth and Patterning of the Cerebellum." *Development (Cambridge, England)* 126 (14): 3089–3100. https://doi.org/10.1242/dev.126.14.3089.

Daigle, I., and C. Li. 1993. "Apl-1, a *Caenorhabditis elegans* Gene Encoding a Protein Related to the Human Beta-Amyloid Protein Precursor." *Proceedings of the National Academy of Sciences of the United States of America* 90 (24): 12045–49. https://doi.org/10.1073/pnas.90.24.12045.

Dal Santo, P., M. A. Logan, A. D. Chisholm, and E. M. Jorgensen. 1999. "The Inositol Trisphosphate Receptor Regulates a 50-Second Behavioral Rhythm in *C. elegans*." *Cell* 98 (6): 757–67. https://doi.org/10.1016/s0092-8674(00)81510-x.

Davies, Andrew G., Jonathan T. Pierce-Shimomura, Hongkyun Kim, Miri K. VanHoven, Tod R. Thiele, Antonello Bonci, Cornelia I. Bargmann, and Steven L. McIntire. 2003. "A Central Role of the BK Potassium Channel in Behavioral Responses to Ethanol in *C. elegans*." *Cell* 115 (6): 655–66. https://doi.org/10.1016/s0092-8674(03)00979-6.

Davis, Lloyd, Inja Radman, Angeliki Goutou, Ailish Tynan, Kieran Baxter, Zhiyan Xi, Jack M. O'Shea, Jason W. Chin, and Sebastian Greiss. 2021. "Precise Optical Control of Gene Expression in C Elegans Using Improved Genetic Code Expansion and Cre Recombinase." *eLife* 10 (August):e67075. https://doi.org/10.7554/eLife.67075.

Davis, M. Wayne, J. Jason Morton, Dana Carroll, and Erik M. Jorgensen. 2008. "Gene Activation Using FLP Recombinase in *C. elegans*." *PLoS Genetics* 4 (3): e1000028. https://doi.org/10.1371/journal.pgen.1000028.

Davis, Paul, Magdalena Zarowiecki, Valerio Arnaboldi, Andrés Becerra, Scott Cain, Juancarlos Chan, Wen J. Chen, et al. 2022. "WormBase in 2022-Data, Processes, and Tools for Analyzing *Caenorhabditis elegans*." *Genetics* 220 (4): iyac003. https://doi.org/10.1093/genetics/iyac003.

Davis, S. J., L. L. Scott, G. Ordemann, A. Philpo, J. Cohn, and J. T. Pierce-Shimomura. 2015. "Putative Calcium-Binding Domains of the *Caenorhabditis elegans* BK Channel Are Dispensable for Intoxication and Ethanol Activation." *Genes, Brain, and Behavior* 14 (6): 454–65. https://doi.org/10.1111/gbb.12229.

Delabar, Jean M., Marco Antônio G. B. Gomes, Marta Fructuoso, Nadège Sarrazin, Nicolas George, Nadia Fleary-Roberts, Hua Sun, et al. 2024. "EGCG-like Non-Competitive Inhibitor of DYRK1A Rescues Cognitive Defect in a down Syndrome Model." *European Journal of Medicinal Chemistry* 265 (February):116098. https://doi.org/10.1016/j.ejmech.2023.116098.

Delabar, Jean M., Julien Lagarde, Marta Fructuoso, Ammara Mohammad, Michel Bottlaender, Eric Doran, Ira Lott, et al. 2023. "Increased Plasma DYRK1A with Aging May Protect against Neurodegenerative Diseases." *Translational Psychiatry* 13 (1): 111. https://doi.org/10.1038/s41398-023-02419-0.

Desbois, Muriel, Steven J. Cook, Scott W. Emmons, and Hannes E. Bülow. 2015. "Directional Trans-Synaptic Labeling of Specific Neuronal Connections in Live Animals." *Genetics* 200 (3): 697–705. https://doi.org/10.1534/genetics.115.177006.

Dickinson, Daniel J., and Bob Goldstein. 2016. "CRISPR-Based Methods for *Caenorhabditis elegans* Genome Engineering." *Genetics* 202 (3): 885–901. https://doi.org/10.1534/genetics.115.182162.

Dossena, Silvia, Charity Nofziger, Emanuele Bernardinelli, Selma Soyal, Wolfgang Patsch, and Markus Paulmichl. 2013. "Use of the Operon Structure of the *C. elegans* Genome as a Tool to Identify Functionally Related Proteins." *Cellular Physiology and Biochemistry: International Journal of Experimental Cellular Physiology, Biochemistry, and Pharmacology* 32 (7): 41–56. https://doi.org/10.1159/000356623.

Down, J. L. 1866. "OBSERVATIONS ON AN ETHNIC CLASSIFICATION OF IDIOTS." *London Hospital Record.*

Duerr, Janet S. 2006. "Immunohistochemistry." In *WormBook: The Online Review of C. elegans Biology [Internet].* WormBook. https://www.ncbi.nlm.nih.gov/books/NBK19743/.

Dupuy, Denis, Qian-Ru Li, Bart Deplancke, Mike Boxem, Tong Hao, Philippe Lamesch, Reynaldo Sequerra, et al. 2004. "A First Version of the *Caenorhabditis elegans* Promoterome." *Genome Research* 14 (10b): 2169–75. https://doi.org/10.1101/gr.2497604.

El Mouridi, Sonia, Faisal Alkhaldi, and Christian Frøkjær-Jensen. 2022. "Modular Safe-Harbor Transgene Insertion for Targeted Single-Copy and Extrachromosomal Array Integration in *Caenorhabditis elegans*." *G3: Genes|Genomes|Genetics* 12 (9): jkac184. https://doi.org/10.1093/g3journal/jkac184.

Ellis, H. M., and H. R. Horvitz. 1986. "Genetic Control of Programmed Cell Death in the Nematode *C. elegans*." *Cell* 44 (6): 817–29. https://doi.org/10.1016/0092-8674(86)90004-8.

Esposito, Giovanni, Elia Di Schiavi, Carmela Bergamasco, and Paolo Bazzicalupo. 2007. "Efficient and Cell Specific Knock-down of Gene Function in Targeted *C. elegans* Neurons." *Gene* 395 (1–2): 170–76. https://doi.org/10.1016/j.gene.2007.03.002.

Evans, Thomas. 2006. "Transformation and Microinjection." *WormBook.* https://doi.org/10.1895/wormbook.1.108.1.

Ewald, Collin Y., Jess N. Landis, Jess Porter Abate, Coleen T. Murphy, and T. Keith Blackwell. 2015. "Dauer-Independent Insulin/IGF-1-Signalling Implicates Collagen Remodelling in Longevity." *Nature* 519 (7541): 97–101. https://doi.org/10.1038/nature14021.

Fang-Yen, Christopher, Christopher V. Gabel, Aravinthan D. T. Samuel, Cornelia I. Bargmann, and Leon Avery. 2012. "Chapter 6 – Laser Microsurgery in *Caenorhabditis elegans*." In *Methods in Cell Biology*, edited by Joel H. Rothman and Andrew Singson, 107:177–206. *Caenorhabditis elegans*: Cell Biology and Physiology. Academic Press. https://doi.org/10.1016/B978-0-12-394620-1.00006-0.

Faunes, Fernando, and Juan Larraín. 2016. "Conservation in the Involvement of Heterochronic Genes and Hormones during Developmental Transitions." *Developmental Biology* 416 (1): 3–17. https://doi.org/10.1016/j.ydbio.2016.06.013.

Fay, David. 2006. "Synthetic and Enhancer Mutations." In *WormBook: The Online Review of C. elegans Biology [Internet].* WormBook. https://www.ncbi.nlm.nih.gov/books/NBK19671/.

Fay, David, and Wade Johnson. 2006. "Suppressor Mutations." In *WormBook: The Online Review of C. elegans Biology [Internet].* WormBook. https://www.ncbi.nlm.nih.gov/books/NBK19712/.

Fay, David S. 2013. "Classical Genetic Methods." In *WormBook: The Online Review of C. elegans Biology [Internet].* WormBook. https://www.ncbi.nlm.nih.gov/books/NBK179229/.

Feinberg, Evan H., Miri K. VanHoven, Andres Bendesky, George Wang, Richard D. Fetter, Kang Shen, and Cornelia I. Bargmann. 2008. "GFP Reconstitution Across Synaptic Partners (GRASP) Defines Cell Contacts and Synapses in Living Nervous Systems." *Neuron* 57 (3): 353–63. https://doi.org/10.1016/j.neuron.2007.11.030.

Ferkey, Denise M., Piali Sengupta, and Noelle D. L'Etoile. 2021. "Chemosensory Signal Transduction in *Caenorhabditis elegans*." *Genetics* 217 (3): iyab004. https://doi.org/10.1093/genetics/iyab004.

Fernandez, Robert W., Kimberly Wei, Erin Y. Wang, Deimante Mikalauskaite, Andrew Olson, Judy Pepper, Nakeirah Christie, et al. 2020. "Cellular Expression and Functional Roles of All 26 Neurotransmitter GPCRs in the *C. elegans* Egg-Laying Circuit." *The Journal of Neuroscience:*

The Official Journal of the Society for Neuroscience 40 (39): 7475–88. https://doi.org/10.1523/JNEUROSCI.1357-20.2020.

Fielenbach, Nicole, and Adam Antebi. 2008. "*C. elegans* Dauer Formation and the Molecular Basis of Plasticity." *Genes & Development* 22 (16): 2149–65. https://doi.org/10.1101/gad.1701508.

Fire, Andrew, SiQun Xu, Mary K. Montgomery, Steven A. Kostas, Samuel E. Driver, and Craig C. Mello. 1998. "Potent and Specific Genetic Interference by Double-Stranded RNA in *Caenorhabditis elegans*." *Nature* 391 (6669): 806–11. https://doi.org/10.1038/35888.

Folkmann, Andrew W., Andrea Putnam, Chiu Fan Lee, and Geraldine Seydoux. 2021. "Regulation of Biomolecular Condensates by Interfacial Protein Clusters." *Science (New York, N.Y.)* 373 (6560): 1218–24. https://doi.org/10.1126/science.abg7071.

Frøkjaer-Jensen, Christian, M. Wayne Davis, Christopher E. Hopkins, Blake J. Newman, Jason M. Thummel, Søren-Peter Olesen, Morten Grunnet, and Erik M. Jorgensen. 2008. "Single-Copy Insertion of Transgenes in *Caenorhabditis elegans*." *Nature Genetics* 40 (11): 1375–83. https://doi.org/10.1038/ng.248.

Frøkjær-Jensen, Christian, M. Wayne Davis, Mihail Sarov, Jon Taylor, Stephane Flibotte, Matthew LaBella, Andrei Pozniakovsky, Donald G. Moerman, and Erik M. Jorgensen. 2014. "Random and Targeted Transgene Insertion in *Caenorhabditis elegans* Using a Modified Mos1 Transposon." *Nature Methods* 11 (5): 529–34. https://doi.org/10.1038/nmeth.2889.

Fujisawa, Kazuko, Jeffrey L. Wrana, and Joseph G. Culotti. 2007. "The Slit Receptor EVA-1 Coactivates a SAX-3/Robo Mediated Guidance Signal in *C. elegans*." *Science (New York, N.Y.)* 317 (5846): 1934–38. https://doi.org/10.1126/science.1144874.

Galati, Domenico F., Kelly D. Sullivan, Andrew T. Pham, Joaquin M. Espinosa, and Chad G. Pearson. 2018. "Trisomy 21 Represses Cilia Formation and Function." *Developmental Cell* 46 (5): 641–650.e6. https://doi.org/10.1016/j.devcel.2018.07.008.

Galbraith, Matthew D., Angela L. Rachubinski, Keith P. Smith, Paula Araya, Katherine A. Waugh, Belinda Enriquez-Estrada, Kayleigh Worek, et al. 2023. "Multidimensional Definition of the Interferonopathy of Down Syndrome and Its Response to JAK Inhibition." *Science Advances* 9 (26): eadg6218. https://doi.org/10.1126/sciadv.adg6218.

Garde, Aastha, Isabel W. Kenny, Laura C. Kelley, Qiuyi Chi, Ayse Sena Mutlu, Meng C. Wang, and David R. Sherwood. 2022. "Localized Glucose Import, Glycolytic Processing, and Mitochondria Generate a Focused ATP Burst to Power Basement-Membrane Invasion." *Developmental Cell* 57 (6): 732–749.e7. https://doi.org/10.1016/j.devcel.2022.02.019.

"Gene Editing and Gene Modulation for Life Sciences." n.d. Accessed February 12, 2024. https://horizondiscovery.com/.

Gerstein, Mark B., Zhi John Lu, Eric L. Van Nostrand, Chao Cheng, Bradley I. Arshinoff, Tao Liu, Kevin Y. Yip, et al. 2010. "Integrative Analysis of the *Caenorhabditis elegans* Genome by the modENCODE Project." *Science* 330 (6012): 1775–87. https://doi.org/10.1126/science.1196914.

Ghasemi, Mehdi, and Robert H. Brown. 2018. "Genetics of Amyotrophic Lateral Sclerosis." *Cold Spring Harbor Perspectives in Medicine* 8 (5): a024125. https://doi.org/10.1101/cshperspect.a024125.

Gilbert, Luke A., Max A. Horlbeck, Britt Adamson, Jacqueline E. Villalta, Yuwen Chen, Evan H. Whitehead, Carla Guimaraes, et al. 2014. "Genome-Scale CRISPR-Mediated Control of Gene Repression and Activation." *Cell* 159 (3): 647–61. https://doi.org/10.1016/j.cell.2014.09.029.

Goodman, M. B., D. H. Hall, L. Avery, and S. R. Lockery. 1998. "Active Currents Regulate Sensitivity and Dynamic Range in *C. elegans* Neurons." *Neuron* 20 (4): 763–72. https://doi.org/10.1016/s0896-6273(00)81014-4.

Goodman, Miriam B., Theodore H. Lindsay, Shawn R. Lockery, and Janet E. Richmond. 2012. "Chapter 14 – Electrophysiological Methods for *Caenorhabditis elegans* Neurobiology." In *Methods in Cell Biology*, edited by Joel H. Rothman and Andrew Singson, 107:409–36. *Caenorhabditis elegans*: Cell Biology and Physiology. Academic Press. https://doi.org/10.1016/B978-0-12-394620-1.00014-X.

Goudeau, Jérôme, Catherine S Sharp, Jonathan Paw, Laura Savy, Manuel D Leonetti, Andrew G York, Dustin L Updike, Cynthia Kenyon, and Maria Ingaramo. 2021. "Split-wrmScarlet and Split-sfGFP: Tools for Faster, Easier Fluorescent Labeling of Endogenous Proteins in *Caenorhabditis elegans*." *Genetics* 217 (4): iyab014. https://doi.org/10.1093/genetics/iyab014.

Greenwald, Iva, and Rhett Kovall. 2013. "Notch Signaling: Genetics and Structure." *WormBook: The Online Review of C. elegans Biology*, January, 1–28. https://doi.org/10.1895/wormbook.1.10.2.

Gu, G, G A Caldwell, and M Chalfie. 1996. "Genetic Interactions Affecting Touch Sensitivity in *Caenorhabditis elegans*." *Proceedings of the National Academy of Sciences of the United States of America* 93 (13): 6577–82.

Guipponi, M., K. Brunschwig, Z. Chamoun, H. S. Scott, K. Shibuya, J. Kudoh, A. L. Delezoide, et al. 2000. "C21orf5, a Novel Human Chromosome 21 Gene, Has a *Caenorhabditis elegans* Ortholog (Pad-1) Required for Embryonic Patterning." *Genomics* 68 (1): 30–40. https://doi.org/10.1006/geno.2000.6250.

Gumienny, Tina L., and Cathy Savage-Dunn. 2013. "TGF-β Signaling in *C. elegans*." *WormBook: The Online Review of C. elegans Biology*, July, 1–34. https://doi.org/10.1895/wormbook.1.22.2.

Guthmueller, Kassandra L., Maggie L. Yoder, and Andrea M. Holgado. 2011. "Determining Genetic Expression Profiles in *C. elegans* Using Microarray and Real-Time PCR." *Journal of Visualized Experiments : JoVE*, no. 53 (July), 2777. https://doi.org/10.3791/2777.

Haley, Ryan, Ying Wang, and Zheng Zhou. 2018. "The Small GTPase RAB-35 Defines a Third Pathway That Is Required for the Recognition and Degradation of Apoptotic Cells." *PLoS Genetics* 14 (8): e1007558. https://doi.org/10.1371/journal.pgen.1007558.

Hammarlund, Marc, Oliver Hobert, David M. Miller, and Nenad Sestan. 2018. "The CeNGEN Project: The Complete Gene Expression Map of an Entire Nervous System." *Neuron* 99 (3): 430–33. https://doi.org/10.1016/j.neuron.2018.07.042.

Harris, Todd W., Erika Hartwieg, H. Robert Horvitz, and Erik M. Jorgensen. 2000. "Mutations in Synaptojanin Disrupt Synaptic Vesicle Recycling." *The Journal of Cell Biology* 150 (3): 589–600.

Hartenstein, Volker. 2006. "The Neuroendocrine System of Invertebrates: A Developmental and Evolutionary Perspective." *The Journal of Endocrinology* 190 (3): 555–70. https://doi.org/10.1677/joe.1.06964.

Hayden, Julia, and Christopher Fang-Yen. 2021. "Cryopreservation of Dehydrated *Caenorhabditis elegans* with Multiple Recoveries Using a Granular Medium." *microPublication Biology* 2021. https://doi.org/10.17912/micropub.biology.000488.

Hedgecock, E. M., J. E. Sulston, and J. N. Thomson. 1983. "Mutations Affecting Programmed Cell Deaths in the Nematode *Caenorhabditis elegans*." *Science (New York, N.Y.)* 220 (4603): 1277–79. https://doi.org/10.1126/science.6857247.

Hedgecock, Edward M., Joseph G. Culotti, and David H. Hall. 1990. "The Unc-5, Unc-6, and Unc-40 Genes Guide Circumferential Migrations of Pioneer Axons and Mesodermal Cells on the Epidermis in *C. elegans*." *Neuron* 4 (1): 61–85. https://doi.org/10.1016/0896-6273(90)90444-K.

Herault, Yann, Jean M. Delabar, Elizabeth M. C. Fisher, Victor L. J. Tybulewicz, Eugene Yu, and Veronique Brault. 2017. "Rodent Models in Down Syndrome Research: Impact and Future Opportunities." *Disease Models & Mechanisms* 10 (10): 1165–86. https://doi.org/10.1242/dmm.029728.

Herculano-Houzel, Suzana, Bruno Mota, and Roberto Lent. 2006. "Cellular Scaling Rules for Rodent Brains." *Proceedings of the National Academy of Sciences* 103 (32): 12138–43. https://doi.org/10.1073/pnas.0604911103.

Herman, Robert K. 1995. "Chapter 6 Mosaic Analysis." In *Methods in Cell Biology*, edited by Henry F. Epstein and Diane C. Shakes, 48:123–46. Cuenorhubditis Elegans: Modern Biologcal Analysis of an Organism. Academic Press. https://doi.org/10.1016/S0091-679X(08)61386-2.

Hobert, Oliver. 2013. "The Neuronal Genome of *Caenorhabditis elegans*." *WormBook*, August, 1–106. https://doi.org/10.1895/wormbook.1.161.1.

———. 2021. "Homeobox Genes and the Specification of Neuronal Identity." *Nature Reviews. Neuroscience* 22 (10): 627–36. https://doi.org/10.1038/s41583-021-00497-x.

Hodgkin, J. 2001. "What Does a Worm Want with 20,000 Genes?" *Genome Biology* 2 (11): COMMENT2008. https://doi.org/10.1186/gb-2001-2-11-comment2008.

Hoerndli, Frédéric J., Angy J. Kallarackal, and Andres V. Maricq. 2015. "Mobile AMPARs Are Required for Synaptic Plasticity." *Channels (Austin, Tex.)* 9 (5): 230–32. https://doi.org/10.1080/19336950.2015.1080931.

Hofschneider, Mark. n.d. "Tiny RNAs That Regulate Gene Function." Lasker Foundation. Accessed February 6, 2024. https://laskerfoundation.org/winners/tiny-rnas-that-regulate-gene-function/.

Horvitz, H. R., and J. E. Sulston. 1980. "Isolation and Genetic Characterization of Cell-Lineage Mutants of the Nematode *Caenorhabditis elegans*." *Genetics* 96 (2): 435–54. https://doi.org/10.1093/genetics/96.2.435.

Horvitz, H. Robert. 2003. "Worms, Life, and Death (Nobel Lecture)." *ChemBioChem* 4 (8): 697–711. https://doi.org/10.1002/cbic.200300614.

Huang, Linda S., and Paul W. Sternberg. 2006. "Genetic Dissection of Developmental Pathways." *WormBook: The Online Review of C. elegans Biology*, June, 1–19. https://doi.org/10.1895/wormbook.1.88.2.

Husson, Steven J., Wagner Steuer Costa, Cornelia Schmitt, and Alexander Gottschalk. 2018. "Keeping Track of Worm Trackers." In *WormBook: The Online Review of C. elegans Biology [Internet]*. WormBook. https://www.ncbi.nlm.nih.gov/books/NBK116078/.

Iliff, Adam J., and X. Z. Shawn Xu. 2020. "*C. elegans*: A Sensible Model for Sensory Biology." *Journal of Neurogenetics* 34 (3–4): 347–50. https://doi.org/10.1080/01677063.2020.1823386.

Inglis, Peter, Ou Guangshuo, Michel Leroux, and Jonathan Scholey. 2006. "The Sensory Cilia of *Caenorhabditis elegans*." *WormBook*. https://doi.org/10.1895/wormbook.1.126.1.

Iyer, Sangeetha, Feba S. Sam, Nina DiPrimio, Graeme Preston, Jan Verheijen, Kausalya Murthy, Zachary Parton, et al. 2019. "Repurposing the Aldose Reductase Inhibitor and Diabetic Neuropathy Drug Epalrestat for the Congenital Disorder of Glycosylation PMM2-CDG." *Disease Models & Mechanisms* 12 (11): dmm040584. https://doi.org/10.1242/dmm.040584.

Jacobs, P. A., A. G. Baikie, W. M. Court Brown, and J. A. Strong. 1959. "The Somatic Chromosomes in Mongolism." *Lancet (London, England)* 1 (7075): 710. https://doi.org/10.1016/s0140-6736(59)91892-6.

James, Gregory, Simon R. Foster, Brian Key, and Annemiek Beverdam. 2013. "The Expression Pattern of EVA1C, a Novel Slit Receptor, Is Consistent with an Axon Guidance Role in the Mouse Nervous System." *PLoS ONE* 8 (9). https://doi.org/10.1371/journal.pone.0074115.

Jang, SoRi, Zhao Xuan, Ross C. Lagoy, Louise M. Jawerth, Ian J. Gonzalez, Milind Singh, Shavanie Prashad, et al. 2021. "Phosphofructokinase Relocalizes into Subcellular Compartments with Liquid-like Properties in Vivo." *Biophysical Journal* 120 (7): 1170–86. https://doi.org/10.1016/j.bpj.2020.08.002.

Jankowsky, Joanna L., and Hui Zheng. 2017. "Practical Considerations for Choosing a Mouse Model of Alzheimer's Disease." *Molecular Neurodegeneration* 12 (December):89. https://doi.org/10.1186/s13024-017-0231-7.

Jeong, Dae-Eun, Yujin Lee, and Seung-Jae V. Lee. 2018. "Western Blot Analysis of *C. elegans* Proteins." *Methods in Molecular Biology (Clifton, N.J.)* 1742:213–25. https://doi.org/10.1007/978-1-4939-7665-2_19.

Ji, Hongfei, Anthony D. Fouad, Shelly Teng, Alice Liu, Pilar Alvarez-Illera, Bowen Yao, Zihao Li, and Christopher Fang-Yen. 2021. "Phase Response Analyses Support a Relaxation Oscillator Model of Locomotor Rhythm Generation in *Caenorhabditis elegans*." *eLife* 10 (September):e69905. https://doi.org/10.7554/eLife.69905.

Ji, Ni, and Alexander van Oudenaarden. 2018. "Single Molecule Fluorescent in Situ Hybridization (smFISH) of *C. elegans* Worms and Embryos." In *WormBook: The Online Review of C. elegans Biology [Internet]*. WormBook. https://www.ncbi.nlm.nih.gov/books/NBK126649/.

Johannesson, Malin, Charlotte Sahlin, Linda Söderberg, Hans Basun, Johanna Fälting, Christer Möller, Olof Zachrisson, et al. 2021. "Elevated Soluble Amyloid Beta Protofibrils in Down Syndrome and Alzheimer's Disease." *Molecular and Cellular Neurosciences* 114 (July):103641. https://doi.org/10.1016/j.mcn.2021.103641.

Jongsma, Elisabeth, Anita Goyala, José Maria Mateos, and Collin Yvès Ewald. 2023. "Removal of Extracellular Human Amyloid Beta Aggregates by Extracellular Proteases in *C. elegans*." *eLife* 12 (September):e83465. https://doi.org/10.7554/eLife.83465.

Jorgensen, Erik M., and Susan E. Mango. 2002. "The Art and Design of Genetic Screens: *Caenorhabditis elegans*." *Nature Reviews Genetics* 3 (5): 356–69. https://doi.org/10.1038/nrg794.

Kage-Nakadai, Eriko, Rieko Imae, Yuji Suehiro, Sawako Yoshina, Sayaka Hori, and Shohei Mitani. 2014. "A Conditional Knockout Toolkit for *Caenorhabditis elegans* Based on the Cre/loxP Recombination." *PloS One* 9 (12): e114680. https://doi.org/10.1371/journal.pone.0114680.

Kage-Nakadai, Eriko, Hiroyuki Kobuna, Osamu Funatsu, Muneyoshi Otori, Keiko Gengyo-Ando, Sawako Yoshina, Sayaka Hori, and Shohei Mitani. 2012. "Single/Low-Copy Integration of Transgenes in *Caenorhabditis elegans* Using an Ultraviolet Trimethylpsoralen Method." *BMC Biotechnology* 12 (January):1. https://doi.org/10.1186/1472-6750-12-1.

Kaletsky, Rachel, Victoria Yao, April Williams, Alexi M. Runnels, Alicja Tadych, Shiyi Zhou, Olga G. Troyanskaya, and Coleen T. Murphy. 2018. "Transcriptome Analysis of Adult *Caenorhabditis elegans* Cells Reveals Tissue-Specific Gene and Isoform Expression." *PLoS Genetics* 14 (8): e1007559. https://doi.org/10.1371/journal.pgen.1007559.

Kano, Takashi, Penelope J. Brockie, Toshihiro Sassa, Hiroyuki Fujimoto, Yasushi Kawahara, Yuichi Iino, Jerry E. Mellem, David M. Madsen, Ryuji Hosono, and Andres V. Maricq. 2008. "Memory in *Caenorhabditis elegans* Is Mediated by NMDA-Type Ionotropic Glutamate Receptors." *Current Biology: CB* 18 (13): 1010–15. https://doi.org/10.1016/j.cub.2008.05.051.

Kemphues, K. J., J. R. Priess, D. G. Morton, and N. S. Cheng. 1988. "Identification of Genes Required for Cytoplasmic Localization in Early *C. elegans* Embryos." *Cell* 52 (3): 311–20. https://doi.org/10.1016/s0092-8674(88)80024-2.

Kenyon, C. 1988. "The Nematode *Caenorhabditis elegans*." *Science (New York, N.Y.)* 240 (4858): 1448–53. https://doi.org/10.1126/science.3287621.

Kepler, Lexis D., Troy A. McDiarmid, and Catharine H. Rankin. 2020. "Habituation in High-Throughput Genetic Model Organisms as a Tool to Investigate the Mechanisms of Neurodevelopmental Disorders." *Neurobiology of Learning and Memory* 171:107208. https://doi.org/10.1016/j.nlm.2020.107208.

Kerr, R., V. Lev-Ram, G. Baird, P. Vincent, R. Y. Tsien, and W. R. Schafer. 2000. "Optical Imaging of Calcium Transients in Neurons and Pharyngeal Muscle of *C. elegans*." *Neuron* 26 (3): 583–94. https://doi.org/10.1016/s0896-6273(00)81196-4.

Kerr, Rex A., Antoine E. Roux, Jérôme Goudeau, and Cynthia Kenyon. 2022. "The *C. elegans* Observatory: High-Throughput Exploration of Behavioral Aging." *Frontiers in Aging* 3:932656. https://doi.org/10.3389/fragi.2022.932656.

Kim, Dennis H., and Jonathan Ewbank. 2018. "Signaling in the Innate Immune Response." *WormBook*, August, 1–35. https://doi.org/10.1895/wormbook.1.83.2.

Kim, Woojin, Ryan S. Underwood, Iva Greenwald, and Daniel D. Shaye. 2018. "OrthoList 2: A New Comparative Genomic Analysis of Human and *Caenorhabditis elegans* Genes." *Genetics* 210 (2): 445–61. https://doi.org/10.1534/genetics.118.301307.

Kimble, J, J Sulston, and J White. 1980. "Regulative Development in the Post-Embryonic Lineages of *Caenorhabditis elegans*." In, 59–68.

Kirienko, Natalia V., Kumaran Mani, and David S. Fay. 2010. "Cancer Models in *Caenorhabditis elegans*." *Developmental Dynamics: An Official Publication of the American Association of Anatomists* 239 (5): 1413–48. https://doi.org/10.1002/dvdy.22247.

Kurz, C. Léopold, and Jonathan J. Ewbank. 2003. "*Caenorhabditis elegans*: An Emerging Genetic Model for the Study of Innate Immunity." *Nature Reviews. Genetics* 4 (5): 380–90. https://doi.org/10.1038/nrg1067.

Kwah, Ji Kent, and Aimee Jaramillo-Lambert. 2023. "Measuring Embryonic Viability and Brood Size in *Caenorhabditis elegans*." *JoVE (Journal of Visualized Experiments)*, no. 192 (February), e65064. https://doi.org/10.3791/65064.

Lai, C. H., C. Y. Chou, L. Y. Ch'ang, C. S. Liu, and W. Lin. 2000. "Identification of Novel Human Genes Evolutionarily Conserved in *Caenorhabditis elegans* by Comparative Proteomics." *Genome Research* 10 (5): 703–13. https://doi.org/10.1101/gr.10.5.703.

Lau, H. L., T. A. Timbers, R. Mahmoud, and C. H. Rankin. 2013. "Genetic Dissection of Memory for Associative and Non-Associative Learning in *Caenorhabditis elegans*." *Genes, Brain, and Behavior* 12 (2): 210–23. https://doi.org/10.1111/j.1601-183X.2012.00863.x.

Lawler, Daniel E., and Dirk R. Albrecht. 2022. "Monitoring Neural Activity during Sleep/Wake Events in Adult *C. elegans* by Automated Sleep Detection and Stimulation." *STAR Protocols* 3 (3): 101532. https://doi.org/10.1016/j.xpro.2022.101532.

Lažetić, Vladimir, Lakshmi E. Batachari, Alistair B. Russell, and Emily R. Troemel. 2023. "Similarities in the Induction of the Intracellular Pathogen Response in *Caenorhabditis elegans* and the Type I Interferon Response in Mammals." *BioEssays: News and Reviews in Molecular, Cellular and Developmental Biology* 45 (11): e2300097. https://doi.org/10.1002/bies.202300097.

Lee, Harksun, Sonia A. Boor, Zoë A. Hilbert, Joshua D. Meisel, Jaeseok Park, Ye Wang, Ryan McKeown, Sylvia E. J. Fischer, Erik C. Andersen, and Dennis H. Kim. 2023. "Genetic Variants That Modify the Neuroendocrine Regulation of Foraging Behavior in *C. elegans*." *bioRxiv: The Preprint Server for Biology*, September, 2023.09.09.556976. https://doi.org/10.1101/2023.09.09.556976.

Lee, Ji E., and Joseph G. Gleeson. 2011. "Cilia in the Nervous System: Linking Cilia Function and Neurodevelopmental Disorders." *Current Opinion in Neurology* 24 (2): 98–105. https://doi.org/10.1097/WCO.0b013e3283444d05.

Lee, Jin Il, Bijaya Kumar Dhakal, Jungsoo Lee, Jaya Bandyopadhyay, Sang Young Jeong, Soo Hyun Eom, Do Han Kim, and Joohong Ahnn. 2003. "The *Caenorhabditis elegans* Homologue of Down Syndrome Critical Region 1, RCN-1, Inhibits Multiple Functions of the Phosphatase Calcineurin." *Journal of Molecular Biology* 328 (1): 147–56. https://doi.org/10.1016/S0022-2836(03)00237-7.

Lee, Kyunghwa, and Douglas S. Portman. 2007. "Neural Sex Modifies the Function of a *C. elegans* Sensory Circuit." *Current Biology: CB* 17 (21): 1858–63. https://doi.org/10.1016/j.cub.2007.10.015.

Lee, R. C., R. L. Feinbaum, and V. Ambros. 1993. "The *C. elegans* Heterochronic Gene Lin-4 Encodes Small RNAs with Antisense Complementarity to Lin-14." *Cell* 75 (5): 843–54. https://doi.org/10.1016/0092-8674(93)90529-y.

Lejeune, J., M. Gautier, and R. Turpin. 1959. "[Study of somatic chromosomes from 9 mongoloid children]." *Comptes Rendus Hebdomadaires Des Seances De l'Academie Des Sciences* 248 (11): 1721–22.

Levenson, Max T., Rio Barrere-Cain, Lisa Truong, Yen-Wei Chen, Karissa Shuck, Blake Panter, Ella Reich, Xia Yang, and Patrick Allard. 2023. "Protocol for Nuclear Dissociation of the Adult *C. elegans* for Single-Nucleus RNA Sequencing and Its Application for Mapping Environmental Responses." *STAR Protocols* 4 (4): 102756. https://doi.org/10.1016/j.xpro.2023.102756.

Li, Dirui, and Ming Wang. 2012. "Construction of a Bicistronic Vector for the Co-Expression of Two Genes in *Caenorhabditis elegans* Using a Newly Identified IRES." *BioTechniques* 52 (3): 173–76. https://doi.org/10.2144/000113821.

Li, Tingting, Xinjian Wang, Zhigang Feng, and Yan Zou. 2022. "Live Imaging of Postembryonic Developmental Processes in *C. elegans*." *STAR Protocols* 3 (2): 101336. https://doi.org/10.1016/j.xpro.2022.101336.

Li, Weixun, Tae-Woo Choi, Joohong Ahnn, and Sun-Kyung Lee. 2016. "Allele-Specific Phenotype Suggests a Possible Stimulatory Activity of RCAN-1 on Calcineurin in *Caenorhabditis elegans*." 2016. https://www.molcells.org/journal/view.html?doi=10.14348/molcells.2016.0222.

Liao, Wen-Wei, Mobin Asri, Jana Ebler, Daniel Doerr, Marina Haukness, Glenn Hickey, Shuangjia Lu, et al. 2023. "A Draft Human Pangenome Reference." *Nature* 617 (7960): 312–24. https://doi.org/10.1038/s41586-023-05896-x.

Liu, Qiang, Philip B. Kidd, May Dobosiewicz, and Cornelia I. Bargmann. 2018a. "*C. elegans* AWA Olfactory Neurons Fire Calcium-Mediated All-or-None Action Potentials." *Cell* 175 (1): 57–70.e17. https://doi.org/10.1016/j.cell.2018.08.018.

Liu, Zheng, Maro J. Kariya, Christopher D. Chute, Amy K. Pribadi, Sarah G. Leinwand, Ada Tong, Kevin P. Curran, et al. 2018b. "Predator-Secreted Sulfolipids Induce Defensive Responses in *C. elegans*." *Nature Communications* 9 (1): 1128. https://doi.org/10.1038/s41467-018-03333-6.

Lockery, Shawn R., and Miriam B. Goodman. 2009. "The Quest for Action Potentials in *C. elegans* Neurons Hits a Plateau." *Nature Neuroscience* 12 (4): 377–78. https://doi.org/10.1038/nn0409-377.

Long, Lijiang, Hong Guo, Di Yao, Kai Xiong, Yongjun Li, Pengpeng Liu, Zuoyan Zhu, and Dong Liu. 2015. "Regulation of Transcriptionally Active Genes via the Catalytically Inactive Cas9 in *C. elegans* and D. Rerio." *Cell Research* 25 (5): 638–41. https://doi.org/10.1038/cr.2015.35.

Lowe, Simon A., Maria M. Usowicz, and James J. L. Hodge. 2019. "Neuronal Overexpression of Alzheimer's Disease and Down's Syndrome Associated DYRK1A/Minibrain Gene Alters Motor Decline, Neurodegeneration and Synaptic Plasticity in Drosophila." *Neurobiology of Disease* 125 (May):107–14. https://doi.org/10.1016/j.nbd.2019.01.017.

Mack, Hildegard I. D., Peichuan Zhang, Bryan R. Fonslow, and John R. Yates. 2017. "The Protein Kinase MBK-1 Contributes to Lifespan Extension in Daf-2 Mutant and Germline-Deficient *Caenorhabditis elegans*." *Aging* 9 (5): 1414–32. https://doi.org/10.18632/aging.101244.

Maggie's Pearl, LLC. 2024. "A Prospective, Randomized, Double-Blind, Placebo-Controlled, Single-Center Study of Oral Epalrestat Therapy in Pediatric Subjects With Phosphomannomutase 2-Congenital Disorder of Glycosylation (PMM2-CDG)." Clinical trial registration NCT04925960. clinicaltrials.gov. https://clinicaltrials.gov/study/NCT04925960.

Mahoney, Timothy R., Shuo Luo, Elaine K. Round, Martin Brauner, Alexander Gottschalk, James H. Thomas, and Michael L. Nonet. 2008. "Intestinal Signaling to GABAergic Neurons Regulates a Rhythmic Behavior in *Caenorhabditis elegans*." *Proceedings of the National Academy of Sciences of the United States of America* 105 (42): 16350–55. https://doi.org/10.1073/pnas.0803617105.

Majeed, Maryam, Haejun Han, Keren Zhang, Wen Xi Cao, Chien-Po Liao, Oliver Hobert, and Hang Lu. 2024. "Toolkits for Detailed and High-Throughput Interrogation of Synapses in *C. elegans*." Edited by Douglas Portman and Albert Cardona. *eLife* 12 (January):RP91775. https://doi.org/10.7554/eLife.91775.

Mango, Susan E. 2007. "The *C. elegans* Pharynx: A Model for Organogenesis." In *WormBook: The Online Review of C. elegans Biology [Internet]*. WormBook. https://www.ncbi.nlm.nih.gov/books/NBK19698/.

Mariol, Marie-Christine, Ludivine Walter, Stéphanie Bellemin, and Kathrin Gieseler. 2013. "A Rapid Protocol for Integrating Extrachromosomal Arrays with High Transmission Rate into the *C. elegans* Genome." *Journal of Visualized Experiments: JoVE*, no. 82 (December), e50773. https://doi.org/10.3791/50773.

McColl, Gawain, Blaine R. Roberts, Tara L. Pukala, Vijaya B. Kenche, Christine M. Roberts, Christopher D. Link, Timothy M. Ryan, et al. 2012. "Utility of an Improved Model of Amyloid-Beta ($A\beta_{1-42}$) Toxicity in *Caenorhabditis elegans* for Drug Screening for Alzheimer's Disease." *Molecular Neurodegeneration* 7 (November):57. https://doi.org/10.1186/1750-1326-7-57.

McGrath, Patrick T., and Ilya Ruvinsky. 2019. "A Primer on Pheromone Signaling in *Caenorhabditis elegans* for Systems Biologists." *Current Opinion in Systems Biology* 13 (February):23–30. https://doi.org/10.1016/j.coisb.2018.08.012.

McIntire, S. L., R. J. Reimer, K. Schuske, R. H. Edwards, and E. M. Jorgensen. 1997. "Identification and Characterization of the Vesicular GABA Transporter." *Nature* 389 (6653): 870–76. https://doi.org/10.1038/39908.

Mellem, Jerry E., Penelope J. Brockie, David M. Madsen, and Andres V. Maricq. 2008. "Action Potentials Contribute to Neuronal Signaling in *C. elegans*." *Nature Neuroscience* 11 (8): 865–67. https://doi.org/10.1038/nn.2131.

Mello, C., and A. Fire. 1995. "DNA Transformation." *Methods in Cell Biology* 48:451–82.

Mills, Holly, Amanda Ortega, Wenjing Law, Vera Hapiak, Philip Summers, Tobias Clark, and Richard Komuniecki. 2016. "Opiates Modulate Noxious Chemical Nociception through a Complex Monoaminergic/Peptidergic Cascade." *The Journal of Neuroscience: The Official*

Journal of the Society for Neuroscience 36 (20): 5498–5508. https://doi.org/10.1523/JNEUROSCI.4520-15.2016.

Mitchison, Hannah M, and Enza Maria Valente. 2017. "Motile and Non-Motile Cilia in Human Pathology: From Function to Phenotypes." *The Journal of Pathology* 241 (2): 294–309. https://doi.org/10.1002/path.4843.

Mondal, Sudip, Evan Hegarty, James J. Sahn, Luisa L. Scott, Sertan Kutal Gökçe, Chris Martin, Navid Ghorashian, et al. 2018. "High-Content Microfluidic Screening Platform Used To Identify σ2R/Tmem97 Binding Ligands That Reduce Age-Dependent Neurodegeneration in *C. elegans* SC_APP Model." *ACS Chemical Neuroscience* 9 (5): 1014–26. https://doi.org/10.1021/acschemneuro.7b00428.

Moreno, Eduardo, Angela McGaughran, Christian Rödelsperger, Manuel Zimmer, and Ralf J. Sommer. 2016. "Oxygen-Induced Social Behaviours in Pristionchus Pacificus Have a Distinct Evolutionary History and Genetic Regulation from *Caenorhabditis elegans*." *Proceedings. Biological Sciences* 283 (1825): 20152263. https://doi.org/10.1098/rspb.2015.2263.

Morrison, G. E., and D. van der Kooy. 2001. "A Mutation in the AMPA-Type Glutamate Receptor, Glr-1, Blocks Olfactory Associative and Nonassociative Learning in *Caenorhabditis elegans*." *Behavioral Neuroscience* 115 (3): 640–49. https://doi.org/10.1037/0735-7044.115.3.640.

Moyer, Anna J., Fabian-Xosé Fernandez, Yicong Li, Donna K. Klinedinst, Liliana D. Florea, Yasuhiro Kazuki, Mitsuo Oshimura, and Roger H. Reeves. 2023. "Overexpression Screen of Chromosome 21 Genes Reveals Modulators of Sonic Hedgehog Signaling Relevant to Down Syndrome." *Disease Models & Mechanisms* 16 (4): dmm049712. https://doi.org/10.1242/dmm.049712.

Moyer, Anna J., Katheleen Gardiner, and Roger H. Reeves. 2020. "All Creatures Great and Small: New Approaches for Understanding Down Syndrome Genetics." *Trends in Genetics: TIG*, October. https://doi.org/10.1016/j.tig.2020.09.017.

Muñoz-Jiménez, Celia, Cristina Ayuso, Agnieszka Dobrzynska, Antonio Torres-Mendéz, Patricia de la Cruz Ruiz, and Peter Askjaer. 2017. "An Efficient FLP-Based Toolkit for Spatiotemporal Control of Gene Expression in *Caenorhabditis elegans*." *Genetics* 206 (4): 1763–78. https://doi.org/10.1534/genetics.117.201012.

Murphy, Coleen T., and Patrick J. Hu. 2013. "Insulin/Insulin-like Growth Factor Signaling in *C. elegans*." *WormBook: The Online Review of C. elegans Biology*, December, 1–43. https://doi.org/10.1895/wormbook.1.164.1.

Nance, Jeremy, and Christian Frøkjær-Jensen. 2019. "The *Caenorhabditis elegans* Transgenic Toolbox." *Genetics* 212 (4): 959–90. https://doi.org/10.1534/genetics.119.301506.

Neumann, Fernanda, Stéphanie Gourdain, Christelle Albac, Alain D. Dekker, Linh Chi Bui, Julien Dairou, Isabelle Schmitz-Afonso, et al. 2018. "DYRK1A Inhibition and Cognitive Rescue in a Down Syndrome Mouse Model Are Induced by New Fluoro-DANDY Derivatives." *Scientific Reports* 8 (1): 2859. https://doi.org/10.1038/s41598-018-20984-z.

Nikoletopoulou, Vassiliki, and Nektarios Tavernarakis. 2014. "Necrotic Cell Death in *Caenorhabditis elegans*." *Methods in Enzymology* 545:127–55. https://doi.org/10.1016/B978-0-12-801430-1.00006-8.

Niu, Longgang, Yan Li, Pengyu Zong, Ping Liu, Yuan Shui, Bojun Chen, and Zhao-Wen Wang. 2020. "Melatonin Promotes Sleep by Activating the BK Channel in *C. elegans*." *Proceedings of the National Academy of Sciences of the United States of America* 117 (40): 25128–37. https://doi.org/10.1073/pnas.2010928117.

Noma, Kentaro, and Yishi Jin. 2018. "Rapid Integration of Multi-Copy Transgenes Using Optogenetic Mutagenesis in *Caenorhabditis elegans*." *G3 (Bethesda, Md.)* 8 (6): 2091–97. https://doi.org/10.1534/g3.118.200158.

Nonet, Michael L. 2020. "Efficient Transgenesis in *Caenorhabditis elegans* Using Flp Recombinase-Mediated Cassette Exchange." *Genetics* 215 (4): 903–21. https://doi.org/10.1534/genetics.120.303388.

Nordquist, Sarah K., Sofia R. Smith, and Jonathan T. Pierce. 2018. "Systematic Functional Characterization of Human 21st Chromosome Orthologs in *Caenorhabditis elegans*." *G3 Genes|Genomes|Genetics* 8 (3): 967–79. https://doi.org/10.1534/g3.118.200019.

Oakes, Mitchell D., Wen Jing Law, Tobias Clark, Bruce A. Bamber, and Richard Komuniecki. 2017. "Cannabinoids Activate Monoaminergic Signaling to Modulate Key *C. elegans* Behaviors." *Journal of Neuroscience* 37 (11): 2859–69. https://doi.org/10.1523/JNEUROSCI.3151-16.2017.

Oberst, A., C. Bender, and D. R. Green. 2008. "Living with Death: The Evolution of the Mitochondrial Pathway of Apoptosis in Animals." *Cell Death and Differentiation* 15 (7): 1139–46. https://doi.org/10.1038/cdd.2008.65.

Oda, Shigekazu, Emi Sato-Ebine, Akinobu Nakamura, Koutarou D. Kimura, and Kazuhiro Aoki. 2023. "Optical Control of Cell Signaling with Red/Far-Red Light-Responsive Optogenetic Tools in *Caenorhabditis elegans*." *ACS Synthetic Biology* 12 (3): 700–708. https://doi.org/10.1021/acssynbio.2c00461.

Offenburger, Sarah-Lena, Elisabeth Jongsma, and Anton Gartner. 2018. "Mutations in *Caenorhabditis elegans* Neuroligin-like Glit-1, the Apoptosis Pathway and the Calcium Chaperone Crt-1 Increase Dopaminergic Neurodegeneration after 6-OHDA Treatment." *PLoS Genetics* 14 (1): e1007106. https://doi.org/10.1371/journal.pgen.1007106.

Olivares, Erick, Eduardo J. Izquierdo, and Randall D. Beer. 2021. "A Neuromechanical Model of Multiple Network Rhythmic Pattern Generators for Forward Locomotion in *C. elegans*." *Frontiers in Computational Neuroscience* 15:572339. https://doi.org/10.3389/fncom.2021.572339.

Olson, Andrew C., Allison M. Butt, Nakeirah T. M. Christie, Ashish Shelar, and Michael R. Koelle. 2023. "Multiple Subthreshold GPCR Signals Combined by the G-Proteins Gαq and Gαs Activate the *Caenorhabditis elegans* Egg-Laying Muscles." *The Journal of Neuroscience: The Official Journal of the Society for Neuroscience* 43 (21): 3789–3806. https://doi.org/10.1523/JNEUROSCI.2301-22.2023.

Onken, Brian, Christine A. Sedore, Anna L. Coleman-Hulbert, David Hall, Erik Johnson, Eleanor Grace Jones, Stephen A. Banse, et al. 2022. "Metformin Treatment of Diverse Caenorhabditis Species Reveals the Importance of Genetic Background in Longevity and Healthspan Extension Outcomes." *Aging Cell* 21 (1): e13488. https://doi.org/10.1111/acel.13488.

Pan, Peng, Zhen Qin, William Sun, Yuxiao Zhou, Shaojia Wang, Pengfei Song, Yong Wang, et al. 2023. "A Spiral Microfluidic Device for Rapid Sorting, Trapping, and Long-Term Live Imaging of *Caenorhabditis elegans* Embryos." *Microsystems & Nanoengineering* 9 (1): 1–11. https://doi.org/10.1038/s41378-023-00485-4.

Parker, Dylan M., Lindsay P. Winkenbach, Annemarie Parker, Sam Boyson, and Erin Osborne Nishimura. 2021. "Improved Methods for Single-Molecule Fluorescence in Situ Hybridization and Immunofluorescence in *Caenorhabditis elegans* Embryos." *Current Protocols* 1 (11): e299. https://doi.org/10.1002/cpz1.299.

Pigazzini, Maria Lucia, and Janine Kirstein. 2020. "In Vivo Quantification of Protein Turnover in Aging *C. elegans* Using Photoconvertible Dendra2." *Journal of Visualized Experiments: JoVE*, no. 160 (June). https://doi.org/10.3791/61196.

Pinan-Lucarré, Bérangère, Haijun Tu, Marie Pierron, Pablo Ibáñez Cruceyra, Hong Zhan, Christian Stigloher, Janet E. Richmond, and Jean-Louis Bessereau. 2014. "*C. elegans* Punctin Specifies Cholinergic versus GABAergic Identity of Postsynaptic Domains." *Nature* 511 (7510): 466–70. https://doi.org/10.1038/nature13313.

Pokala, Navin, Qiang Liu, Andrew Gordus, and Cornelia I. Bargmann. 2014. "Inducible and Titratable Silencing of *Caenorhabditis elegans* Neurons in Vivo with Histamine-Gated Chloride Channels." *Proceedings of the National Academy of Sciences of the United States of America* 111 (7): 2770–75. https://doi.org/10.1073/pnas.1400615111.

Porta-de-la-Riva, Montserrat, Laura Fontrodona, Alberto Villanueva, and Julián Cerón. 2012. "Basic *Caenorhabditis elegans* Methods: Synchronization and Observation." *Journal of Visualized Experiments : JoVE*, no. 64 (June), 4019. https://doi.org/10.3791/4019.

Praitis, V., E. Casey, D. Collar, and J. Austin. 2001. "Creation of Low-Copy Integrated Transgenic Lines in *Caenorhabditis elegans*." *Genetics* 157 (3): 1217–26. https://doi.org/10.1093/genetics/157.3.1217.

Priyadarshini, Monika, Julie Zhouli Ni, Amhed M. Vargas-Velazquez, Sam Guoping Gu, and Christian Frøkjær-Jensen. 2022. "Reprogramming the piRNA Pathway for Multiplexed and Transgenerational Gene Silencing in *C. elegans*." *Nature Methods* 19 (2): 187–94. https://doi.org/10.1038/s41592-021-01369-z.

Prömel, Simone, Franziska Fiedler, Claudia Binder, Jana Winkler, Torsten Schöneberg, and Doreen Thor. 2016. "Deciphering and Modulating G Protein Signalling in *C. elegans* Using the DREADD Technology." *Scientific Reports* 6 (1): 28901. https://doi.org/10.1038/srep28901.

Qiao, L., J. L. Lissemore, P. Shu, A. Smardon, M. B. Gelber, and E. M. Maine. 1995. "Enhancers of Glp-1, a Gene Required for Cell-Signaling in *Caenorhabditis elegans*, Define a Set of Genes Required for Germline Development." *Genetics* 141 (2): 551–69. https://doi.org/10.1093/genetics/141.2.551.

Rahmani, Aelon, and Yee Lian Chew. 2021. "Investigating the Molecular Mechanisms of Learning and Memory Using *Caenorhabditis elegans*." *Journal of Neurochemistry* 159 (3): 417–51. https://doi.org/10.1111/jnc.15510.

Raich, William B, Celine Moorman, Clay O Lacefield, Jonah Lehrer, Dusan Bartsch, Ronald H A Plasterk, Eric R Kandel, and Oliver Hobert. 2003. "Characterization of *Caenorhabditis elegans* Homologs of the Down Syndrome Candidate Gene DYRK1A." *Genetics* 163 (2): 571–80. https://doi.org/10.1093/genetics/163.2.571.

Raizen, David M., John E. Zimmerman, Matthew H. Maycock, Uyen D. Ta, Young-jai You, Meera V. Sundaram, and Allan I. Pack. 2008. "Lethargus Is a *Caenorhabditis elegans* Sleep-like State." *Nature* 451 (7178): 569–72. https://doi.org/10.1038/nature06535.

Randi, Francesco, Anuj K. Sharma, Sophie Dvali, and Andrew M. Leifer. 2023. "Neural Signal Propagation Atlas of *Caenorhabditis elegans*." *Nature* 623 (7986): 406–14. https://doi.org/10.1038/s41586-023-06683-4.

Rankin, C. H., C. D. Beck, and C. M. Chiba. 1990. "*Caenorhabditis elegans*: A New Model System for the Study of Learning and Memory." *Behavioural Brain Research* 37 (1): 89–92. https://doi.org/10.1016/0166-4328(90)90074-o.

Rawsthorne, Helena, Fernando Calahorro, Lindy Holden-Dye, Vincent O'Connor, and James Dillon. 2021. "Investigating Autism Associated Genes in *C. elegans* Reveals Candidates with a Role in Social Behaviour." *PloS One* 16 (5): e0243121. https://doi.org/10.1371/journal.pone.0243121.

Reynolds, John J., Louise S. Bicknell, Paula Carroll, Martin R. Higgs, Ranad Shaheen, Jennie E. Murray, Dimitrios K. Papadopoulos, et al. 2017. "Mutations in DONSON Disrupt Replication Fork Stability and Cause Microcephalic Dwarfism." *Nature Genetics* 49 (4): 537–49. https://doi.org/10.1038/ng.3790.

Richmond, Janet E. 2006. "Electrophysiological Recordings from the Neuromuscular Junction of *C. elegans*." *WormBook: The Online Review of C. elegans Biology*, October, 1–8. https://doi.org/10.1895/wormbook.1.112.1.

Ripoll-Sánchez, Lidia, Jan Watteyne, HaoSheng Sun, Robert Fernandez, Seth R. Taylor, Alexis Weinreb, Barry L. Bentley, et al. 2023. "The Neuropeptidergic Connectome of *C. elegans*." *Neuron* 111 (22): 3570–3589.e5. https://doi.org/10.1016/j.neuron.2023.09.043.

Rueda, Noemí, Jesús Flórez, and Carmen Martínez-Cué. 2013. "Apoptosis in Down's Syndrome: Lessons from Studies of Human and Mouse Models." *Apoptosis: An International Journal on Programmed Cell Death* 18 (2): 121–34. https://doi.org/10.1007/s10495-012-0785-3.

Saeki, S., M. Yamamoto, and Y. Iino. 2001. "Plasticity of Chemotaxis Revealed by Paired Presentation of a Chemoattractant and Starvation in the Nematode *Caenorhabditis elegans*." *Journal of Experimental Biology* 204 (10): 1757–64.

Sae-Lee, Wisath, Luisa L. Scott, Lotti Brose, Aliyah J. Encarnacion, Ted Shi, Pragati Kore, Lashaun O. Oyibo, Congxi Ye, Susan K. Rozmiarek, and Jonathan T. Pierce. 2020. "APP-

Induced Patterned Neurodegeneration Is Exacerbated by APOE4 in *Caenorhabditis elegans*." *G3 (Bethesda, Md.)* 10 (8): 2851–61. https://doi.org/10.1534/g3.120.401486.

Sammi, Shreesh Raj, Laura E. Jameson, Kendra D. Conrow, Maxwell C. K. Leung, and Jason R. Cannon. 2022. "*Caenorhabditis elegans* Neurotoxicity Testing: Novel Applications in the Adverse Outcome Pathway Framework." *Frontiers in Toxicology* 4 (March):826488. https://doi.org/10.3389/ftox.2022.826488.

Sanchez, S, Briana Syed, Brooke Frohock, and Jon Pierce. 2024. "Survival of the Fittest Dosage: Restricted Natural Distribution of Gene Expression Level Predicts Duplication Sensitivity."

Sarov, Mihail, John I. Murray, Kristin Schanze, Andrei Pozniakovski, Wei Niu, Karolin Angermann, Susanne Hasse, et al. 2012. "A Genome-Scale Resource for in Vivo Tag-Based Protein Function Exploration in *C. elegans*." *Cell* 150 (4): 855–66. https://doi.org/10.1016/j.cell.2012.08.001.

Sasakura, Hiroyuki, and Ikue Mori. 2013. "Behavioral Plasticity, Learning, and Memory in *C. elegans*." *Current Opinion in Neurobiology* 23 (1): 92–99. https://doi.org/10.1016/j.conb.2012.09.005.

Sawa, Hitoshi, and Hendrik C. Korswagen. 2013. "Wnt Signaling in *C. elegans*." *WormBook: The Online Review of C. elegans Biology*, December, 1–30. https://doi.org/10.1895/wormbook.1.7.2.

Schweinsberg, Peter J., and Barth D. Grant. 2013. "*C. elegans* Gene Transformation by Microparticle Bombardment." *WormBook: The Online Review of C. elegans Biology*, December, 1–10. https://doi.org/10.1895/wormbook.1.166.1.

Scott, Euan, Adam Hudson, Emily Feist, Fernando Calahorro, James Dillon, Raissa de Freitas, Matthew Wand, Liliane Schoofs, Vincent O'Connor, and Lindy Holden-Dye. 2017. "An Oxytocin-Dependent Social Interaction between Larvae and Adult *C. elegans*." *Scientific Reports* 7 (1): 10122. https://doi.org/10.1038/s41598-017-09350-7.

Scott, Luisa L., James J. Sahn, Antonio Ferragud, Rachel C. Yen, Praveen N. Satarasinghe, Michael D. Wood, Timothy R. Hodges, et al. 2018. "Small Molecule Modulators of σ2R/Tmem97 Reduce Alcohol Withdrawal-Induced Behaviors." *Neuropsychopharmacology: Official Publication of the American College of Neuropsychopharmacology* 43 (9): 1867–75. https://doi.org/10.1038/s41386-018-0067-z.

Sen, Ilke, Alan Kavšek, and Christian G. Riedel. 2021. "Chromatin Immunoprecipitation and Sequencing (ChIP-Seq) Optimized for Application in *Caenorhabditis elegans*." *Current Protocols* 1 (7): e187. https://doi.org/10.1002/cpz1.187.

Shaye, Daniel D., and Iva Greenwald. 2011. "OrthoList: A Compendium of *C. elegans* Genes with Human Orthologs." *PloS One* 6 (5): e20085. https://doi.org/10.1371/journal.pone.0020085.

Shekari, Arman, and Margaret Fahnestock. 2021. "Chapter 13 – Cholinergic Neurodegeneration in Alzheimer Disease Mouse Models." In *Handbook of Clinical Neurology*, edited by Dick F. Swaab, Ruud M. Buijs, Felix Kreier, Paul J. Lucassen, and Ahmad Salehi, 182:191–209. The Human Hypothalamus: Neuropsychiatric Disorders. Elsevier. https://doi.org/10.1016/B978-0-12-819973-2.00013-7.

Sheu, Shu-Hsien, Srigokul Upadhyayula, Vincent Dupuy, Song Pang, Fei Deng, Jinxia Wan, Deepika Walpita, et al. 2022. "A Serotonergic Axon-Cilium Synapse Drives Nuclear Signaling to Alter Chromatin Accessibility." *Cell* 185 (18): 3390–3407.e18. https://doi.org/10.1016/j.cell.2022.07.026.

Siew, Wei-Hong, Kai-Leng Tan, Maryam Abbaspour Babaei, Pike-See Cheah, and King-Hwa Ling. 2013. "MicroRNAs and Intellectual Disability (ID) in Down Syndrome, X-Linked ID, and Fragile X Syndrome." *Frontiers in Cellular Neuroscience* 7 (April):41. https://doi.org/10.3389/fncel.2013.00041.

Silva-García, Carlos G., Anne Lanjuin, Caroline Heintz, Sneha Dutta, Nicole M. Clark, and William B. Mair. 2019. "Single-Copy Knock-In Loci for Defined Gene Expression in *Caenorhabditis elegans*." *G3 (Bethesda, Md.)* 9 (7): 2195–98. https://doi.org/10.1534/g3.119.400314.

Singh, Jogender. 2021. "Harnessing the Power of Genetics: Fast Forward Genetics in *Caenorhabditis elegans*." *Molecular Genetics and Genomics: MGG* 296 (1): 1–20. https://doi.org/10.1007/s00438-020-01721-6.

Sönnichsen, B., L. B. Koski, A. Walsh, P. Marschall, B. Neumann, M. Brehm, A.-M. Alleaume, et al. 2005. "Full-Genome RNAi Profiling of Early Embryogenesis in *Caenorhabditis elegans*." *Nature* 434 (7032): 462–69. https://doi.org/10.1038/nature03353.

Soutschek, Michael, and Gerhard Schratt. 2023. "Non-Coding RNA in the Wiring and Remodeling of Neural Circuits." *Neuron* 111 (14): 2140–54. https://doi.org/10.1016/j.neuron.2023.04.031.

Spies, Jan, and Henrik Bringmann. 2018. "Automated Detection and Manipulation of Sleep in *C. elegans* Reveals Depolarization of a Sleep-Active Neuron during Mechanical Stimulation-Induced Sleep Deprivation." *Scientific Reports* 8 (1): 9732. https://doi.org/10.1038/s41598-018-28095-5.

Statzer, Cyril, Jin Meng, Richard Venz, Monet Bland, Stacey Robida-Stubbs, Krina Patel, Dunja Petrovic, et al. 2022. "ATF-4 and Hydrogen Sulfide Signalling Mediate Longevity in Response to Inhibition of Translation or mTORC1." *Nature Communications* 13 (1): 967. https://doi.org/10.1038/s41467-022-28599-9.

Steiner, Florian A., Paul B. Talbert, Sivakanthan Kasinathan, Roger B. Deal, and Steven Henikoff. 2012. "Cell-Type-Specific Nuclei Purification from Whole Animals for Genome-Wide Expression and Chromatin Profiling." *Genome Research* 22 (4): 766–77. https://doi.org/10.1101/gr.131748.111.

Sterken, Mark G., L. Basten Snoek, Jan E. Kammenga, and Erik C. Andersen. 2015. "The Laboratory Domestication of *Caenorhabditis elegans*." *Trends in Genetics: TIG* 31 (5): 224–31. https://doi.org/10.1016/j.tig.2015.02.009.

Stevenson, Zachary C., Megan J. Moerdyk-Schauwecker, Brennen Jamison, and Patrick C. Phillips. 2020. "Rapid Self-Selecting and Clone-Free Integration of Transgenes into Engineered CRISPR Safe Harbor Locations in *Caenorhabditis elegans*." *G3 (Bethesda, Md.)* 10 (10): 3775–82. https://doi.org/10.1534/g3.120.401400.

Stiernagle, Theresa. 2006. "Maintenance of *C. elegans*." In *WormBook: The Online Review of C. elegans Biology [Internet]*. WormBook. https://www.ncbi.nlm.nih.gov/books/NBK19649/.

Stinchcomb, D. T., J. E. Shaw, S. H. Carr, and D. Hirsh. 1985. "Extrachromosomal DNA Transformation of *Caenorhabditis elegans*." *Molecular and Cellular Biology* 5 (12): 3484–96. https://doi.org/10.1128/mcb.5.12.3484-3496.1985.

Sulston, J., M. Dew, and S. Brenner. 1975. "Dopaminergic Neurons in the Nematode *Caenorhabditis elegans*." *The Journal of Comparative Neurology* 163 (2): 215–26. https://doi.org/10.1002/cne.901630207.

Susoy, Vladislav, Wesley Hung, Daniel Witvliet, Joshua E. Whitener, Min Wu, Core Francisco Park, Brett J. Graham, Mei Zhen, Vivek Venkatachalam, and Aravinthan D. T. Samuel. 2021. "Natural Sensory Context Drives Diverse Brain-Wide Activity during *C. elegans* Mating." *Cell* 184 (20): 5122–5137.e17. https://doi.org/10.1016/j.cell.2021.08.024.

Sutphin, George L., Grant Backer, Susan Sheehan, Shannon Bean, Caroline Corban, Teresa Liu, Marjolein J. Peters, et al. 2017. "*Caenorhabditis elegans* Orthologs of Human Genes Differentially Expressed with Age Are Enriched for Determinants of Longevity." *Aging Cell* 16 (4): 672–82. https://doi.org/10.1111/acel.12595.

Taylor, Seth R., Gabriel Santpere, Alexis Weinreb, Alec Barrett, Molly B. Reilly, Chuan Xu, Erdem Varol, et al. 2021. "Molecular Topography of an Entire Nervous System." *Cell* 184 (16): 4329–4347.e23. https://doi.org/10.1016/j.cell.2021.06.023.

Teo, Emelyne, Sudharshan Ravi, Diogo Barardo, Hyung-Seok Kim, Sheng Fong, Amaury Cazenave-Gassiot, Tsze Yin Tan, et al. 2019. "Metabolic Stress Is a Primary Pathogenic Event in Transgenic *Caenorhabditis elegans* Expressing Pan-Neuronal Human Amyloid Beta." *eLife* 8 (October):e50069. https://doi.org/10.7554/eLife.50069.

Thapliyal, Shruti, and Kavita Babu. 2018. "*C. elegans* Locomotion: Finding Balance in Imbalance." In *Biochemical and Biophysical Roles of Cell Surface Molecules*, edited by Kausik Chattopadhyay and Subhash C. Basu, 185–96. Advances in Experimental Medicine and Biology. Singapore: Springer. https://doi.org/10.1007/978-981-13-3065-0_14.

Thompson, Owen, Mark Edgley, Pnina Strasbourger, Stephane Flibotte, Brent Ewing, Ryan Adair, Vinci Au, et al. 2013. "The Million Mutation Project: A New Approach to Genetics in *Caenorhabditis elegans*." *Genome Research* 23 (10): 1749–62. https://doi.org/10.1101/gr.157651.113.

Tian, Lin, S. Andrew Hires, Tianyi Mao, Daniel Huber, M. Eugenia Chiappe, Sreekanth H. Chalasani, Leopoldo Petreanu, et al. 2009. "Imaging Neural Activity in Worms, Flies and Mice with Improved GCaMP Calcium Indicators." *Nature Methods* 6 (12): 875–81. https://doi.org/10.1038/nmeth.1398.

Timmons, L., and A. Fire. 1998. "Specific Interference by Ingested dsRNA." *Nature* 395 (6705): 854. https://doi.org/10.1038/27579.

Tong, Yong-Guang, and Thomas R. Bürglin. 2010. "Conditions for Dye-Filling of Sensory Neurons in *Caenorhabditis elegans*." *Journal of Neuroscience Methods* 188 (1): 58–61. https://doi.org/10.1016/j.jneumeth.2010.02.003.

Tramutola, Antonella, Gilda Pupo, Fabio Di Domenico, Eugenio Barone, Andrea Arena, Chiara Lanzillotta, Diede Brokeaart, et al. 2016. "Activation of P53 in Down Syndrome and in the Ts65Dn Mouse Brain Is Associated with a Pro-Apoptotic Phenotype." *Journal of Alzheimer's Disease: JAD* 52 (1): 359–71. https://doi.org/10.3233/JAD-151105.

Treinin, M., B. Gillo, L. Liebman, and M. Chalfie. 1998. "Two Functionally Dependent Acetylcholine Subunits Are Encoded in a Single *Caenorhabditis elegans* Operon." *Proceedings of the National Academy of Sciences of the United States of America* 95 (26): 15492–95. https://doi.org/10.1073/pnas.95.26.15492.

Trent, C., N. Tsuing, and H. R. Horvitz. 1983. "Egg-Laying Defective Mutants of the Nematode Caenorhabditis Elegans." *Genetics* 104 (4): 619–47. https://doi.org/10.1093/genetics/104.4.619.

Trojanowski, Nicholas F., and David M. Raizen. 2016. "Call It Worm Sleep." *Trends in Neurosciences* 39 (2): 54–62. https://doi.org/10.1016/j.tins.2015.12.005.

Tsien, Roger. 2007. "Roger Tsien: Bringing Color to Cell Biology. A Passion for Chemistry, Color, and Conversations with Cells Is What Drives Roger Tsien. Interview by Ruth Williams." *The Journal of Cell Biology* 179 (1): 6–8. https://doi.org/10.1083/jcb.1791pi.

Tsukada, Yuki, and Ikue Mori. 2021. "Optogenetics in *Caenorhabditis elegans*." In *Optogenetics: Light-Sensing Proteins and Their Applications in Neuroscience and Beyond*, edited by Hiromu Yawo, Hideki Kandori, Amane Koizumi, and Ryoichiro Kageyama, 321–34. Advances in Experimental Medicine and Biology. Singapore: Springer. https://doi.org/10.1007/978-981-15-8763-4_20.

Tursun, Baris, Luisa Cochella, Inés Carrera, and Oliver Hobert. 2009. "A Toolkit and Robust Pipeline for the Generation of Fosmid-Based Reporter Genes in *C. elegans*." *PloS One* 4 (3): e4625. https://doi.org/10.1371/journal.pone.0004625.

Ullrich, Milena, Vanessa Liang, Yee Lian Chew, Samuel Banister, Xiaomin Song, Thiri Zaw, Hong Lam, et al. 2014. "Bio-Orthogonal Labeling as a Tool to Visualize and Identify Newly Synthesized Proteins in *Caenorhabditis elegans*." *Nature Protocols* 9 (9): 2237–55. https://doi.org/10.1038/nprot.2014.150.

Uzel, Kerem, Saul Kato, and Manuel Zimmer. 2022. "A Set of Hub Neurons and Non-Local Connectivity Features Support Global Brain Dynamics in *C. elegans*." *Current Biology: CB* 32 (16): 3443–3459.e8. https://doi.org/10.1016/j.cub.2022.06.039.

Van Robays, J. 2016. "John Langdon Down (1828 – 1896)." *Facts, Views & Vision in ObGyn* 8 (2): 131–36.

Van Sinay, Elien, Olivier Mirabeau, Geert Depuydt, Matthias Boris Van Hiel, Katleen Peymen, Jan Watteyne, Sven Zels, Liliane Schoofs, and Isabel Beets. 2017. "Evolutionarily Conserved TRH Neuropeptide Pathway Regulates Growth in *Caenorhabditis elegans*." *Proceedings of the National Academy of Sciences of the United States of America* 114 (20): E4065–74. https://doi.org/10.1073/pnas.1617392114.

Vavouri, Tanya, Klaudia Walter, Walter R. Gilks, Ben Lehner, and Greg Elgar. 2007. "Parallel Evolution of Conserved Non-Coding Elements That Target a Common Set of Developmental Regulatory Genes from Worms to Humans." *Genome Biology* 8 (2): R15. https://doi.org/10.1186/gb-2007-8-2-r15.

Veerapathiran, Sapthaswaran, and Thorsten Wohland. 2018. "Fluorescence Techniques in Developmental Biology." *Journal of Biosciences* 43 (3): 541–53.

Veland, Iben R., Aashir Awan, Lotte B. Pedersen, Bradley K. Yoder, and Søren T. Christensen. 2009. "Primary Cilia and Signaling Pathways in Mammalian Development, Health and Disease." *Nephron. Physiology* 111 (3): p 39–53. https://doi.org/10.1159/000208212.

Vidal-Gadea, Andrés, Stephen Topper, Layla Young, Ashley Crisp, Leah Kressin, Erin Elbel, Thomas Maples, et al. 2011. "*Caenorhabditis elegans* Selects Distinct Crawling and Swimming Gaits via Dopamine and Serotonin." *Proceedings of the National Academy of Sciences of the United States of America* 108 (42): 17504–9. https://doi.org/10.1073/pnas.1108673108.

Voutev, Roumen, and E. Jane Albert Hubbard. 2008. "A 'FLP-Out' System for Controlled Gene Expression in *Caenorhabditis elegans*." *Genetics* 180 (1): 103–19. https://doi.org/10.1534/genetics.108.090274.

Wade, Nicholas. 2019. "Sydney Brenner, a Decipherer of the Genetic Code, Is Dead at 92." *The New York Times*, April 5, 2019, sec. Obituaries. https://www.nytimes.com/2019/04/05/obituaries/sydney-brenner-dead.html.

Waggoner, L. E., G. T. Zhou, R. W. Schafer, and W. R. Schafer. 1998. "Control of Alternative Behavioral States by Serotonin in *Caenorhabditis elegans*." *Neuron* 21 (1): 203–14. https://doi.org/10.1016/s0896-6273(00)80527-9.

Wang, Baiping, Hongmei Li, Sena A. Mutlu, Devon A. Bowser, Michael J. Moore, Meng C. Wang, and Hui Zheng. 2017a. "The Amyloid Precursor Protein Is a Conserved Receptor for Slit to Mediate Axon Guidance." *eNeuro* 4 (3). https://doi.org/10.1523/ENEURO.0185-17.2017.

Wang, Han, Jonathan Liu, Shahla Gharib, Cynthia M. Chai, Erich M. Schwarz, Navin Pokala, and Paul W. Sternberg. 2017b. "cGAL, a Temperature-Robust GAL4-UAS System for *Caenorhabditis elegans*." *Nature Methods* 14 (2): 145–48. https://doi.org/10.1038/nmeth.4109.

Wang, M., and P. W. Sternberg. 2001. "Pattern Formation during *C. elegans* Vulval Induction." *Current Topics in Developmental Biology* 51:189–220. https://doi.org/10.1016/s0070-2153(01)51006-6.

Wang, Zheng, and David R. Sherwood. 2011. "Dissection of Genetic Pathways in *C. elegans*." *Methods in Cell Biology* 106:113–57. https://doi.org/10.1016/B978-0-12-544172-8.00005-0.

Way, J. C., and M. Chalfie. 1988. "Mec-3, a Homeobox-Containing Gene That Specifies Differentiation of the Touch Receptor Neurons in *C. elegans*." *Cell* 54 (1): 5–16. https://doi.org/10.1016/0092-8674(88)90174-2.

Weissman, Tamily A., and Y. Albert Pan. 2015. "Brainbow: New Resources and Emerging Biological Applications for Multicolor Genetic Labeling and Analysis." *Genetics* 199 (2): 293–306. https://doi.org/10.1534/genetics.114.172510.

White, JG, Eileen Southgate, JN Thomson, and S Brenner. 1976. "The Structure of the Ventral Nerve Cord of *Caenorhabditis elegans*." *Philosophical Transactions of the Royal Society of London. B, Biological Sciences* 275 (938): 327–48. https://doi.org/10.1098/rstb.1976.0086.

White, J. G., E. Southgate, J. N. Thomson, and S. Brenner. 1986. "The Structure of the Nervous System of the Nematode *Caenorhabditis elegans*." *Philosophical Transactions of the Royal Society of London. Series B, Biological Sciences* 314 (1165): 1–340. https://doi.org/10.1098/rstb.1986.0056.

Widmayer, Samuel J, Kathryn S Evans, Stefan Zdraljevic, and Erik C Andersen. 2022. "Evaluating the Power and Limitations of Genome-Wide Association Studies in *Caenorhabditis elegans*." *G3 Genes|Genomes|Genetics* 12 (7): jkac114. https://doi.org/10.1093/g3journal/jkac114.

Wightman, B., I. Ha, and G. Ruvkun. 1993. "Posttranscriptional Regulation of the Heterochronic Gene Lin-14 by Lin-4 Mediates Temporal Pattern Formation in *C. elegans*." *Cell* 75 (5): 855–62. https://doi.org/10.1016/0092-8674(93)90530-4.

Woglar, Alexander, Kei Yamaya, Baptiste Roelens, Alistair Boettiger, Simone Köhler, and Anne M. Villeneuve. 2020. "Quantitative Cytogenetics Reveals Molecular Stoichiometry and Longitudinal Organization of Meiotic Chromosome Axes and Loops." *PLoS Biology* 18 (8): e3000817. https://doi.org/10.1371/journal.pbio.3000817.

Wolfe, Aaron D., John N. Koberstein, Chadwick B. Smith, Melissa L. Stewart, Ian J. Gonzalez, Marc Hammarlund, Anthony A. Hyman, Philip J. S. Stork, Richard H. Goodman, and Daniel A. Colón-Ramos. 2024. "Local and Dynamic Regulation of Neuronal Glycolysis in Vivo." *Proceedings of the National Academy of Sciences* 121 (3): e2314699121. https://doi.org/10.1073/pnas.2314699121.

Wolinsky, E., and J. Way. 1990. "The Behavioral Genetics of *Caenorhabditis elegans*." *Behavior Genetics* 20 (2): 169–89. https://doi.org/10.1007/BF01067789.

Woods, John O, Ulf Martin Singh-Blom, Jon M Laurent, Kriston L McGary, and Edward M Marcotte. 2013. "Prediction of Gene–Phenotype Associations in Humans, Mice, and Plants Using Phenologs." *BMC Bioinformatics* 14 (June):203. https://doi.org/10.1186/1471-2105-14-203.

Wu, Yi-Chun, Xiaochen Wang, and Ding Xue. 2012. "Chapter 10 – Methods for Studying Programmed Cell Death in *C. elegans*." In *Methods in Cell Biology*, edited by Joel H. Rothman and Andrew Singson, 107:295–320. *Caenorhabditis elegans*: Cell Biology and Physiology. Academic Press. https://doi.org/10.1016/B978-0-12-394620-1.00010-2.

Xia, Yisui, Remi Sonneville, Michael Jenkyn-Bedford, Liqin Ji, Constance Alabert, Ye Hong, Joseph T. P. Yeeles, and Karim P. M. Labib. 2023. "DNSN-1 Recruits GINS for CMG Helicase Assembly during DNA Replication Initiation in *Caenorhabditis elegans*." *Science (New York, N.Y.)* 381 (6664): eadi4932. https://doi.org/10.1126/science.adi4932.

Xu, Qin, Zherui Liang, and Yadong Huang. 2024. "APOE4 Homozygosity Is a New Genetic Form of Alzheimer's Disease." *Nature Medicine*, May. https://doi.org/10.1038/s41591-024-02923-w.

Yemini, Eviatar, Albert Lin, Amin Nejatbakhsh, Erdem Varol, Ruoxi Sun, Gonzalo E. Mena, Aravinthan D.T. Samuel, Liam Paninski, Vivek Venkatachalam, and Oliver Hobert. 2021. "NeuroPAL: A Multicolor Atlas for Whole-Brain Neuronal Identification in *C. elegans*." *Cell* 184 (1): 272–288.e11. https://doi.org/10.1016/j.cell.2020.12.012.

Yi, Bitna, James J. Sahn, Pooneh Memar Ardestani, Andrew K. Evans, Luisa Scott, Jessica Z. Chan, Sangeetha Iyer, et al. 2017. "Small Molecule Modulator of Sigma 2 Receptor Is Neuroprotective and Reduces Cognitive Deficits and Neuro-Inflammation in Experimental Models of Alzheimer's Disease." *Journal of Neurochemistry* 140 (4): 561–75. https://doi.org/10.1111/jnc.13917.

Yochem, John, and Robert K. Herman. 2003. "Investigating *C. elegans* Development through Mosaic Analysis." *Development* 130 (20): 4761–68. https://doi.org/10.1242/dev.00701.

Yoshina, Sawako, Yuji Suehiro, Eriko Kage-Nakadai, and Shohei Mitani. 2015. "Locus-Specific Integration of Extrachromosomal Transgenes in *C. elegans* with the CRISPR/Cas9 System." *Biochemistry and Biophysics Reports* 5 (December):70–76. https://doi.org/10.1016/j.bbrep.2015.11.017.

Yu, Hui, Boanerges Aleman-Meza, Shahla Gharib, Marta K. Labocha, Christopher J. Cronin, Paul W. Sternberg, and Weiwei Zhong. 2013. "Systematic Profiling of *Caenorhabditis elegans* Locomotive Behaviors Reveals Additional Components in G-Protein Gαq Signaling." *Proceedings of the National Academy of Sciences of the United States of America* 110 (29): 11940–45. https://doi.org/10.1073/pnas.1310468110.

Zhang, Gaotian, Jake D Mostad, and Erik C Andersen. 2021. "Natural Variation in Fecundity Is Correlated with Species-Wide Levels of Divergence in *Caenorhabditis elegans*." *G3: Genes|Genomes|Genetics* 11 (8): jkab168. https://doi.org/10.1093/g3journal/jkab168.

Zhang, Gaotian, Nicole M. Roberto, Daehan Lee, Steffen R. Hahnel, and Erik C. Andersen. 2022. "The Impact of Species-Wide Gene Expression Variation on *Caenorhabditis elegans* Complex Traits." *Nature Communications* 13 (1): 3462. https://doi.org/10.1038/s41467-022-31208-4.

Zhang, Liangyu, Jordan D. Ward, Ze Cheng, and Abby F. Dernburg. 2015. "The Auxin-Inducible Degradation (AID) System Enables Versatile Conditional Protein Depletion in *C. elegans*." *Development (Cambridge, England)* 142 (24): 4374–84. https://doi.org/10.1242/dev.129635.

Zhao, Yu, Camila Coelho, Amanda L. Hughes, Luciana Lazar-Stefanita, Sandy Yang, Aaron N. Brooks, Roy S. K. Walker, et al. 2023. "Debugging and Consolidating Multiple Synthetic Chromosomes Reveals Combinatorial Genetic Interactions." *Cell* 186 (24): 5220–5236.e16. https://doi.org/10.1016/j.cell.2023.09.025.

Zheng, Hui, and Edward H. Koo. 2006. “The Amyloid Precursor Protein: Beyond Amyloid.” *Molecular Neurodegeneration* 1 (1): 5. https://doi.org/10.1186/1750-1326-1-5.

Zhou, Xin, and Jean-Louis Bessereau. 2019. “Molecular Architecture of Genetically-Tractable GABA Synapses in *C. elegans*.” *Frontiers in Molecular Neuroscience* 12:304. https://doi.org/10.3389/fnmol.2019.00304.

Zhu, Ping Jun, Sanjeev Khatiwada, Ya Cui, Lucas C. Reineke, Sean W. Dooling, Jean J. Kim, Wei Li, Peter Walter, and Mauro Costa-Mattioli. 2019. “Activation of the ISR Mediates the Behavioral and Neurophysiological Abnormalities in Down Syndrome.” *Science (New York, N.Y.)* 366 (6467): 843–49. https://doi.org/10.1126/science.aaw5185.

Using *Drosophila* to Study the Mechanisms That Underlie Down Syndrome

Bing Ye

Abstract Down syndrome (DS) is a chromosomal disorder caused by an extra copy of most or all of human chromosome 21 (Hsa21), which contains over 200 protein-coding genes and more than 400 non-protein-coding genes. This trisomy results in the overexpression of most of these genes, presenting a challenge in identifying those responsible for specific symptoms of DS. Moreover, some DS symptoms may result from interactions among multiple genes. *Drosophila melanogaster*, with its short lifecycle and capacity for easy genetic manipulation, is a valuable model for such research. Approximately half of the protein-coding genes on Hsa21 have *Drosophila* homologs, enabling the study of these genes in a simpler genetic model. This chapter outlines key *Drosophila* techniques used to explore DS mechanisms, reviews significant research contributions, and discusses the limitations of using *Drosophila* models to understand DS pathogenesis.

Keywords Down syndrome · *Drosophila* · Genetic models · Overexpression · Neurodevelopment · Heart development · *Dscam* · *Dyrk1a* · *DSCR1* · Therapeutic treatment

1 Introduction

Down syndrome (DS) is a chromosomal disorder caused by the presence of an extra copy of most or all human chromosome 21 (Hsa21). Hsa21 is the smallest autosome in humans, but it contains a broad genetic repertoire with over 200 protein-coding genes and more than 400 non-protein-coding genes (Frankish et al. 2019; Hattori et al. 2000). A major challenge in DS research is pinpointing the genes responsible

B. Ye (✉)
Life Sciences Institute, University of Michigan–Ann Arbor, Ann Arbor, MI, USA

Department of Cell and Developmental Biology, University of Michigan, Ann Arbor, MI, USA
e-mail: bingye@umich.edu

B. Ye, R. H. Reeves (eds.), *Genetic Models of Down Syndrome*,
https://doi.org/10.1007/978-3-031-78611-2_5

for distinct medical conditions. Trisomy of the whole Hsa21 or a fragment of it leads to an extra copy of Hsa21 genes, resulting in heightened expression of most of these genes (Pelleri et al. 2018). From a gene expression standpoint, DS is predominantly an overexpression disorder. It is crucial to delve into the pathology at the level of individual gene expression. Recognizing genes that directly contribute to the unique symptoms of DS is instrumental in devising targeted therapeutic strategies. Furthermore, there is a possibility that certain symptoms arise from the combined effect of multiple genes. For instance, the combined overexpression of both the *Down syndrome cell adhesion molecules (Dscam)* and *Collagen Type VI Alpha 2 Chain (Col6A2)* congenital heart disease in mice, while the overexpression of either gene on its own only has milder effects on heart health (Grossman et al. 2011). Testing each gene independently, let alone considering interactions among Hsa21 genes, for potential contributions to DS-associated symptoms is a labor-intensive endeavor. Many methodologies suitable for single-gene disorders fall short when applied to DS. Consequently, our insight into the molecular mechanisms tied to specific DS-associated diseases remains rudimentary.

The genetics of *Drosophila melanogaster*, together with its short lifecycle, allows for easy manipulation of genes, including knockdown, overexpression, and mutagenesis. Approximately 47% of protein-coding genes on Hsa21 have *Drosophila* homologs (Table 1), allowing the study of these genes in a simplified context. Culturing and experimenting with *Drosophila* are much cheaper and faster compared to mammalian models. However, the chromosomal arrangement of genes is different between *Drosophila* and mammals. There has not been a chromosomal model of DS in *Drosophila*. Instead, *Drosophila* studies on DS have focused on individual genes or small groups of genes on Hsa21 or their homologs.

In this chapter, I will explain the major techniques in *Drosophila* that are particularly useful for studying mechanisms underlying DS and present the major studies in *Drosophila* that have led to mechanistic insights into the pathogenesis of DS, followed by a discussion of the limitations of *Drosophila* systems for studying DS mechanisms.

2 Comparisons of Organ Systems Between *Drosophila* and Human

Many organ systems of *Drosophila melanogaster* are analogous to those of human. As such, they are often used to model human diseases in the studies of the underlying mechanisms. For those organ systems that do not have *Drosophila* counterparts, researchers have successfully used *Drosophila* to model the molecular pathways. For example, the *Drosophila* eye development was a fruitful model for key discoveries in Ras signaling (Wassarman et al. 1995) and Tor signaling (Miron and Sonenberg 2001).

Table 1 *Drosophila* homologs of Hsa21 genes

Human gene symbol	*Drosophila* gene symbol	*Drosophila* gene annotation number
HSPA13	*Hsc70-4*	CG4264
SAMSN1	*SKIP*	CG31163
USP25	*Puf*	CG5794
BTG3	*Tob*	CG9214
C21orf91	*Rhol*	CG9366
NCAM2	*Fas2*	CG3665
MRPL39	*mRpL*	CG17166
ATP5J	*ATPsynCF6*	CG4412
APP	*Appl*	CG7727
N6AMT1		CG9960
LTN1	*Ltn1*	CG32210
RWDD2B		CG30338
USP16	*Usp16-45*	CG4165
CCT8	*CCT8*	CG8258
MAP3K7CL	*Tak1*	CG18492
BACH1	*cnc*	CG43286
GRIK1		CG3822
TIAM1	*sif*	CG34418
SOD1	*Sod1*	CG11793
SCAF4		CG4266
HUNK	*AMPKalpha*	CG3051
MRAP	*rod*	CG1569
SYNJ1	*Synj*	CG6562
PAXBP1		CG1965
OLIG2	*Oli*	CG5545
OLIG1	*Oli*	CG5545
TMEM50B		CG15012
DNAJC28		CG43322
GART	*ade3*	CG31628
SON		CG8273
DONSON	*hd*	CG2669
CRYZL1		CG16935
ITSN1	*Dap160*	CG1099
ATP5O	*ATPsynO*	CG4307
MRPS6	*mRpS6*	CG15016
SLC5A3	*smvt*	CG2191
KCNE2	*Gel*	CG1106
RCAN1	*sra*	CG6072
SETD4		CG33230
CBR1		CG11200
CBR3		CG40486

(continued)

Table 1 (continued)

Human gene symbol	*Drosophila* gene symbol	*Drosophila* gene annotation number
DOPEY2		CG15099
MORC3		CG12693
SIM2	*sim*	CG7771
HLCS	*Hcs*	CG14670
PIGP	*PIG-P*	CG14550
TTC3	*Hop*	CG1594
VPS26C	*CG4074*	CG4074
DYRK1A	*mnb*	CG42273
KCNJ6	*Irk1*	CG44159
ERG	*Ets65A*	CG7018
ETS2	*pnt*	CG17077
PSMG1	*Cyp6u1*	CG3567
BRWD1	*BRWD3*	CG31132
WRB		CG32022
SH3BGR	*Sh3beta*	CG8582
B3GALT5		CG8668
PCP4	*igl*	CG18285
DSCAM	*Dscam1*	CG17800
BACE2	*Bace*	CG13095
MX1	*shi*	CG18102
TMPRSS2	*Corin*	CG2105
C2CD2		CG10737
ZNF295	*kel*	CG7210
ABCG1		CG5853
UBASH3A		CG13604
SLC37A1	*MFS16*	CG10069
PDE9A	*Pde9*	CG42276
WDR4	*wuho*	CG15897
NDUFV3	*NdufV3*	CG11752
CBS	*Cbs*	CG1753
U2AF1	*U2af38*	CG3582
CRYAA	*l(2)efl*	CG4533
SNF1LK	*Sik2*	CG4290
RRP1B	*Nnp-1*	CG12396
PDXK	*Pdxk*	CG34455
RRP1	*Nnp-1*	CG12396
AGPAT3	*CG4729*	CG4729
TMEM1	*SIDL*	CG6623
PWP2		CG12325
ICOSLG	*robo2*	CG5481
AIRE	*Mi-2*	CG8103
PFKL	*Pfk*	CG4001

(continued)

Table 1 (continued)

Human gene symbol	*Drosophila* gene symbol	*Drosophila* gene annotation number
C21orf2	*CG14995*	CG14995
TRPM2	*Trpm*	CG44240
UBE2G2	*Ubc7*	CG4443
SUMO3	*smt3*	CG4494
PTTG1IP	*px*	CG4444
SLX9	*CG16753*	CG16753
ADARB1	*Adar*	CG12598
POFUT2	*O-fut2*	CG14789
SLC19A1		CG6574
PCBP3	*mub*	CG7437
COL6A1	*Col4a1*	CG4145
COL6A2	*vkg*	CG16858
MCM3AP	*xmas-2*	CG32562
PCNT	*plp*	CG33957
DIP2A	*DIP2*	CG7020
S100B	*CanB2*	CG11217
PRMT2	*Art1*	CG6554

The homologs are based on the annotations in Flybase.org (Ozturk-Colak et al. 2024), as of July 2024. The human gene list is based on Korbel et al. (2009), which listed 213 protein-coding genes

When using *Drosophila* organ systems to model diseases in human equivalents, it is important to keep in mind that the analogous organs usually have structural and physiological differences. The simplicity of *Drosophila* organ systems, while beneficial for studies on molecular and cellular mechanisms, may not capture the complexity of human tissues. Thus, while *Drosophila* offers a convenient and efficient platform for genetic research, results require subsequent validation in more complex models to confirm their relevance to human diseases.

I will focus on the nervous system and the heart, two organ systems that have contributed the most to DS research in *Drosophila* so far. For introductions to other organ systems, such as fat body and kidney-like structure, the readers can read review articles previously published (Ugur et al. 2016).

2.1 The Nervous System

Overall organization Like human, the nervous system in *Drosophila* consists of the central nervous system (CNS) and the peripheral nervous system (PNS). Moreover, the *Drosophila* CNS is divided into the brain and the fly's equivalent of spinal cord, called ventral nerve cord (VNC). The anatomy of the fly's brain and VNC is very different from that of humans, so there are no direct counterparts of the human CNS regions in *Drosophila*. However, the genetic, cellular, neurophysiological, and neurochemical properties are mostly conserved.

Neuron morphology Each neuron typically forms an input compartment, the dendrites, and an output compartment, the axon. Dendrites and axons are different in electrical excitability, morphology, microtubule orientation, and distribution of specific molecules and organelles (Caceres et al. 2012; Craig and Banker 1994). Establishing and maintaining such a cellular polarity are key to a neuron's function. Moreover, the axon of a neuron is guided to its target region, often traveling a long distance from the cell body to connect to neurons downstream. *Drosophila* neurons have contributed significantly to our understanding of neuron morphology, including dendrite and axon development. Readers can learn about these in previously published reviews (Caceres et al. 2012; Grueber et al. 2005; Jan and Jan 2003; Tessier-Lavigne and Goodman 1996).

Based on the number of dendrites connected to the cell body, neurons can be grouped into three major morphological types: unipolar, bipolar, and multipolar (Cajal 1995). Although all three types are present in different species, the majority of invertebrate neurons are unipolar, while that of vertebrate neurons are multipolar (Grueber et al. 2005; Laurent 1999; Strausfeld 1976). Unipolar neurons extend a single process from the cell body and then bifurcate into dendrites and an axon. The unipolar morphology facilitates synapse formation away from the cell body, possibly serving as an alternative to neuronal migration (Wang et al. 2015), which is common in the vertebrate CNS (Hatten 1999) but rare in the insect CNS (Harris 2001).

Both multipolar and bipolar neurons are present in the *Drosophila* PNS (Orgogozo and Grueber 2005). The multipolar dendritic arborization (da) neurons in the *Drosophila* larva is a widely used model for studying dendrite development (Grueber et al. 2005).

Synapses Synapses are specialized junctions through which neurons communicate with each other. They play a crucial role in transmitting and computing information across the neural network. A typical synapse consists of a presynaptic area equipped with neurotransmitter release mechanisms, which aligns precisely with a postsynaptic area for receiving and processing these chemical signals. Most of the *Drosophila* research on synapses are done on the larval neuromuscular junction (NMJs) (Prokop and Meinertzhagen 2006).

The *Drosophila* NMJs uses glutamate as the neurotransmitter (Jan and Jan 1976) and are thus different from the vertebrate NMJs, which use acetylcholine as the transmitter (Eccles 1976). *Drosophila* NMJs exhibit stereotypic morphology, which facilities the studies of the morphology of synaptic terminals and boutons (Johansen et al. 1989). The larval NMJs also exhibit synaptic plasticity (Davis and Muller 2015; Keshishian et al. 1994; Van Vactor and Sigrist 2017). The *Drosophila* NMJ system has contributed much to our understanding of the molecular and cellular mechanisms that underlie synaptic development and plasticity (Collins and DiAntonio 2007; Menon et al. 2013).

2.2 The Heart

The heart of *Drosophila*, termed the dorsal vessel, is simpler than the human heart, comprising a single muscular tube closed at the posterior end and extending from the abdomen to the thorax (Lehmacher et al. 2012). Despite its simplicity, the fly heart shares some similarities with the human heart (Bier and Bodmer 2004; Choma et al. 2011), such as division into four chambers separated by valve-like structures that regulate the flow of hemolymph, the insect equivalent of blood. The dorsal vessel's aorta is composed of cardiomyocytes that transport hemolymph toward the head, where it disperses into the body cavity.

The *Drosophila* heart has been useful for exploring genetic factors involved in heart development. For example, *Drosophila* research discovered the "*tinman*" gene as a key regulator of heart development (Bodmer 1993). It was later found that the mammalian homolog of tinman, called Nkx2-5, is also essential for heart development (Lyons et al. 1995; Tanaka et al. 1999). Mutations in Nkx2-5 are linked to congenital heart disease and sudden cardiac arrest in humans. The *Drosophila* heart was used to screen candidate genes involved in the congenital heart defects (CHDs) caused by trisomy 21 (Grossman et al. 2011). This study led to the discovery that multiple Hsa21 genes collectively contribute to the severity of heart disease, which was further confirmed in mice. See more about this in section "*Overexpression of Dscam and COL6A2 Cooperatively Causes Congenital Heart Defects: From Drosophila to Mouse Models*."

3 Hsa21 Genes Conserved in *Drosophila*

Among the approximately 200 protein-coding genes on Hsa21 (Hattori et al. 2000; Korbel et al. 2009), the 100 in Table 1 have homologs in *Drosophila*.

4 Technical Approaches for Studying DS Genes in *Drosophila*

Drosophila melanogaster stands out as an organism with one of the largest repertoires of genetic tools, partly due to unique features of this species and the long history of *Drosophila* genetics. It is impossible to cover the breadth and depth of the genetic approaches in *Drosophila* in this chapter, so I focus on the most commonly used ones for studying DS genes. Readers interested in learning more about the genetic techniques and tools in *Drosophila* are encouraged to read books and review articles dedicated to those topics (Dahmann 2016; Senturk and Bellen 2018; Sullivan et al. 2000; Zhang et al. 2010).

4.1 The GAL4-UAS System

The GAL4-UAS system is a powerful technique in *Drosophila* research, allowing for the precise expression of genes in select cells, at specific life stages, or both (Brand and Perrimon 1993; Jenett et al. 2012). Originally identified in yeast, the primary components of this system are the GAL4 transcription factor and an enhancer called upstream activation sequence (UAS) (Fig. 1). GAL4 is a transcriptional activator that, when it binds to the UAS enhancer, initiates the transcription of the DNA sequence downstream of UAS. When using the GAL4-UAS system, two transgenes, typically in two separate transgenic fly lines, are needed. In the first line, the GAL4 cDNA is placed downstream of a tissue-specific promoter/enhancer (Fig. 1a). This ensures that GAL4 is only produced in certain cells or at specific times in the fly's life. This line is often referred to as the "driver" line. In the second line, the gene of interest (to be studied or manipulated) is placed downstream of the UAS enhancer. This line is dubbed the "responder" or "UAS" line. When a GAL4 driver fly is mated with a responder fly (carrying UAS-gene of interest), the offspring inherit both the GAL4 and the UAS-gene of interest. Consequently, in cells

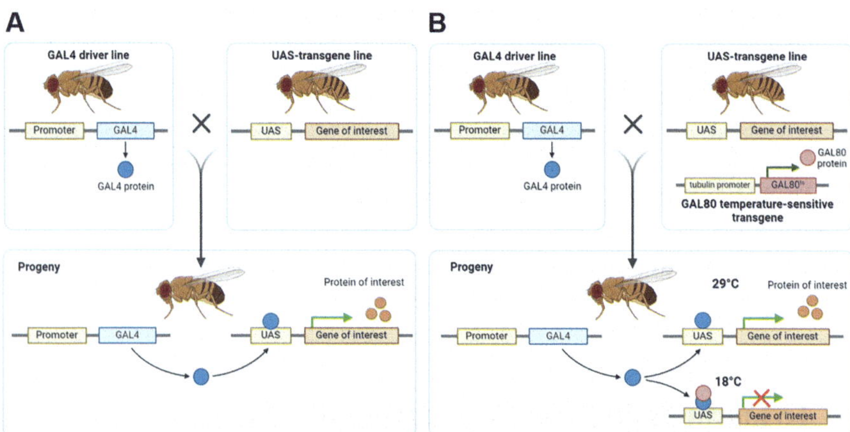

Fig. 1 The GAL4-UAS system allows spatiotemporal control of transgene expression in *Drosophila*. (**a**) A fly expressing the yeast transcription factor GAL4 under a cell-type-specific promoter is mated with another fly carrying one or more transgenes for expressing gene(s) of interest with the UAS enhancer. The progeny inherits both the GAL4 and the UAS-gene of interest. In these flies, the GAL4 proteins are expressed in the desired cell types and then act on the UAS enhancer to express the proteins of interest. In addition to overexpressing transgenes, the GAL4-UAS system can be used to express UAS-transgenes for RNAi. (**b**) The yeast repressor protein GAL80 binds to and inhibits the GAL4 protein. A transgene for expressing GAL80 under a ubiquitous promoter (e.g., tubulin promoter) is integrated into the genome of either the GAL4 driver line or the UAS-transgene line (the latter in the example shown). In the example in (**b**), a temperature-sensitive mutant allele of GAL80, called GAL80^{ts}, is used. GAL80^{ts} is functional (i.e., it inhibits GAL4) at a permissive temperature (usually 18 °C) but is inactive at 29 °C. By controlling the environmental temperature, the UAS-transgene can be expressed at specific time in the fly's life. (This figure is created with Biorender)

where GAL4 is expressed by the tissue-specific promoter, GAL4 binds to the UAS, which in turn prompts the expression of the gene of interest.

The GAL4-UAS system offers several advantages. One of its standout features is facilitating cell-type-specific expression. By using tissue-specific GAL4 driver lines, researchers can dictate the expression location of any UAS-transgenes within the fly. This specificity paves the way for in-depth studies on how specific genes influence particular tissues or organs. There are catalogs of GAL4 and UAS flies to order from public stock centers or request from investigators.

For conditions like DS, the GAL4-UAS system is particularly useful. Since DS is primarily an overexpression condition, researchers can generate UAS-transgenes to express genes on Hsa21 or their *Drosophila* homologs. Concurrently, GAL4 driver lines can be selected to express the DS-related genes in cells that are relevant to conditions that co-occur with DS. When studying the effect of overexpressing more than one Hsa21 genes, two or more UAS transgenes can be used. While I am not aware of a record of the maximal number of UAS-transgenes that have been used in a fly, most studies use 1–3 UAS-transgenes.

It is difficult to provide a general estimation of the expression level of the UAS-transgenes because many factors affect the expression level (Pfeiffer et al. 2010, 2012). These factors include the 5′ and 3′ untranslated regions (UTRs) flanking the coding cDNA, the number of UAS enhancers in the transgene, the number of UAS-transgenes in the flies to be studied, and the temperature in which the flies are raised. It is thus important to quantify the expression level change that is caused by the GAL-UAS overexpression in key experiments of a study. In our experience with UAS-*Dscam* transgenes that carry different 5′ or 3′ UTRs, some led to mild increases in Dscam protein levels in neurons (<1-fold), which caused observable overgrowth of axon terminals (unpublished results) (Kim et al. 2013). It is a common practice to raise the flies at 18 °C to lower GAL4-UAS expression (Duffy 2002), though the developmental time of *Drosophila* is roughly doubled at 18 °C, compared to the optimal temperature of 25 °C (Kuntz and Eisen 2014). Researchers can also make UAS-transgenes that express the gene of interest at various levels by using DNA constructs containing different numbers of the UAS enhancer (Pfeiffer et al. 2010, 2012).

In addition to overexpressing transgenes, the GAL4-UAS system can be used to express either long double-strand RNAs (dsRNAs) or short hairpin RNAs (shRNA) (Perkins et al. 2015), which can knock down specific genes via RNA interference (RNAi). This use of GAL4-UAS to knock down gene expression provides an efficient approach for studying the loss of gene function in a cell-type- and time-specific manner.

4.2 Temporal Control of Transgene Expression

When incorporated with elements like temperature-sensitive control or drug-responsive triggers, the GAL4-UAS system allows for the precise temporal control of gene expression (McGuire et al. 2004). Researchers can designate the timing of transgene expression at a specific stage of the fly's life.

The yeast protein GAL80 binds to and inhibits the GAL4 protein, thus regulating the expression of genes within the GAL4-UAS system. Transgenes expressing GAL80 have been used to control the timing of UAS-transgene expression in *Drosophila*. To temporally control gene expression in the GAL4-UAS system, a temperature-sensitive mutant allele of GAL80, called $GAL80^{ts}$, can be used (McGuire et al. 2003) (Fig. 1b). This allele of GAL80 is functional (i.e., it inhibits GAL4) at a permissive temperature (usually 18 °C) but becomes inactive at a restrictive temperature (usually 29 °C). By controlling the environment, researchers can precisely determine when the UAS-transgene is expressed. For instance, keeping the flies at 18 °C will ensure that the gene remains off, and shifting them to 29 °C will turn the gene on.

The GeneSwitch system is a drug-triggered tool that enables controlled gene expression in specific tissues or at a particular time (Mao et al. 2004; Osterwalder et al. 2001). In this system GAL4 is fused with the ligand-binding domain of the human progesterone receptor. This fusion protein, known as GeneSwitch GAL4, remains inactive until the presence of a specific steroid hormone, usually the synthetic steroid RU486 (mifepristone). By feeding flies or applying to cells RU486 at desired times, researchers can activate the GeneSwitch GAL4 driver in a temporally controlled manner. When activated, GeneSwitch GAL4 binds to the UAS placed upstream of the cDNA of a gene of interest, driving its expression.

The $GAL80^{ts}$ and GeneSwitch systems are particularly useful in developmental studies where gene expression needs to be tightly controlled at specific stages, or in experiments where the researcher needs to activate or deactivate a gene to study its effects at particular times (e.g., degeneration in aging flies).

4.3 Loss-of-Function Studies in Specific Types of Cells

To study gene functions, we often need to eliminate or reduce the function of a gene in specific cell types or even a subpopulation of cells within a type. For example, to determine whether a gene functions in the cells that exhibit a phenotype (i.e., cell-autonomous function), we need to eliminate or reduce the function of a gene only in these cells but not in the cells surrounding the phenotype-exhibiting cells. Although cell-type specific loss-of-function studies can be achieved by RNAi, RNAi usually does not lead to a complete loss of the gene function. It is often advantageous to study loss of function with genetic mutants that are null for genomic DNAs, mRNAs, or proteins. In mice, this is typically achieved through conditional knockout via the Cre/LoxP system. In *Drosophila*, the Genetic Mosaic with a Repressible Cell Marker (MARCM) is a commonly used approach for generating single-cells or a group of cells that are homozygous mutant for a gene (Lee and Luo 1999, 2001).

MARCM can generate genetic mosaics, which are organisms composed of cells with different genotypes (Fig. 2a). It allows researchers to investigate the function of specific genes in a subpopulation of cells within an otherwise genetically normal organism. MARCM combines the GAL4-UAS system with a cell marker and the GAL80 repressor to achieve targeted gene expression or loss of function in specific

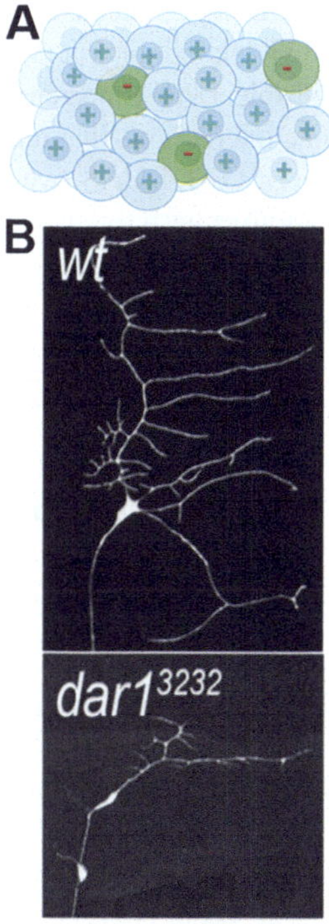

Fig. 2 The MARCM technique. (**a**) A schematic drawing that shows genetic mosaicism in which homozygous mutant cells ("−") are intermixed with wild-type or heterozygous cells ("+"). In MARCM, the homozygous mutant cells are labeled by a cell marker, such as GFP. (**b**) Examples of genetic mosaics generated by MARCM. The MARCM mosaic clone is labeled by GFP. Shown are class I da neurons in the *Drosophila* larva. Top: a wild-type (wt) neuron; bottom: a neuron that is mutant for the gene *dar1* (*dar1*^{3232}). The dendritic arbor is reduced in the mutant neuron. Scale bar, 20 μm. (Modified from Wang et al. (2015))

cells. In MARCM, when mitotic recombination mediated by the FLP/FRT recombinase system occurs in a subset of cells (or even a single cell), some of the progeny cells lose the GAL80 transgene that is placed on the same chromosomal arm as the mutation. As a result, the GAL4 activates the expression of a UAS-linked marker gene, typically a fluorescent protein. This selective expression allows the visualization of the cells where the recombination event occurred and thus where GAL80 is inactivated. The positively labeled cells can be designed to be homozygous for a mutation, while the other sibling cells are heterozygotes or wild-type (Fig. 2). Thus, by combining this system with mutations, researchers can study the impact of specific genetic changes in only the marked cells.

MARCM can also be used to overexpress transgenes of interest in single cells or a group of cells, if a UAS-transgene—instead of a mutation—is incorporated into the system.

4.4 Genetic Screens

Drosophila systems are often conducive to efficient genetic screening. While traditional genetic mutants are commonly used in such screens, the GAL4-UAS system is particularly valuable for identifying genes that may contribute to DS pathogenesis, because overexpression of Hsa21 genes is a fundamental contributor to DS-related disorders. The flexibility of the GAL4-UAS system to accommodate multiple UAS-transgenes in a single fly enhances its efficacy in assessing gene–gene interactions that may play a role in DS pathogenesis. Furthermore, the integration of GAL4-UAS overexpression strategies for Hsa21 genes or their *Drosophila* homologs with the use of genetic mutants or RNAi empowers researchers to delineate genetic pathways involving specific Hsa21 genes.

5 The Role of *DSCAM* in Down Syndrome

5.1 **Drosophila** *Studies Led to the Discovery of the Contribution of Dscam to Excessive GABAergic Synapses in DS Mouse Models*

DSCAM was initially discovered as a Hsa21 gene encoding a cell adhesion molecule in the immunoglobulin superfamily (Yamakawa et al. 1998). The studies of the *Drosophila* homolog of *DSCAM*, termed *Dscam* (Schmucker and Chen 2009), provided the guidance for investigations of *Dscam* in mouse models of DS and led to the discovery of its contribution to excessive GABAergic synapses in mouse models (Hergenreder et al. 2024).

The *Drosophila Dscam* contains various splicing targets to create 38,016 variants (Schmucker et al. 2000), which allow the dendritic or axonal branches of a neuron to recognize each other (Hattori et al. 2008; Matthews et al. 2007; Zhu et al. 2011; Zipursky and Sanes 2010). Loss of *Dscam* function in *Drosophila* leads to the crossing of dendritic branches that belong to the same neuron (Hughes et al. 2007; Matthews et al. 2007; Soba et al. 2007). Mammalian *DSCAM* genes lack the splicing diversity; instead, increased DSCAM levels have been observed in the brains of individuals with DS and those of mouse DS models (Alves-Sampaio et al. 2010; Liu et al. 2023; Saito et al. 2000), raising the possibility that inappropriate DSCAM levels may lead to medical conditions.

The importance of Dscam levels in neuron development was established by studies conducted in the class IV dendritic arborization (C4da) neurons of the *Drosophila* larva (Kim et al. 2013), which are a type of somatosensory neurons. Kim et al. discovered that the expression level of Dscam determines the size of the presynaptic arbor of a C4da neuron and demonstrated that this function of *Dscam* is independent of its splicing diversity (Kim et al. 2013). Overexpression of *Dscam* in C4da neurons increased their presynaptic terminals without increasing dendritic growth (Fig. 3a). Conversely, loss of *Dscam* reduced the size of presynaptic terminals in C4da neurons. These studies benefited from genetic mosaic techniques that allow labeling single cells in *Drosophila* (Fig. 2) and the fact that the presynaptic terminal morphology of fly neurons is stereotypical and easy to quantify. In mammals, the sizes of presynaptic terminals are typically much larger than those in *Drosophila*, and axons are usually projected to a longer distance, making the evaluation of

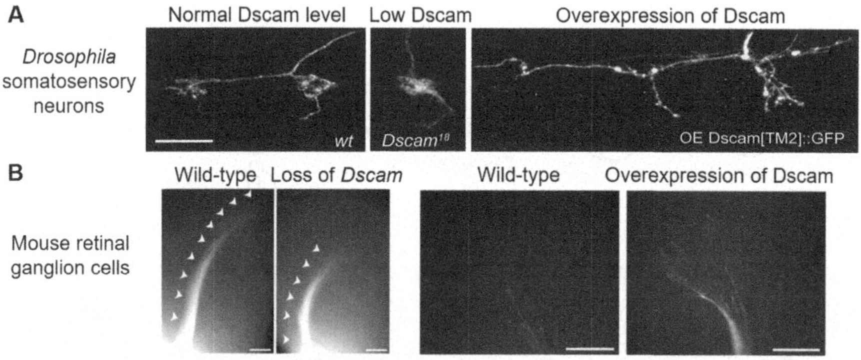

Fig. 3 The Dscam level determines the size of axon terminals. (**a**) Dscam expression levels determine the sizes of presynaptic terminals in *Drosophila* somatosensory neurons. Presynaptic terminal length of *Drosophila* C4da neurons are determined by Dscam levels. Low Dscam levels (*Dscam*^{18}) decrease presynaptic terminal length in comparison to wild-type (wt) neurons, while overexpression of Dscam increases presynaptic terminal length. The MARCM technique was used to generate the GFP+ C4da neurons that were homozygous for *Dscam*^{18} or expressing UAS-Dscam. Scale bar: 10 μm. (**b**) In mice, loss of *DSCAM* impairs, while overexpression promotes, the axonal growth in retinal ganglion cells. Shown are optic tracts of E13.5 (loss of *Dscam*), E14.5 (Dscam overexpression), and their respective wild-type littermates. Scale bars: 250 μm. (Panel **a** is adapted from Kim et al. (2013) and Panel **b** is from Bruce et al. (2017), both of which are with permission from the publisher)

presynaptic terminals harder. Later studies found that this function of Dscam is conserved in mammals. In mice, axon growth of retinal ganglion cells (RGCs) is impaired in *Dscam* loss-of-function mutants (Bruce et al. 2017) (Fig. 3b). Conversely, Dscam overexpression leads to overgrowth of RGC axons. These studies in *Drosophila* and mice raised the possibility that the excessive DSCAM proteins in DS are involved in the pathogenesis of neuronal defects in DS.

While evaluating presynaptic terminals is generally difficult in mammals, advanced genetic labeling techniques have made it possible to study the axon terminals of a single neuron in some types of neurons, particularly some GABAergic neurons. Importantly, many studies have shown an excessive number of GABAergic synapses in DS brains (Costa and Grybko 2005; Deidda et al. 2014; Fernandez et al. 2007; Kleschevnikov et al. 2012a, b). It is thus not surprising that the first study testing the contribution of *Dscam* to presynaptic development was conducted on GABAergic neurons.

This study was carried out in the Ts65Dn mouse (see chapter "Mouse Models of Down Syndrome", Folz et al.). Previous findings have indicated that these mice exhibit enhanced GABAergic inhibition that compromises cognitive abilities, an effect that is reversible by GABAergic antagonists (Braudeau et al. 2011a, b; Colas et al. 2013; Fernandez et al. 2007; Martinez-Cue et al. 2013). The number and size of GABAergic synapses are increased in the cerebral cortex of Ts65Dn mice (Belichenko et al. 2009; Kurt et al. 2000; Nosheny et al. 2015), but the underlying molecular mechanism was unknown.

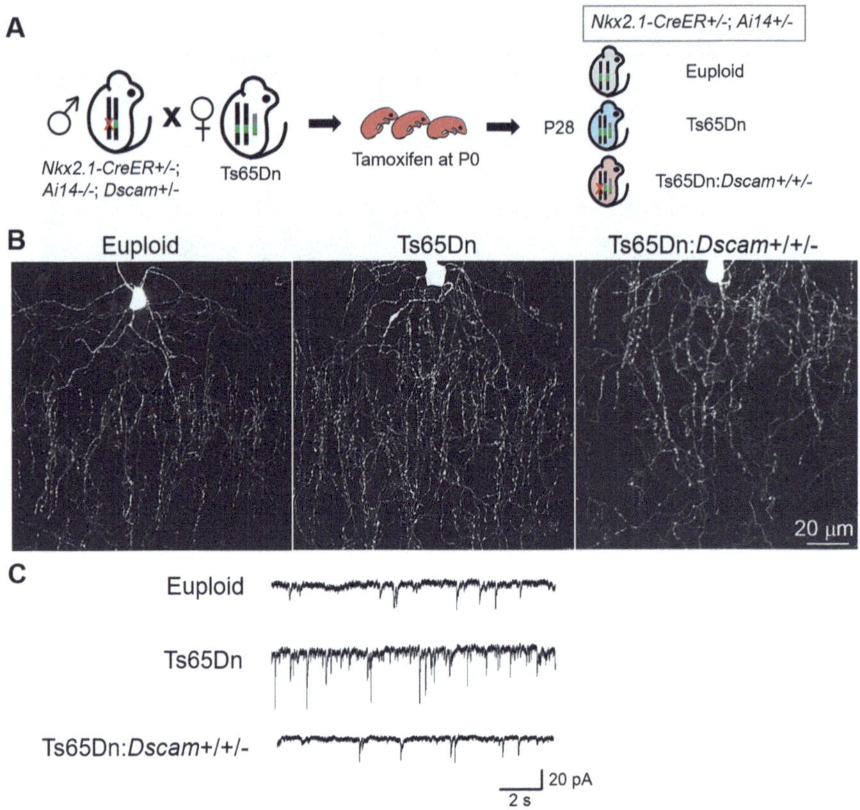

Fig. 4 The Ts65Dn mouse model of DS displays an increase in the lengths of presynaptic terminals of the GABAergic chandelier cells in comparison to euploid mice (**b**), while normalizing *Dscam* gene dosage to two copies in the trisomy mice (Ts65Dn:*Dscam*+/+/−) (**a**) eliminates the overgrowth of chandelier presynaptic terminals (**b**). (**c**) The excessive GABAergic synapses in trisomy mice lead to an increase in the frequency of mIPSCs, which is eliminated in trisomy mice that carry only two copies of the *Dscam* gene (Ts65Dn:*Dscam*+/+/−). (Adapted from Liu et al. (2023))

By breeding Ts65Dn mice with heterozygous loss-of-function *Dscam* mutants, offspring were generated that have two functional *Dscam* genes in the trisomic background alongside with the euploid and trisomic littermates (Liu et al. 2023) (Fig. 4a). This allows for an examination of the role of *Dscam* in the phenotypes observed in the trisomic mice. The Dscam protein level in the *Dscam* dosage-corrected mice was comparable to that seen in wild-type mice. The presynaptic terminals of the chandelier GABAergic neurons, which are a type of parvalbumin-expressing neurons, are overgrown (Fig. 4b). Moreover, the number and size of GABAergic synapses formed by basket cells (another major type of parvalbumin-expressing neurons) are increased in the cerebral cortices of Ts65Dn mice (Liu et al. 2023). The *Dscam* dosage-corrected trisomy mice had normal presynaptic terminal length of chandelier GABAergic neuron as well as normal number and size of GABAergic synapses. Moreover, patch-clamp recording of the pyramidal neurons in the cerebral cortex showed that these mice had normal miniature inhibitory postsynaptic current (mIPSC) frequency (Fig. 4c), while the Ts65Dn mice showed an increase in mIPSC frequency, underscoring a presynaptic role of Dscam in GABAergic synaptic transmission. Using a Cre/Lox-based conditional correction of *Dscam* dosage in GABAergic neurons, it was found that these effects of *Dscam* on GABAergic synapses require the functions of *Dscam* within GABAergic neurons.

In summary, basic studies of the *Drosophila* homolog of *Dscam* laid the foundation for discovering the contribution of *Dscam* to GABAergic phenotypes in mouse models of DS.

5.2 *A Third Copy of* Dscam *Causes Axonal Targeting Errors and Neuromuscular Junction Defects in* Drosophila

In most *Drosophila* studies on DS, overexpression of Hsa21 genes or their fly homologs has been achieved using the GAL4-UAS system. However, two studies illustrated the use of a *Drosophila* strain harboring a third copy of the Hsa21 homolog in its full-length form.

A transgenesis platform was used to facilitate the insertion of bacterial artificial chromosomes (BAC) into the *Drosophila* genome (Venken et al. 2006). One of the strains, *Dscam^{BAC}*, harbors the complete genomic locus of *Dscam*, ensuring that the additional *Dscam* transgene is expressed under the control of its native promoter. This results in elevated Dscam protein levels (Cvetkovska et al. 2013). When these flies are crossed with a wild-type strain, the resulting progenies possess three copies of the *Dscam* gene, referred to as *DscamX3*. Cvetkovska et al. (2013) demonstrated that *DscamX3* flies exhibit errors in axonal targeting (Cvetkovska et al. 2013).

To investigate synaptic function of *DscamX3*, electrophysiological recordings were performed at the neuromuscular junction (NMJ) of *Drosophila* larvae (Lowe et al. 2018), a model system for studying glutamatergic excitatory synapses. In *DscamX3* larvae, a reduction was observed in the amplitude of the miniature excitatory junction potentials (mEJPs), which reflect the postsynaptic depolarization triggered by single synaptic vesicles. Concurrently, there was an increase in the number of vesicles released by motor neurons. Despite these alterations at the synaptic level, the amplitude of the evoked EJPs remained unchanged. These electrophysiological findings support the notion that the presence of an additional copy of *Dscam* exerts significant effects on synaptic transmission and plasticity.

6 *Drosophila* Studies of the Roles of Dyrk1A in Down Syndrome

Dual specificity tyrosine-phosphorylation-regulated kinase 1A (Dyrk1A) is a protein kinase that has been extensively characterized for its pivotal roles in neurodevelopmental deficits associated with DS. The *Dyrk1A* gene was initially identified and studied in *Drosophila*, where it is referred to as *minibrain* (*Mnb*) and is known to regulate neurogenesis (Tejedor et al. 1995). *Drosophila* mutants lacking *Mnb* exhibit reduced brain size and impaired neural functions. Mammalian *Dyrk1A* genes were subsequently discovered through their homology to *Mnb* (Guimera et al. 1996; Kentrup et al. 1996; Song et al. 1996). Dyrk1A proteins are distributed in both the cell nucleus and the cytoplasm (Martí et al. 2003), and it phosphorylates numerous nuclear and cytoplasmic proteins (Park et al. 2009). In addition to its expression in the nervous system, Dyrk1A is found in various tissues and has been implicated in a multitude of medical conditions, including the brain disorders and cardiovascular defects associated with DS (Ahn et al. 2006; Altafaj et al. 2001; Arron et al. 2006; Atas-Ozcan et al. 2021; Lana-Elola et al. 2024; Smith et al. 1997).

6.1 The Effects of Dyrk1A/Mnb Overexpression on Synaptic Development and Functions

Transgenic mice overexpressing Dyrk1A exhibit altered synaptic plasticity as well as impairments in learning and memory (Ahn et al. 2006; Altafaj et al. 2001; Branchi et al. 2004). The *Drosophila* larval NMJ has provided insights into the synaptic functions of Mnb and the underlying molecular pathways. Dyrk1A and Mnb localize to the presynaptic terminals of NMJs in mice and *Drosophila*, respectively (Arque et al. 2013; Chen et al. 2014); Dyrk1A is transported to the cell bodies through axons (Arque et al. 2013). Overexpression of Mnb in motor neurons of *Drosophila* larvae specifically increases the number of one of the two bouton types at the NMJ (the 1b type), which is the primary bouton type involved in synaptic plasticity (Lowe et al. 2019). Moreover, it increases the frequency—though decreases the amplitude—of spontaneously mEJPs, which is consistent with the morphological changes in 1b boutons. Furthermore, Mnb overexpression in motor neurons reduces the recovery of EJP from frequency-dependent depression, suggesting that Mnb overexpression slows replenishment of the readily releasable pool of vesicles at the NMJ. These effects of Mnb are mediated in part by synaptojanin (Synj), which is encoded by the fly homolog of another Hsa21 gene. Mnb phosphorylates Synj, following neuronal stimulation; this phosphorylation facilitates rapid endocytosis of synaptic vesicles and contributes to synaptic growth (Chen et al. 2014).

Drosophila studies further highlighted the coordination between Mnb and calcineurin (CaN) in regulating Synj, demonstrating their critical roles in balancing different endocytic pathways at the NMJ under varying synaptic demands (Peng et al. 2021) (see below).

Overexpression of Mnb in *Drosophila* also accelerates age-dependent neurodegeneration, leads to motor decline, and reduces lifespan (Lowe et al. 2019), aligning with previous studies that linked Dyrk1A overexpression to neurodegeneration, including in Alzheimer's disease (AD) associated with DS (García-Cerro et al. 2017; Watson-Scales et al. 2018; Wegiel et al. 2008).

These molecular and physiological insights gleaned from *Drosophila* studies offer insights for in-depth mechanistic investigations of motor and synaptic alterations in DS models across various species, including mammals.

6.2 The Effect of Dyrk1A/Mnb Overexpression on Dendrite Development

Dendritic branching and length are first increased in DS infants (Becker et al. 1986; Prinz et al. 1997), followed by marked reductions in older DS children (Becker et al. 1986; Kaufmann and Moser 2000). Cultures of embryonic cortical neurons from *Dyrk1A* transgenic mice show reduced dendritic arborization after 2 days in vitro (2 DIV) (Lepagnol-Bestel et al. 2009). Overexpressing *Dyrk1A* in wild-type young cortical neurons (1–2 DIV) also leads to similar changes in culture. It should be noted that the dendrites are in their nascent form at 2 DIV. Thus, whether these observations in cultured mouse neurons are relevant to human is unclear.

In *Drosophila, Mnb* regulates microtubule dynamics in larval sensory neurons (Ori-McKenney et al. 2016). Mnb directly binds to tubulin, independent of its kinase activity, and phosphorylates β-tubulin at serine 172, leading to an inhibition of microtubule polymerization. This inhibition of microtubule growth is also observed for the mammalian Dyrk1A (Ori-McKenney et al. 2016). Overexpression of Mnb with a UAS-transgene or a Mnb duplication line shortened the terminal branches of dendrites of the class III da neurons (Fig. 5), a type of mechanosensory neurons in *Drosophila* larvae. These findings suggest that an excessive amount of Dyrk1A might reduce dendrite development by directly affecting microtubule growth.

6.3 The Effect of Dyrk1A/Mnb Overexpression on Neurogenesis

The *Drosophila Mnb* mutants exhibit smaller optic lobes and central brain (Tejedor et al. 1995). The morphogenic neuroepithelial layers that give rise to the optic lobes are disrupted in *Mnb* mutants, which probably results in reduced proliferation of neuroblasts. This led to the idea that *Mnb*, and possibly its homologs (*Dyrk1A*), regulates neurogenesis. Indeed, overexpression of Dyk1A phosphorylates p53, inducing $p21^{CIP1}$,

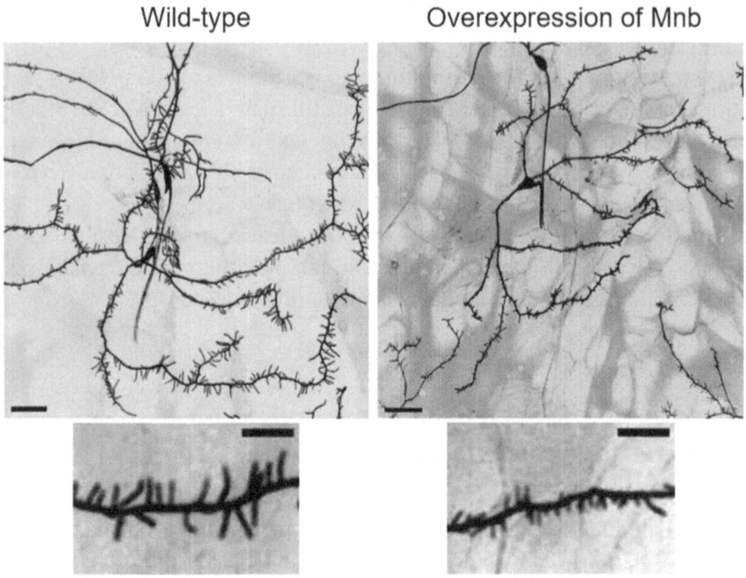

Fig. 5 Overexpression of Mnb reduces the length of the terminal branches of class III da neurons in *Drosophila* larvae. The bottom panels are enlarged pictures of selected dendritic branches, showing the difference in terminal branch length between the two genotypes. (Adapted from Ori-McKenney et al. (2016) with permission from the publisher)

and consequently impairs G1/G0–S phase transition in a cell line of immortalized rat hippocampal progenitors (Park et al. 2010). Overexpression of Dyrk1A also promotes the nuclear export of cyclin D1 in developing neocortices of mice, which may lead to cyclin D1 degradation and neuronal proliferation (Yabut et al. 2010).

7 *Drosophila* Studies on the Role of DSCR1 in Down Syndrome

The *Down syndrome critical region 1* (*DSCR1*) gene, also termed *RCAN1* (Regulator of Calcineurin 1), is overexpressed in brains of individuals with DS (Fuentes et al. 2000; Sun et al. 2011). *DSCR1* encodes a protein that inhibits calcineurin (CaN), a serine/threonine phosphatase regulated by calcium and calmodulin (Fuentes et al. 2000). The *Drosophila* homolog of *DSCR1* is known as *Nebula* (*Nla*), a gene that plays a crucial role in learning and memory. Research in *Drosophila* revealed that both a deficiency and an excess of Nla result in learning impairments (Chang et al. 2003), a phenomenon later corroborated in mouse models where alterations in DSCR1 levels were found to adversely affect memory (Dierssen et al. 2011; Hoeffer et al. 2007; Martin et al. 2012).

Further investigations using *Drosophila* have illuminated a significant role of Nla in maintaining mitochondrial structure and function (Chang and Min 2005). The Nla protein localizes to the mitochondria, with changes in its abundance affecting various mitochondrial parameters. These include enzyme activities, DNA content, and the number and size of the mitochondria themselves. These observations in *Drosophila* raise the intriguing hypothesis that DSCR1 overexpression might lead to mitochondrial structure and function, which are frequently observed in DS models.

8 Using *Drosophila* to Study the Effects of Overexpressing Multiple Hsa21 Genes

A strength of *Drosophila* systems for studying DS is that overexpressing two or more transgenes in vivo is straightforward. A cell-type-specific GAL4 can be used to drive the expression of multiple UAS-transgenes, each for a different Hsa21 gene. Moreover, by using GAL80^{ts} (Fig. 1b), one can control the timing of the UAS-transgene expression and thus restrict the overexpression to a specific stage in a fly's life.

It is also relatively easy to perform epistasis analysis in which overexpression of a Hsa21 gene is combined with a loss-of-function mutant or a UAS-transgene of another gene, allowing us to study gene interactions that lead to specific phenotypes.

These strengths have led to studies that provided important insights into the molecular mechanisms underlying DS phenotypes.

8.1 Overexpression of Dscam and COL6A2 Cooperatively Causes Congenital Heart Defects: From Drosophila *to Mouse Models*

DS is a leading cause of congenital heart defects (CHDs) (Benhaourech et al. 2016). Approximately half of individuals with DS experience CHDs, many requiring surgical interventions for correction (Benhaourech et al. 2016; Bergström et al. 2016; Freeman et al. 1998). Research involving individuals with DS who have segmental trisomies has identified a critical region at the distal end of Hsa21 that correlates with CHD incidence (Barlow et al. 2001). Within this region, *Dscam* is notably expressed during heart development, suggesting a potential association between elevated levels of Dscam and CHDs in DS. CHDs in DS have been observed to involve increased adhesion between cardiac fibroblasts (Wright et al. 1984), a process that could be influenced by the role of Dscam in cell adhesion.

The effects of overexpressing certain Hsa21 genes on heart function in *Drosophila* models were assessed with parameters including basal heart rate, response to stress-induced heart failure, and recovery post-failure (Grossman et al. 2011) (Fig. 6a).

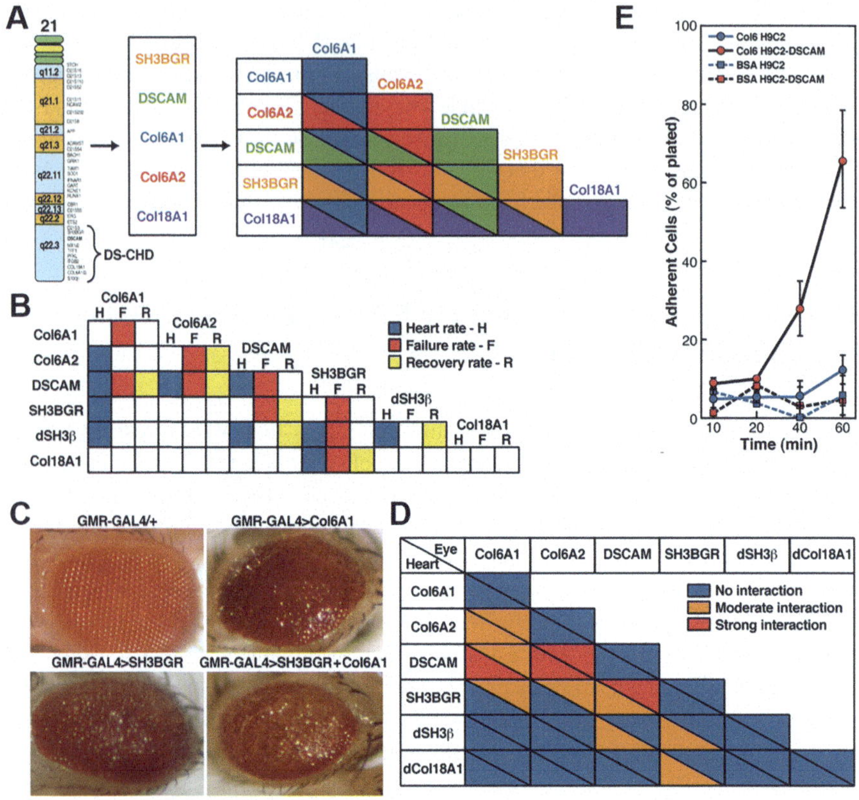

Fig. 6 (**a**) A schematic depiction of the steps of testing candidate CHD genes from CHD-associated Hsa21 genes identified through human genetic mapping. The candidate genes were screened by overexpressing them (with the GAL4-UAS system) in the Drosophila heart individually and in all possible pair-wise combinations. (**b**) A summary of the interactions between CHD candidate genes in the fly heart. H: Heart rate; F: failure rate; R: recovery rate. Colored squares indicate significant difference in heart phenotypes relative to control flies that do not have the UAS-CHD-gene. (**c**) Examples of a genetic interaction between the SH3BGR and COL6A1 in the fly eye. The GMR-GAL4 driver drive UAS-transgene expression in the fly eye. (**d**) A summary of the genetic interactions between CHD candidate genes in the fly heart (lower triangles in each interaction box) and eye (upper triangles in each interaction box). The combination of Dscam and COL6A2 led to the strongest interactions in both the heart and the eye. (**e**) Cell adhesion assay in the H9C2 cardiomyocyte cell line. A stable cell line of H9C2 cardiomyocyte cells expressing Dscam (H9C2-DSCAM) or parental H9C2 cells were grown on BSA or COL6 coated plates. The percentage of adherent cells on the plates were counted at the indicated time points. Dscam and COL6 synergistically enhances cell adhesion. (Adapted from Grossman et al. (2011))

While the overexpression of various genes, such as *Dscam*, *COL6A1*, *COL6A2*, *COL18A1*, and *SH3BGR*, individually affected cardiac dysfunction under stress, none exacerbated all three measured parameters (Fig. 6b). This led researchers to explore whether these genes might collectively contribute to heart disease severity. Significantly, co-expressing the combination of *Dscam* and *COL6A2* or *Dscam* and

SH3BGR led to strong genetic interactions and markedly worsened heart failure rates (Grossman et al. 2011) (Fig. 6c, d).

Further analysis in mouse models, with separate lines overexpressing *Dscam* or *COL6A2*, showed no detrimental effects on heart health or survival (Grossman et al. 2011). However, simultaneous overexpression of *Dscam* and *COL6A2* led to over 50% mortality and significant cardiac anomalies in survivors. This co-overexpression notably increased cellular adhesion beyond the effects of each gene alone.

The precise mechanism by which altered cell adhesion influences cardiac structure remains to be elucidated. Results from in vitro cell adhesion assays show that the combination of Dscam and COL6A2 synergistically enhances cell adhesion (Fig. 6e), which might cause severe defects in heart development or function.

Taken together, these findings highlight the synergistic effects of Dscam and COL6A2 overexpression in inducing CHDs, underscoring the critical need to examine the collective influence of Hsa21 genes on DS-associated conditions. This work also demonstrates the utility of leveraging multiple model systems to unravel the genetic intricacies contributing to DS phenotypes.

8.2 Interactions Between Dscam *and Amyloid Precursor Protein genes*

There is a strong association between DS and Alzheimer's disease (AD) (Lai and Williams 1989; Wisniewski et al. 1985). The AD-associated neurodegeneration is, in a significant part, caused by the extra copy of the gene encoding the amyloid precursor protein (APP), which is located on Hsa21. Like its mutations (Chartier-Harlin et al. 1991; Murrell et al. 1991), duplications of *APP* are risk factors for familial AD (Kasuga et al. 2009; Rovelet-Lecrux et al. 2006; Sleegers et al. 2006). An individual with partial trisomy 21 missing the extra copy of *APP* did not develop early onset of AD (Doran et al. 2017). Overexpression of the human APP in *Drosophila* causes neurodegeneration and learning and memory deficits in a conditioned courtship suppression assay (Chakraborty et al. 2011; Greeve et al. 2004); that in mice impairs the cognitive abilities in the Morris water maze and in an object recognition task (Simon et al. 2009). Increased expression of APP in the DS mouse model Ts65Dn causes the degeneration of basal forebrain cholinergic neurons (Salehi et al. 2006).

In transgenic mice that overexpress APP, the level of Dscam protein is elevated in the cerebral cortex (Jia et al. 2011). Consistent with this finding, overexpression of APP in *Drosophila* neurons increases Dscam expression (Pizzano et al. 2023). The *Drosophila* homolog of *APP*, termed *Amyloid precursor protein-like* (*Appl*), requires *Dscam* to promote axon terminal growth in larval sensory neurons. The Appl-induced increase in Dscam protein expression is through a post-transcriptional mechanism that is yet to be identified.

8.3 *Dyrk1A/Mnb, CaN, and Synj1/Synj Interact to Modulate Synaptic Endocytosis*

Both *Dyrk1A* and *Synj1* (*Synaptojanin1*) genes are located on Hsa21. A series of studies in *Drosophila* have elucidated that the Dyrk1A homolog Mnb kinase and the calcium/calmodulin-dependent phosphatase CaN orchestrate the activity of the Synj1 homolog Synj, a phosphoinositol phosphatase, to modulate endocytosis in accordance with the level of synaptic transmission at the NMJ (Chen et al. 2014; Geng et al. 2016; Peng et al. 2021). Synaptic vesicles are retrieved and recycled through distinct mechanisms that vary with the intensity of neural activity. Clathrin-mediated endocytosis (CME) predominates during low to moderate activity, retrieving synaptic vesicles for recycling. In contrast, during high-frequency neuronal firing, activity-dependent bulk endocytosis (ADBE) becomes dominant, efficiently recycling large portions of the synaptic membrane to replenish the reserve pool vesicles depleted by such activity.

Mnb is critical for efficient CME by facilitating synaptic vesicle retrieval during moderate neuronal activities; it also suppresses ADBE (Fig. 7). Overexpression of Mnb enhances CME but reduces ADBE (Peng et al. 2021). Mnb facilitates CME by phosphorylating Synj (Chen et al. 2014; Geng et al. 2016). This phosphorylation enhances the 5′-phosphoinositol phosphatase activity of Synj, affecting its interaction with proteins like endophilin A and thereby possibly accelerating clathrin disassembly for CME.

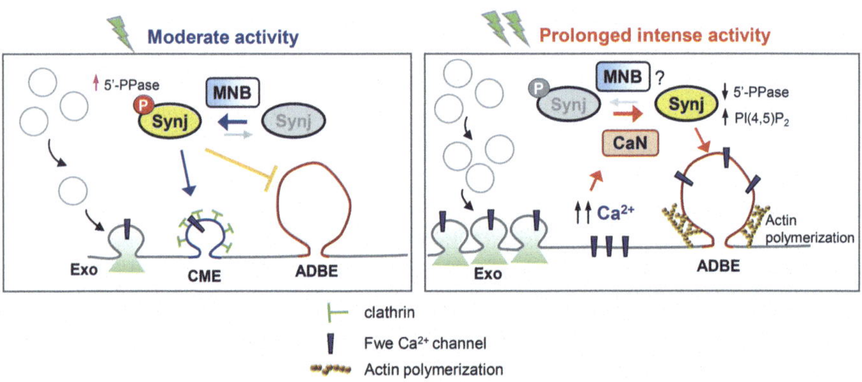

Fig. 7 The Mnb-CaN-Synj pathway in the regulation of synaptic vesicle endocytosis at the *Drosophila* NMJ. Mnb and CaN regulate Synj's activity to modulate endocytosis across different synaptic transmission levels. During moderate motor neuron activity, Mnb phosphorylates Synj, enhancing CME while suppressing ADBE. In contrast, prolonged intense activity raises intracellular Ca^{2+} levels, activating CaN, which dephosphorylates Synj. This reduction in Synj's 5′-PPase activity leads to PI(4,5)P2 accumulation and actin polymerization, promoting ADBE. It remains unclear whether Mnb is inactivated during these intense activities. (The figure is from Peng et al. (2021) with permission from the publisher)

Phosphorylation of Synj inversely affects ADBE (Fig. 7). During intense neuronal activity, CaN dephosphorylates Synj at Serine 1029, reducing Synj's 5′-PPase activity, and consequently activates ADBE (Geng et al. 2016). This dephosphorylation by CaN is crucial for facilitating efficient ADBE, contrasting the role of Mnb. Overexpressing a dephosphorylated form of Synj elevates ADBE even in the context of reduced CaN activity, highlighting the critical function of CaN-mediated dephosphorylation of Synj in ADBE regulation. This dynamic interplay between Mnb, CaN, and Synj fine-tunes synaptic vesicle recycling, demonstrating a complex regulatory network responsive to different levels of neuronal stimulation to maintain reliable communication.

8.4 Upregulation of Itsn1/Dap160, Synj1/Synj, and DSCR1/ Nla Together Leads to Altered Synaptic Functions

A study on the effects of overexpressing multiple Hsa21 homologs on synaptic morphology and endocytosis showed that while overexpressing *Drosophila* homologs of *Itsn1* (*Dap160*), *Synj1* (*Synj*), and *DSCR1* (*Nla*) individually led to changes in synaptic morphology, it was necessary to overexpress the three genes together to observe impaired vesicle recycling and reduced locomotion (Chang and Min 2009). These three genes interact with each other in a complex way. Dap160 is a scaffold protein required for synaptic growth and vesicle endocytosis, and is enriched at the synaptic terminals (Koh et al. 2004; Marie et al. 2004). Dap160 overexpression reduces the Synj level at the NMJ but increases it in the brain, possibly by affecting the transport of Synj from cell bodies to the axonal terminals (Chang and Min 2009). Nla overexpression suppresses CaN activity, reducing the 5′-phosphatase activity of Synj—which mitigates the adverse effects of Synj overexpression at the NMJ.

8.5 DSCR1/Nla Overexpression Has Distinct Effects on APP-Induced Memory Defects at Different Stages of Development

DSCR1 is overexpressed in individuals with DS as well as those with AD, suggesting a potential link between this gene and neurodegenerative processes (Cook et al. 2005; Ermak et al. 2001; Sun et al. 2011). Investigations in mammalian cells have uncovered a complex relationship between DSCR1, cell death, and survival. In cultured cells, DSCR1 upregulation paradoxically promotes both cell death and survival, adding layers of complexity to its role in cellular function and disease (Ermak et al. 2002; Porta et al. 2007; Sun et al. 2011, 2014; Wu et al. 2015).

The *Drosophila* homolog of *DSCR1*, *Nla*, has been shown to counteract the neurodegenerative effects and axonal transport defects induced by APP overexpression (Shaw and Chang 2013). Interestingly, while overexpression of Nla initially mitigates APP-induced memory defects, it accelerates memory decline and mitochondrial dysfunction in older flies (Shaw et al. 2015). These effects are mediated through the regulation of CaN activity, as pharmacological inhibition of CaN protects aged flies from APP-induced memory loss. Surprisingly, acute overexpression of Nla in aged flies also rescues APP-induced memory loss (Shaw et al. 2015). In this study the GeneSwitch-GAL4 system was employed (see sect. 4.2), allowing for temporal control over APP and Nla transgene expression in the mushroom bodies of adult flies through the administration of RU486 (Mao et al. 2004). This experiment showed that transient APP overexpression in the mushroom bodies of adult flies resulted in memory defects. However, simultaneous overexpression of APP and Nla, regardless of the age of the flies, ameliorated the memory deficits induced by APP. This contrasts with the observation that chronic co-overexpression of these genes predominantly impairs memory in aged flies. These findings suggest that the age of the flies does not solely account for Nla's inability to rescue APP phenotypes in older specimens. Instead, it appears that short-term Nla overexpression is protective against APP-induced memory loss, whereas long-term overexpression has detrimental effects.

Through these *Drosophila* studies, we gain a nuanced understanding of the age-dependent, bidirectional effects of DSCR1 overexpression, shedding light on the intricate mechanisms underlying age-related memory impairments associated with AD in DS.

8.6 DS and Fragile X Syndrome

Studies in *Drosophila* provided insights into the molecular links between DS and fragile X syndrome (FXS) (Chang et al. 2013). DS and FXS are the two most common genetic causes of intellectual disability. FXS is caused by loss-of-function mutations of the gene encoding Fragile X Mental Retardation Protein (FMRP). These mutations are characterized by expansions of the CGG trinucleotide repeats in the 5′ UTR of the *FMRP* gene (also called *fragile X mental retardation 1* or *fmr1*) (Santoro et al. 2012). FMRP represses mRNA translation.

Two independent studies in rodents found that FMRP binds to *Dscam* mRNAs (Brown et al. 2001; Darnell et al. 2011). *Drosophila* studies confirmed this binding and found that the fly homolog of FMRP represses *Dscam* translation. In *Drosophila*, loss of *FMRP* leads to excessive growth of presynaptic terminals (Kim et al. 2013; Zhang et al. 2001), a phenotype that is also caused by overexpression of Dscam in the somatosensory neurons (Kim et al. 2013). Both loss of *FMRP* and gain of *Dscam* require the Abelson tyrosine kinase (Abl) to cause the overgrowth of presynaptic terminals (Sterne et al. 2015), further supporting that they function in the same molecular pathway.

9 Using *Drosophila* to Discover Pharmacological Strategies for Therapeutic Treatments

Compounds identified to mitigate the effects of DS-linked gene overexpression in *Drosophila* serve as potential therapeutic agents for treating DS. Several studies utilizing *Drosophila* have validated the therapeutic potential of some chemical compounds.

9.1 *Targeting the Dscam-Abl Signaling*

Abl overexpression has been shown to induce a significant increase in presynaptic terminals in larval C4da neurons (Sterne et al. 2015), similar to the effects observed with Dscam overexpression (Kim et al. 2013). The constitutively active form of Abl, BCR-Abl, but not the kinase-inactive mutant, causes a marked enhancement of presynaptic terminal growth in these neurons (Sterne et al. 2015), indicating that Abl's regulatory effect on axonal overgrowth is dependent on its kinase activity. On the other hand, loss of *Abl* resulted in only a slight reduction in presynaptic terminals, but it completely inhibited the overgrowth caused by Dscam overexpression (Kim et al. 2013). This epistasis test confirmed that Abl is required for the Dscam-induced overgrowth of presynaptic terminals. Dscam overexpression leads to increased levels of active Abl in both cell cultures and neural tissues, suggesting that Dscam activates Abl (Fig. 8a).

These findings led to the hypothesis that neurological disorders resulting from DSCAM overexpression might be alleviated by inhibiting Abl activity. This hypothesis was tested using nilotinib and bafetinib, two FDA-approved Abl inhibitors (Druker et al. 2001; Hebron et al. 2013; Weisberg et al. 2005), which successfully reduced the axon terminal overgrowth in *Drosophila* larvae overexpressing Dscam in C4da neurons (Sterne et al. 2015) (Fig. 8b–d).

9.2 *Targeting Dyrk1A*

In a *Drosophila* study, the Dyrk1A inhibitor PST-001 was found to reduce the phosphorylation of human Tau at serine 262 in flies overexpressing human Tau (Zhu et al. 2022). This inhibitor also diminished neuronal degeneration in a fly photoreceptor model overexpressing human Tau, human Amyloid-β42, or Mnb (the *Drosophila* homolog of Dyrk1A), and it extended the lifespan of these flies. This study suggests the potential of Dyrk1A inhibition as a therapeutic strategy for DS-associated AD.

A structure-based informatics study followed by a cell-based assay was conducted to search for novel Dyrk1A inhibitors (Choi et al. 2021). This study

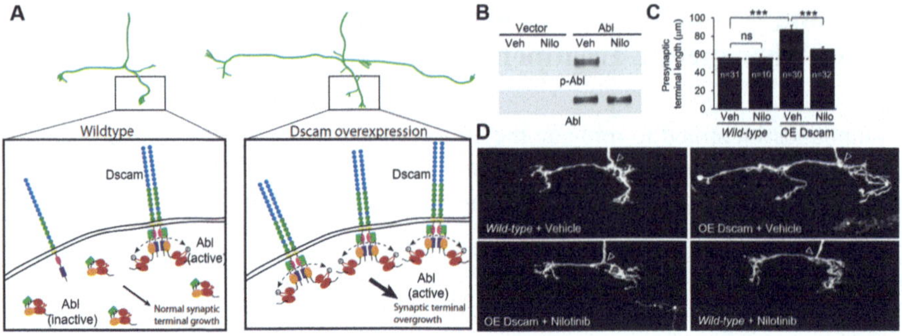

Fig. 8 Pharmacological inhibition of Abl mitigates the neuronal defects caused by increased Dscam expression in vivo. (**a**) Increasing Dscam levels activates Abl (as indicated by p-Abl levels) through its cytoplasmic domain, leading to the overgrowth of presynaptic terminals in *Drosophila* neurons. (**b**) Nilotinib (Nilo) inhibits *Drosophila* Abl kinase in cultured Schneider-2 cells transfected with either a Myc-vector plasmid ("Vector) or a plasmid for expression Myc-tagged Abl. The cells were treated with either vehicle (DMSO) or 5 μM nilotinib. Total lysates were subjected to western blot analysis with phospho-Y412-Abl (p-Abl) (top) and Myc antibodies (bottom). (**c**) Quantification of the presynaptic terminal length of the indicated genotypes and drug treatment. Sample number is shown inside each bar. (**d**) Nilotinib treatment mitigates presynaptic arbor enlargement caused by Dscam overexpression (OE Dscam). Nilotinib treatment alone does not affect presynaptic terminal growth. The arrowhead in each panel points to the location where an axon elaborates the presynaptic terminal arbor. The MARCM technique was used to generate and visualize single presynaptic terminals of mutant C4da neurons. Scale bar: 10 μm. (Panel **a** is adapted from Hergenreder et al. (2024) with permission from the publisher. Panel **b–d** are adapted from Sterne et al. (2015))

discovered aristolactam BIII, an alkaloid from plants, as a potent inhibitor of Dyrk1A. Overexpression of *Mnb* in the fly wings shortened the lateral vein 5 (L5) (Degoutin et al. 2013). Feeding larvae and adults aristolactam BIII for 2 weeks reduced the wing defect. Feeding larvae aristolactam BIII during development also reduced the rough eye phenotype caused by the coexpression of *Mnb* and human Tau. Lastly, feeding parent flies with aristolactam BIII rescued the neurogenesis defects caused by overexpression of Mnb. These *Drosophila* studies provided in vivo support of the therapeutic potential of aristolactam BIII for diseases caused by Dyrk1A hyperactivity. Additional studies in Dyrk1A transgenic mice showed that aristolactam BIII reduces defects in locomotion and those in an open-field test for exploratory behavior.

9.3 Targeting the DSCR1-CaN Signaling

DSCR1 encodes a protein that inhibits CaN, a phosphatase important for learning and memory. In aged flies, chronic DSCR1 overexpression exacerbates memory loss induced by APP, whereas acute overexpression mitigates this memory loss (Mao et al. 2004; Shaw et al. 2015). This raised the question of whether acute CaN inhibition could also ameliorate APP-induced memory deficits. Pharmacological inhibition of CaN was achieved using cyclosporin A or FK506, which significantly alleviated memory loss in aged flies overexpressing APP. Additionally, acute CaN inhibition enhanced short-term memory in aged flies, even in the absence of APP overexpression. This suggests that these drugs may counteract the increased CaN activity observed in aged wild-type flies (Shaw et al. 2015).

10 Limitations of *Drosophila* Systems for Studying DS Mechanisms

While *Drosophila* systems offer genetic manipulations of single cells and tissues at desired time points of the animal life, efficiency in testing gene interactions, and unbiased genetic screens for identifying molecular mechanisms, there are several notable weaknesses in this system for studying DS mechanisms.

First, the *Drosophila* system has not been used to study the effects of supernumerary chromosomes in DS and may not be a good system for this line of research. Accumulating evidence suggests that in addition to the changes in Hsa21 gene dosage, the supernumerary chromosome influences genome-wide gene expression (see chapter "Impact of Cre/LoxP-Mediated Chromosome Engineering Technology on Down Syndrome Research", Xing et al.). Given the chromosomal differences between *Drosophila* and humans, creating a full or even partial chromosomal model of DS in fruit flies is challenging. To date, such a model does not exist. However, although rare, extra chromosomes do occur in some *Drosophila* strains (Bauerly et al. 2014). It will be interesting to find out if *Drosophila* can be used to study the effects of the supernumerary chromosome in DS.

Second, while there is substantial genetic homology between humans and *Drosophila*, the arrangement of genes on chromosomes is different. Humans have 23 pairs of chromosomes, whereas *Drosophila* has only four pairs. This difference can affect how genes interact with each other, which might not always mirror human genetics.

Third, because of the lack of a chromosomal model, research on DS using *Drosophila* primarily focuses on individual Hsa21 genes or their interactive partners, without considering the genetic context of trisomy 21. While this approach yields valuable insights into the functions of specific genes, it may not fully encompass the complex gene interactions that underlie DS symptoms. In *Drosophila* studies, the main strategy to investigate a gene's role in DS involves overexpressing the Hsa21 gene of interest. However, this technique does not allow for the exploration of how this gene functions in the presence of additional copies of other Hsa21 genes. By contrast, in mouse trisomy models of DS, the role of an individual gene can be elucidated by normalizing its copy number to two within the trisomic context. This method maintains the triplication of other genes, facilitating the identification of the unique contribution of the specific gene in question to DS pathogenesis.

Despite their limitations, *Drosophila* models are valuable for understanding the pathogenesis of DS and developing therapeutic approaches, as demonstrated in many human conditions. The effectiveness of these models hinges on using cellular or physiological processes that mirror human biology. Using such models, researchers can discover gene functions and interactions, generating hypotheses about the molecular and cellular mechanisms underlying DS. These hypotheses can then be tested in mammalian models, which are more closely related to humans. Integrating *Drosophila* models with other systems can accelerate our understanding and treatment of DS.

Acknowledgments The author thanks Drs. Karen Chang, Summer Thyme, Jonathan Pierce, and Roger H. Reeves for suggestions on earlier versions of the manuscript. The Down syndrome research in the Ye laboratory is funded by the National Institute of Health of the United States.

References

Ahn, K.J., Jeong, H.K., Choi, H.S., Ryoo, S.R., Kim, Y.J., Goo, J.S., Choi, S.Y., Han, J.S., Ha, I., and Song, W.J. (2006). DYRK1A BAC transgenic mice show altered synaptic plasticity with learning and memory defects. Neurobiology of disease *22*, 463–472.

Altafaj, X., Dierssen, M., Baamonde, C., Marti, E., Visa, J., Guimera, J., Oset, M., Gonzalez, J.R., Florez, J., Fillat, C., *et al.* (2001). Neurodevelopmental delay, motor abnormalities and cognitive deficits in transgenic mice overexpressing Dyrk1A (minibrain), a murine model of Down's syndrome. Hum Mol Genet *10*, 1915–1923.

Alves-Sampaio, A., Troca-Marin, J.A., and Montesinos, M.L. (2010). NMDA-mediated regulation of DSCAM dendritic local translation is lost in a mouse model of Down's syndrome. J Neurosci *30*, 13537–13548.

Arque, G., Casanovas, A., and Dierssen, M. (2013). Dyrk1A Is Dynamically Expressed on Subsets of Motor Neurons and in the Neuromuscular Junction: Possible Role in Down Syndrome. *Plos One 8(1), e54285.*

Arron, J.R., Winslow, M.M., Polleri, A., Chang, C.P., Wu, H., Gao, X., Neilson, J.R., Chen, L., Heit, J.J., Kim, S.K., *et al.* (2006). NFAT dysregulation by increased dosage of DSCR1 and DYRK1A on chromosome 21. Nature *441*, 595–600.

Atas-Ozcan, H., Brault, V., Duchon, A., and Herault, Y. (2021). Dyrk1a from Gene Function in Development and Physiology to Dosage Correction across Life Span in Down Syndrome. *Genes (Basel) 12(11), 1833.*

Barlow, G.M., Chen, X.N., Shi, Z.Y., Lyons, G.E., Kurnit, D.M., Celle, L., Spinner, N.B., Zackai, E., Pettenati, M.J., Van Riper, A.J., *et al.* (2001). Down syndrome congenital heart disease: a narrowed region and a candidate gene. Genetics in medicine : official journal of the American College of Medical Genetics *3*, 91–101.

Bauerly, E., Hughes, S.E., Vietti, D.R., Miller, D.E., McDowell, W., and Hawley, R.S. (2014). Discovery of supernumerary B chromosomes in Drosophila melanogaster. Genetics *196*, 1007–1016.

Becker, L.E., Armstrong, D.L., and Chan, F. (1986). Dendritic atrophy in children with Down's syndrome. Ann Neurol *20*, 520–526.

Belichenko, P.V., Kleschevnikov, A.M., Masliah, E., Wu, C., Takimoto-Kimura, R., Salehi, A., and Mobley, W.C. (2009). Excitatory-inhibitory relationship in the fascia dentata in the Ts65Dn mouse model of Down syndrome. The Journal of comparative neurology *512*, 453–466.

Benhaourech, S., Drighil, A., and Hammiri, A.E. (2016). Congenital heart disease and Down syndrome: various aspects of a confirmed association. Cardiovasc J Afr *27*, 287–290.

Bergström, S., Carr, H., Petersson, G., Stephansson, O., Bonamy, A.K., Dahlström, A., Halvorsen, C.P., and Johansson, S. (2016). Trends in Congenital Heart Defects in Infants With Down Syndrome. *Pediatrics 138 (1): e20160123.*

Bier, E., and Bodmer, R. (2004). Drosophila, an emerging model for cardiac disease. Gene *342*, 1–11.

Bodmer, R. (1993). The Gene Tinman Is Required for Specification of the Heart and Visceral Muscles in Drosophila. Development *118*, 719–729.

Branchi, I., Bichler, Z., Minghetti, L., Delabar, J.M., Malchiodi-Albedi, F., Gonzalez, M.C., Chettouh, Z., Nicolini, A., Chabert, C., Smith, D.J., *et al.* (2004). Transgenic mouse in vivo library of human Down syndrome critical region 1: Association between DYRK1A overexpression, brain development abnormalities, and cell cycle protein alteration. Journal of neuropathology and experimental neurology *63*, 429–440.

Brand, A.H., and Perrimon, N. (1993). Targeted gene expression as a means of altering cell fates and generating dominant phenotypes. Development *118*, 401–415.

Braudeau, J., Dauphinot, L., Duchon, A., Loistron, A., Dodd, R.H., Herault, Y., Delatour, B., and Potier, A.C. (2011a). Chronic Treatment with a Promnesiant GABA-A a5-Selective Inverse Agonist Increases Immediate Early Genes Expression during Memory Processing in Mice and Rectifies Their Expression Levels in a Down Syndrome Mouse Model. Advances in *Pharmacological Sciences 2011, 153218.*

Braudeau, J., Delatour, B., Duchon, A., Pereira, P.L., Dauphinot, L., de Chaumont, F., Olivo-Marin, J.C., Dodd, R.H., Herault, Y., and Potier, M.C. (2011b). Specific targeting of the GABA-A receptor alpha5 subtype by a selective inverse agonist restores cognitive deficits in Down syndrome mice. J Psychopharmacol *25*, 1030–1042.

Brown, V., Jin, P., Ceman, S., Darnell, J.C., O'Donnell, W.T., Tenenbaum, S.A., Jin, X., Feng, Y., Wilkinson, K.D., Keene, J.D., *et al.* (2001). Microarray identification of FMRP-associated brain mRNAs and altered mRNA translational profiles in fragile X syndrome. Cell *107*, 477–487.

Bruce, F.M., Brown, S., Smith, J.N., Fuerst, P.G., and Erskine, L. (2017). DSCAM promotes axon fasciculation and growth in the developing optic pathway. Proceedings of the National Academy of Sciences of the United States of America *114*, 1702–1707.

Caceres, A., Ye, B., and Dotti, C.G. (2012). Neuronal polarity: demarcation, growth and commitment. Current opinion in cell biology *24*, 547–553.

Cajal, R. (1995). Histology of the nervous system of man and vertebrates (Oxford University Press, USA).

Chakraborty, R., Vepuri, V., Mhatre, S.D., Paddock, B.E., Miller, S., Michelson, S.J., Delvadia, R., Desai, A., Vinokur, M., Melicharek, D.J., *et al.* (2011). Characterization of a Drosophila Alzheimer's disease model: pharmacological rescue of cognitive defects. PLoS One *6*, e20799.

Chang, K.T., and Min, K.T. (2005). homolog of Down syndrome critical region 1 is critical for mitochondrial function. Nature Neuroscience *8*, 1577–1585.

Chang, K.T., and Min, K.T. (2009). Upregulation of three Drosophila homologs of human chromosome 21 genes alters synaptic function: implications for Down syndrome. Proceedings of the National Academy of Sciences of the United States of America *106*, 17117–17122.

Chang, K.T., Ro, H., Wang, W., and Min, K.T. (2013). Meeting at the crossroads: common mechanisms in Fragile X and Down syndrome. Trends in neurosciences *36*, 685–694.

Chang, K.T., Shi, Y.J., and Min, K.T. (2003). The Drosophila homolog of Down's syndrome critical region 1 gene regulates learning: implications for mental retardation. Proceedings of the National Academy of Sciences of the United States of America *100*, 15794–15799.

Chartier-Harlin, M.C., Crawford, F., Houlden, H., Warren, A., Hughes, D., Fidani, L., Goate, A., Rossor, M., Roques, P., Hardy, J., *et al.* (1991). Early-onset Alzheimer's disease caused by mutations at codon 717 of the beta-amyloid precursor protein gene. Nature *353*, 844–846.

Chen, C.K., Bregere, C., Paluch, J., Lu, J.F., Dickman, D.K., and Chang, K.T. (2014). Activity-dependent facilitation of Synaptojanin and synaptic vesicle recycling by the Minibrain kinase. Nature communications *5*, 4246.

Choi, M., Kim, A.K., Ham, Y., Lee, J.Y., Kim, D., Yang, A., Jo, M.J., Yoon, E., Heo, J.N., Han, S.B., *et al.* (2021). Aristolactam BIII, a naturally derived DYRK1A inhibitor, rescues Down syndrome-related phenotypes. Phytomedicine *92*, 153695.

Choma, M.A., Suter, M.J., Vakoc, B.J., Bouma, B.E., and Tearney, G.J. (2011). Physiological homology between Drosophila melanogaster and vertebrate cardiovascular systems. Disease models & mechanisms *4*, 411–420.

Colas, D., Chuluun, B., Warrier, D., Blank, M., Wetmore, D.Z., Buckmaster, P., Garner, C.C., and Heller, H.C. (2013). Short-term treatment with the GABAA receptor antagonist pentylenetetrazole produces a sustained pro-cognitive benefit in a mouse model of Down's syndrome. Brit J Pharmacol *169*, 963–973.

Collins, C.A., and DiAntonio, A. (2007). Synaptic development: insights from Drosophila. Current opinion in neurobiology *17*, 35–42.

Cook, C.N., Hejna, M.J., Magnuson, D.J., and Lee, J.M. (2005). Expression of calcipressin1, an inhibitor of the phosphatase calcineurin, is altered with aging and Alzheimer's disease. Journal of Alzheimers Disease *8*, 63–73.

Costa, A.C., and Grybko, M.J. (2005). Deficits in hippocampal CA1 LTP induced by TBS but not HFS in the Ts65Dn mouse: a model of Down syndrome. Neuroscience letters *382*, 317–322.

Craig, A.M., and Banker, G. (1994). Neuronal polarity. Annual review of neuroscience *17*, 267–310.

Cvetkovska, V., Hibbert, A.D., Emran, F., and Chen, B.E. (2013). Overexpression of Down syndrome cell adhesion molecule impairs precise synaptic targeting. Nat Neurosci *16*, 677–682.

Dahmann, C. (2016). Drosophila: Methods and Protocols, 2nd edn (Totowa, NJ, USA: Humana Press).

Darnell, J.C., Van Driesche, S.J., Zhang, C., Hung, K.Y., Mele, A., Fraser, C.E., Stone, E.F., Chen, C., Fak, J.J., Chi, S.W., *et al.* (2011). FMRP stalls ribosomal translocation on mRNAs linked to synaptic function and autism. Cell *146*, 247–261.

Davis, G.W., and Muller, M. (2015). Homeostatic control of presynaptic neurotransmitter release. Annu Rev Physiol *77*, 251–270.

Degoutin, J.L., Milton, C.C., Yu, E., Tipping, M., Bosveld, F., Yang, L., Bellaiche, Y., Veraksa, A., and Harvey, K.F. (2013). Riquiqui and minibrain are regulators of the hippo pathway downstream of Dachsous. Nat Cell Biol *15*, 1176–1185.

Deidda, G., Bozarth, I.F., and Cancedda, L. (2014). Modulation of GABAergic transmission in development and neurodevelopmental disorders: investigating physiology and pathology to gain therapeutic perspectives. Front Cell Neurosci *8*, 119.

Dierssen, M., Arque, G., McDonald, J., Andreu, N., Martinez-Cue, C., Florez, J., and Fillat, C. (2011). Behavioral characterization of a mouse model overexpressing DSCR1/ RCAN1. PLoS One *6*, e17010.

Doran, E., Keator, D., Head, E., Phelan, M.J., Kim, R., Totoiu, M., Barrio, J.R., Small, G.W., Potkin, S.G., and Lott, I.T. (2017). Down Syndrome, Partial Trisomy 21, and Absence of Alzheimer's Disease: The Role of APP. J Alzheimers Dis *56*, 459–470.

Druker, B.J., Talpaz, M., Resta, D.J., Peng, B., Buchdunger, E., Ford, J.M., Lydon, N.B., Kantarjian, H., Capdeville, R., Ohno-Jones, S., *et al.* (2001). Efficacy and safety of a specific inhibitor of the BCR-ABL tyrosine kinase in chronic myeloid leukemia. N Engl J Med *344*, 1031–1037.

Duffy, J.B. (2002). GAL4 system in Drosophila: a fly geneticist's Swiss army knife. Genesis *34*, 1–15.

Eccles, J. (1976). From electrical to chemical transmission in the central nervous system. Notes Rec R Soc Lond *30*, 219–230.

Ermak, G., Harris, C.D., and Davies, K.J. (2002). The DSCR1 (Adapt78) isoform 1 protein calcipressin 1 inhibits calcineurin and protects against acute calcium-mediated stress damage, including transient oxidative stress. Faseb J *16*, 814–824.

Ermak, G., Morgan, T.E., and Davies, K.J.A. (2001). Chronic overexpression of the calcineurin inhibitory gene is associated with Alzheimer's disease. Journal of Biological Chemistry *276*, 38787–38794.

Fernandez, F., Morishita, W., Zuniga, E., Nguyen, J., Blank, M., Malenka, R.C., and Garner, C.C. (2007). Pharmacotherapy for cognitive impairment in a mouse model of Down syndrome. Nat Neurosci *10*, 411–413.

Frankish, A., Diekhans, M., Ferreira, A.M., Johnson, R., Jungreis, I., Loveland, J., Mudge, J.M., Sisu, C., Wright, J., Armstrong, J., *et al.* (2019). GENCODE reference annotation for the human and mouse genomes. Nucleic acids research *47*, D766–D773.

Freeman, S.B., Taft, L.F., Dooley, K.J., Allran, K., Sherman, S.L., Hassold, T.J., Khoury, M.J., and Saker, D.M. (1998). Population-based study of congenital heart defects in Down syndrome. Am J Med Genet *80*, 213–217.

Fuentes, J.J., Genescà, L., Kingsbury, T.J., Cunningham, K.W., Pérez-Riba, M., Estivill, X., and de la Luna, S. (2000). DSCR1, overexpressed in Down syndrome, is an inhibitor of calcineurin-mediated signaling pathways. Human Molecular Genetics *9*, 1681–1690.

García-Cerro, S., Rueda, N., Vidal, V., Lantigua, S., and Martínez-Cué, C. (2017). Normalizing the gene dosage of Dyrk1A in a mouse model of Down syndrome rescues several Alzheimer's disease phenotypes. Neurobiology of disease *106*, 76–88.

Geng, J., Wang, L., Lee, J.Y., Chen, C.K., and Chang, K.T. (2016). Phosphorylation of Synaptojanin Differentially Regulates Endocytosis of Functionally Distinct Synaptic Vesicle Pools. J Neurosci *36*, 8882–8894.

Greeve, I., Kretzschmar, D., Tschape, J.A., Beyn, A., Brellinger, C., Schweizer, M., Nitsch, R.M., and Reifegerste, R. (2004). Age-dependent neurodegeneration and Alzheimer-amyloid plaque formation in transgenic Drosophila. J Neurosci *24*, 3899–3906.

Grossman, T.R., Gamliel, A., Wessells, R.J., Taghli-Lamallem, O., Jepsen, K., Ocorr, K., Korenberg, J.R., Peterson, K.L., Rosenfeld, M.G., Bodmer, R., *et al.* (2011). Over-Expression of DSCAM and COL6A2 Cooperatively Generates Congenital Heart Defects. *PLoS Genet 7(11): e1002344.*

Grueber, W.B., Yang, C.H., Ye, B., and Jan, Y.N. (2005). The development of neuronal morphology in insects. Curr Biol *15*, R730–738.

Guimera, J., Casas, C., Pucharcos, C., Solans, A., Domenech, A., Planas, A.M., Ashley, J., Lovett, M., Estivill, X., and Pritchard, M.A. (1996). A human homologue of Drosophila minibrain (MNB) is expressed in the neuronal regions affected in Down syndrome and maps to the critical region. Human Molecular Genetics *5*, 1305–1310.

Harris, W.A. (2001). Temporal coordinates: the genes that fix cell fate with birth order. Developmental cell *1*, 313–314.

Hatten, M.E. (1999). Central nervous system neuronal migration. Annual review of neuroscience *22*, 511–539.

Hattori, D., Millard, S.S., Wojtowicz, W.M., and Zipursky, S.L. (2008). Dscam-mediated cell recognition regulates neural circuit formation. Annual review of cell and developmental biology *24*, 597–620.

Hattori, M., Fujiyama, A., Taylor, T.D., Watanabe, H., Yada, T., Park, H.S., Toyoda, A., Ishii, K., Totoki, Y., Choi, D.K., *et al.* (2000). The DNA sequence of human chromosome 21. Nature *405*, 311–319.

Hebron, M.L., Lonskaya, I., and Moussa, C.E. (2013). Nilotinib reverses loss of dopamine neurons and improves motor behavior via autophagic degradation of alpha-synuclein in Parkinson's disease models. Hum Mol Genet *22*, 3315–3328.

Hergenreder, T., Yang, T., and Ye, B. (2024). The role of Down syndrome cell adhesion molecule in Down syndrome. Med Rev *4*, 31–41.

Hoeffer, C.A., Dey, A., Sachan, N., Wong, H., Patterson, R.J., Shelton, J.M., Richardson, J.A., Klann, E., and Rothermel, B.A. (2007). The Down syndrome critical region protein RCAN1 regulates long-term potentiation and memory via inhibition of phosphatase signaling. J Neurosci *27*, 13161–13172.

Hughes, M.E., Bortnick, R., Tsubouchi, A., Baumer, P., Kondo, M., Uemura, T., and Schmucker, D. (2007). Homophilic Dscam interactions control complex dendrite morphogenesis. Neuron *54*, 417–427.

Jan, L.Y., and Jan, Y.N. (1976). L-glutamate as an excitatory transmitter at the Drosophila larval neuromuscular junction. The Journal of physiology *262*, 215–236.

Jan, Y.N., and Jan, L.Y. (2003). The control of dendrite development. Neuron *40*, 229–242.

Jenett, A., Rubin, G.M., Ngo, T.T., Shepherd, D., Murphy, C., Dionne, H., Pfeiffer, B.D., Cavallaro, A., Hall, D., Jeter, J., *et al.* (2012). A GAL4-driver line resource for Drosophila neurobiology. Cell Rep *2*, 991–1001.

Jia, Y.L., Jing, L.J., Li, J.Y., Lu, J.J., Han, R., Wang, S.Y., Peng, T., and Jia, Y.J. (2011). Expression and significance of DSCAM in the cerebral cortex of APP transgenic mice. Neuroscience letters *491*, 153–157.

Johansen, J., Halpern, M.E., Johansen, K.M., and Keshishian, H. (1989). Stereotypic morphology of glutamatergic synapses on identified muscle cells of Drosophila larvae. J Neurosci *9*, 710–725.

Kasuga, K., Shimohata, T., Nishimura, A., Shiga, A., Mizuguchi, T., Tokunaga, J., Ohno, T., Miyashita, A., Kuwano, R., Matsumoto, N., *et al.* (2009). Identification of independent APP locus duplication in Japanese patients with early-onset Alzheimer disease. Journal of Neurology, Neurosurgery, and Psychiatry *80*, 1050–1052.

Kaufmann, W.E., and Moser, H.W. (2000). Dendritic anomalies in disorders associated with mental retardation. Cereb Cortex *10*, 981–991.

Kentrup, H., Becker, W., Heukelbach, J., Wilmes, A., Schurmann, A., Huppertz, C., Kainulainen, H., and Joost, H.G. (1996). Dyrk, a dual specificity protein kinase with unique structural features whose activity is dependent on tyrosine residues between subdomains VII and VIII. Journal of Biological Chemistry *271*, 3488–3495.

Keshishian, H., Chang, T.N., and Jarecki, J. (1994). Precision and plasticity during Drosophila neuromuscular development. Faseb J *8*, 731–737.

Kim, J.H., Wang, X., Coolon, R., and Ye, B. (2013). Dscam expression levels determine presynaptic arbor sizes in Drosophila sensory neurons. Neuron *78*, 827–838.

Kleschevnikov, A.M., Belichenko, P.V., Faizi, M., Jacobs, L.F., Htun, K., Shamloo, M., and Mobley, W.C. (2012a). Deficits in cognition and synaptic plasticity in a mouse model of Down syndrome ameliorated by GABAB receptor antagonists. J Neurosci *32*, 9217–9227.

Kleschevnikov, A.M., Belichenko, P.V., Gall, J., George, L., Nosheny, R., Maloney, M.T., Salehi, A., and Mobley, W.C. (2012b). Increased efficiency of the GABAA and GABAB receptor-mediated neurotransmission in the Ts65Dn mouse model of Down syndrome. Neurobiology of disease *45*, 683–691.

Koh, T.W., Verstreken, P., and Bellen, H.J. (2004). Dap160/intersectin acts as a stabilizing scaffold required for synaptic development and vesicle endocytosis. Neuron *43*, 193–205.

Korbel, J.O., Tirosh-Wagner, T., Urban, A.E., Chen, X.N., Kasowski, M., Dai, L., Grubert, F., Erdman, C., Gao, M.C., Lange, K., *et al.* (2009). The genetic architecture of Down syndrome phenotypes revealed by high-resolution analysis of human segmental trisomies. Proceedings of the National Academy of Sciences of the United States of America *106*, 12031–12036.

Kuntz, S.G., and Eisen, M.B. (2014). Drosophila embryogenesis scales uniformly across temperature in developmentally diverse species. PLoS Genet *10*, e1004293.

Kurt, M.A., Davies, D.C., Kidd, M., Dierssen, M., and Florez, J. (2000). Synaptic deficit in the temporal cortex of partial trisomy 16 (Ts65Dn) mice. Brain Res *858*, 191–197.

Lai, F., and Williams, R.S. (1989). A prospective study of Alzheimer disease in Down syndrome. Arch Neurol *46*, 849–853.

Lana-Elola, E., Aoidi, R., Llorian, M., Gibbins, D., Buechsenschuetz, C., Bussi, C., Flynn, H., Gilmore, T., Watson-Scales, S., Haugsten Hansen, M., *et al.* (2024). Increased dosage of DYRK1A leads to congenital heart defects in a mouse model of Down syndrome. Sci Transl Med *16*, eadd6883.

Laurent, G. (1999). Dendritic processing in invertebrates: a link to function. In Dendrites, G. Stuart, Spruston, N., and Hausser M., ed. (Oxford University Press), pp. 290–297.

Lee, T., and Luo, L. (1999). Mosaic analysis with a repressible cell marker for studies of gene function in neuronal morphogenesis. Neuron *22*, 451–461.

Lee, T., and Luo, L. (2001). Mosaic analysis with a repressible cell marker (MARCM) for Drosophila neural development. Trends in Neurosciences *24*, 251–254.

Lehmacher, C., Abeln, B., and Paululat, A. (2012). The ultrastructure of Drosophila heart cells. Arthropod Struct Dev *41*, 459–474.

Lepagnol-Bestel, A.M., Zvara, A., Maussion, G., Quignon, F., Ngimbous, B., Ramoz, N., Imbeaud, S., Loe-Mie, Y., Benihoud, K., Agier, N., *et al.* (2009). DYRK1A interacts with the REST/NRSF-SWI/SNF chromatin remodelling complex to deregulate gene clusters involved in the neuronal phenotypic traits of Down syndrome. Hum Mol Genet *18*, 1405–1414.

Liu, H., Caballero-Floran, R.N., Hergenreder, T., Yang, T., Hull, J.M., Pan, G., Li, R., Veling, M.W., Isom, L.L., Kwan, K.Y., *et al.* (2023). DSCAM gene triplication causes excessive GABAergic synapses in the neocortex in Down syndrome mouse models. PLoS Biol *21*, e3002078.

Lowe, S.A., Hodge, J.J.L., and Usowicz, M.M. (2018). A third copy of the Down syndrome cell adhesion molecule (Dscam) causes synaptic and locomotor dysfunction in Drosophila. Neurobiology of Disease *110*, 93–101.

Lowe, S.A., Usowicz, M.M., and Hodge, J.J.L. (2019). Neuronal overexpression of Alzheimer's disease and Down's syndrome associated DYRK1A/minibrain gene alters motor decline, neurodegeneration and synaptic plasticity in Drosophila. Neurobiology of Disease *125*, 107–114.

Lyons, I., Parsons, L.M., Hartley, L., Li, R., Andrews, J.E., Robb, L., and Harvey, R.P. (1995). Myogenic and morphogenetic defects in the heart tubes of murine embryos lacking the homeo box gene Nkx2-5. Genes Dev *9*, 1654–1666.

Mao, Z., Roman, G., Zong, L., and Davis, R.L. (2004). Pharmacogenetic rescue in time and space of the rutabaga memory impairment by using Gene-Switch. Proceedings of the National Academy of Sciences of the United States of America *101*, 198–203.

Marie, B., Sweeney, S.T., Poskanzer, K.E., Roos, J., Kelly, R.B., and Davis, G.W. (2004). Dap160/intersectin scaffolds the periactive zone to achieve high-fidelity endocytosis and normal synaptic growth. Neuron *43*, 207–219.

Martí, E., Altafaj, X., Dierssen, M., de la Luna, S., Fotaki, V., Alvarez, M., Pérez-Riba, M., Ferrer, I., and Estivill, X. (2003). Dyrk1A expression pattern supports specific roles of this kinase in the adult central nervous system. Brain Research *964*, 250–263.

Martin, K.R., Corlett, A., Dubach, D., Mustafa, T., Coleman, H.A., Parkington, H.C., Merson, T.D., Bourne, J.A., Porta, S., Arbonés, M.L., *et al.* (2012). Over-expression of RCAN1 causes Down syndrome-like hippocampal deficits that alter learning and memory. Human Molecular Genetics *21*, 3025–3041.

Martinez-Cue, C., Martinez, P., Rueda, N., Vidal, R., Garcia, S., Vidal, V., Corrales, A., Montero, J.A., Pazos, A., Florez, J., *et al.* (2013). Reducing GABAA alpha5 receptor-mediated inhibition rescues functional and neuromorphological deficits in a mouse model of down syndrome. J Neurosci *33*, 3953–3966.

Matthews, B.J., Kim, M.E., Flanagan, J.J., Hattori, D., Clemens, J.C., Zipursky, S.L., and Grueber, W.B. (2007). Dendrite Self-Avoidance Is Controlled by Dscam. Cell *129*, 593–604.

McGuire, S.E., Le, P.T., Osborn, A.J., Matsumoto, K., and Davis, R.L. (2003). Spatiotemporal rescue of memory dysfunction in Drosophila. Science (New York, NY) *302*, 1765–1768.

McGuire, S.E., Roman, G., and Davis, R.L. (2004). Gene expression systems in Drosophila: a synthesis of time and space. Trends in Genetics: TIG *20*, 384–391.

Menon, K.P., Carrillo, R.A., and Zinn, K. (2013). Development and plasticity of the Drosophila larval neuromuscular junction. Wiley Interdiscip Rev Dev Biol *2*, 647–670.

Miron, M., and Sonenberg, N. (2001). Regulation of translation via TOR signaling: insights from Drosophila melanogaster. J Nutr *131*, 2988S–2993S.

Murrell, J., Farlow, M., Ghetti, B., and Benson, M.D. (1991). A mutation in the amyloid precursor protein associated with hereditary Alzheimer's disease. Science (New York, NY) *254*, 97–99.

Nosheny, R.L., Belichenko, P.V., Busse, B.L., Weissmiller, A.M., Dang, V., Das, D., Fahimi, A., Salehi, A., Smith, S.J., and Mobley, W.C. (2015). Increased cortical synaptic activation of TrkB and downstream signaling markers in a mouse model of Down Syndrome. Neurobiology of Disease *77*, 173–190.

Orgogozo, V., and Grueber, W.B. (2005). FlyPNS, a database of the Drosophila embryonic and larval peripheral nervous system. BMC Dev Biol *5*, 4.

Ori-McKenney, K.M., McKenney, R.J., Huang, H.H., Li, T., Meltzer, S., Jan, L.Y., Vale, R.D., Wiita, A.P., and Jan, Y.N. (2016). Phosphorylation of beta-Tubulin by the Down Syndrome Kinase, Minibrain/DYRK1a, Regulates Microtubule Dynamics and Dendrite Morphogenesis. Neuron *90*, 551–563.

Osterwalder, T., Yoon, K.S., White, B.H., and Keshishian, H. (2001). A conditional tissue-specific transgene expression system using inducible GAL4. Proceedings of the National Academy of Sciences of the United States of America *98*, 12596–12601.

Ozturk-Colak, A., Marygold, S.J., Antonazzo, G., Attrill, H., Goutte-Gattat, D., Jenkins, V.K., Matthews, B.B., Millburn, G., Dos Santos, G., Tabone, C.J., *et al.* (2024). FlyBase: updates to the Drosophila genes and genomes database. *Genetics 227(1), iyad211.*

Park, J., Oh, Y., Yoo, L., Jung, M.S., Song, W.J., Lee, S.H., Seo, H., and Chung, K.C. (2010). Dyrk1A phosphorylates p53 and inhibits proliferation of embryonic neuronal cells. J Biol Chem *285*, 31895–31906.

Park, J., Song, W.J., and Chung, K.C. (2009). Function and regulation of Dyrk1A: towards understanding Down syndrome. Cell Mol Life Sci *66*, 3235–3240.

Pelleri, M.C., Cattani, C., Vitale, L., Antonaros, F., Strippoli, P., Locatelli, C., Cocchi, G., Piovesan, A., and Caracausi, M. (2018). Integrated Quantitative Transcriptome Maps of Human Trisomy 21 Tissues and Cells. Front Genet *9*, 125.

Peng, Y.J., Geng, J.H., Wu, Y., Pinales, C., Langen, J., Chang, Y.C., Buser, C., and Chang, K.T. (2021). Minibrain kinase and calcineurin coordinate activity-dependent bulk endocytosis through synaptojanin. *Journal of Cell Biology 220(12): e202011028.*

Perkins, L.A., Holderbaum, L., Tao, R., Hu, Y.H., Sopko, R., McCall, K., Donghui, Y.Z., Flockhart, I., Binari, R., Shim, H.S., *et al.* (2015). The Transgenic RNAi Project at Harvard Medical School: Resources and Validation. Genetics *201*, 843–U868.

Pfeiffer, B.D., Ngo, T.T.B., Hibbard, K.L., Murphy, C., Jenett, A., Truman, J.W., and Rubin, G.M. (2010). Refinement of Tools for Targeted Gene Expression in Drosophila. Genetics *186*, 735–U488.

Pfeiffer, B.D., Truman, J.W., and Rubin, G.M. (2012). Using translational enhancers to increase transgene expression in Drosophila. Proceedings of the National Academy of Sciences of the United States of America *109*, 6626–6631.

Pizzano, S., Sterne, G.R., Veling, M.W., Xu, L.A., Hergenreder, T., and Ye, B. (2023). The Drosophila homolog of APP promotes Dscam expression to drive axon terminal growth, revealing interaction between Down syndrome genes. *Disease Models & Mechanisms 16(9): dmm049725.*

Porta, S., Serra, S.A., Huch, M., Valverde, M.A., Llorens, F., Estivill, X., Arbones, M.L., and Marti, E. (2007). RCAN1 (DSCR1) increases neuronal susceptibility to oxidative stress: a potential pathogenic process in neurodegeneration. Hum Mol Genet *16*, 1039–1050.

Prinz, M., Prinz, B., and Schulz, E. (1997). The growth of non-pyramidal neurons in the primary motor cortex of man: a Golgi study. Histol Histopathol *12*, 895–900.

Prokop, A., and Meinertzhagen, I.A. (2006). Development and structure of synaptic contacts in Drosophila. Seminars in Cell & Developmental Biology *17*, 20–30.

Rovelet-Lecrux, A., Hannequin, D., Raux, G., Le Meur, N., Laquerriere, A., Vital, A., Dumanchin, C., Feuillette, S., Brice, A., Vercelletto, M., *et al.* (2006). APP locus duplication causes autosomal dominant early-onset Alzheimer disease with cerebral amyloid angiopathy. Nature Genetics *38*, 24–26.

Saito, Y., Oka, A., Mizuguchi, M., Motonaga, K., Mori, Y., Becker, L.E., Arima, K., Miyauchi, J., and Takashima, S. (2000). The developmental and aging changes of Down's syndrome cell adhesion molecule expression in normal and Down's syndrome brains. Acta Neuropathol *100*, 654–664.

Salehi, A., Delcroix, J.D., Belichenko, P.V., Zhan, K., Wu, C., Valletta, J.S., Takimoto-Kimura, R., Kleschevnikov, A.M., Sambamurti, K., Chung, P.P., *et al.* (2006). Increased App expression in a mouse model of Down's syndrome disrupts NGF transport and causes cholinergic neuron degeneration. Neuron *51*, 29–42.

Santoro, M.R., Bray, S.M., and Warren, S.T. (2012). Molecular mechanisms of fragile X syndrome: a twenty-year perspective. Annu Rev Pathol *7*, 219–245.

Schmucker, D., and Chen, B. (2009). Dscam and DSCAM: complex genes in simple animals, complex animals yet simple genes. Genes Dev *23*, 147–156.

Schmucker, D., Clemens, J.C., Shu, H., Worby, C.A., Xiao, J., Muda, M., Dixon, J.E., and Zipursky, S.L. (2000). Drosophila Dscam is an axon guidance receptor exhibiting extraordinary molecular diversity. Cell *101*, 671–684.

Senturk, M., and Bellen, H.J. (2018). Genetic strategies to tackle neurological diseases in fruit flies. Current Opinion in Neurobiology *50*, 24–32.

Shaw, J.L., and Chang, K.T. (2013). Nebula/DSCR1 upregulation delays neurodegeneration and protects against APP-induced axonal transport defects by restoring calcineurin and GSK-3beta signaling. PLoS Genet *9*, e1003792.

Shaw, J.L., Zhang, S., and Chang, K.T. (2015). Bidirectional Regulation of Amyloid Precursor Protein-Induced Memory Defects by Nebula/DSCR1: A Protein Upregulated in Alzheimer's Disease and Down Syndrome. J Neurosci *35*, 11374–11383.

Simon, A.M., Schiapparelli, L., Salazar-Colocho, P., Cuadrado-Tejedor, M., Escribano, L., Lopez de Maturana, R., Del Rio, J., Perez-Mediavilla, A., and Frechilla, D. (2009). Overexpression of wild-type human APP in mice causes cognitive deficits and pathological features unrelated to Abeta levels. Neurobiology of Disease *33*, 369–378.

Sleegers, K., Brouwers, N., Gijselinck, I., Theuns, J., Goossens, D., Wauters, J., Del-Favero, J., Cruts, M., Duijn, C.M.V., and Broeckhoven, C.V. (2006). APP duplication is sufficient to cause early onset Alzheimer's dementia with cerebral amyloid angiopathy. Brain *129*, 2977–2983.

Smith, D.J., Stevens, M.E., Sudanagunta, S.P., Bronson, R.T., Makhinson, M., Watabe, A.M., O'Dell, T.J., Fung, J., Weier, H.U., Cheng, J.F., *et al.* (1997). Functional screening of 2 Mb of human chromosome 21q22.2 in transgenic mice implicates minibrain in learning defects associated with Down syndrome. Nature Genetics *16*, 28–36.

Soba, P., Zhu, S., Emoto, K., Younger, S., Yang, S.J., Yu, H.H., Lee, T., Jan, L.Y., and Jan, Y.N. (2007). Drosophila sensory neurons require Dscam for dendritic self-avoidance and proper dendritic field organization. Neuron *54*, 403–416.

Song, W.J., Sternberg, L.R., KastenSportes, C., VanKeuren, M.L., Chung, S.H., Slack, A.C., Miller, D.E., Glover, T.W., Chiang, P.W., Lou, L.D., *et al.* (1996). Isolation of human and murine homologues of the Drosophila minibrain gene: Human homologue maps to 21q22.2 in the down syndrome "critical region". Genomics *38*, 331–339.

Sterne, G.R., Kim, J.H., and Ye, B. (2015). Dysregulated Dscam levels act through Abelson tyrosine kinase to enlarge presynaptic arbors. eLife *4*, e05196.

Strausfeld, N.J. (1976). Atlas of an Insect Brain (Berlin, Heidelberg, New York: Springer-Verlag).

Sullivan, W., Ashburner, M., and Hawley, R.S. (2000). Drosophila protocols (Cold Spring Harbor, N.Y.: Cold Spring Harbor Laboratory Press).

Sun, X., Wu, Y., Herculano, B., and Song, W. (2014). RCAN1 overexpression exacerbates calcium overloading-induced neuronal apoptosis. PLoS One *9*, e95471.

Sun, X.L., Wu, Y.L., Chen, B., Zhang, Z.H., Zhou, W.H., Tong, Y.G., Yuan, J.Y., Xia, K., Gronemeyer, H., Flavell, R.A., *et al.* (2011). Regulator of Calcineurin 1 (RCAN1) Facilitates Neuronal Apoptosis through Caspase-3 Activation. Journal of Biological Chemistry *286*, 9049–9062.

Tanaka, M., Chen, Z., Bartunkova, S., Yamasaki, N., and Izumo, S. (1999). The cardiac homeobox gene Csx/Nkx2.5 lies genetically upstream of multiple genes essential for heart development. Development *126*, 1269–1280.

Tejedor, F., Zhu, X.R., Kaltenbach, E., Ackermann, A., Baumann, A., Canal, I., Heisenberg, M., Fischbach, K.F., and Pongs, O. (1995). minibrain: a new protein kinase family involved in postembryonic neurogenesis in Drosophila. Neuron *14*, 287–301.

Tessier-Lavigne, M., and Goodman, C.S. (1996). The molecular biology of axon guidance. Science (New York, NY) *274*, 1123–1133.

Ugur, B., Chen, K., and Bellen, H.J. (2016). Drosophila tools and assays for the study of human diseases. Disease Models & Mechanisms *9*, 235–244.

Van Vactor, D., and Sigrist, S.J. (2017). Presynaptic morphogenesis, active zone organization and structural plasticity in Drosophila. Current Opinion in Neurobiology *43*, 119–129.

Venken, K.J.T., He, Y.C., Hoskins, R.A., and Bellen, H.J. (2006). P[acman]: A BAC transgenic platform for targeted insertion of large DNA fragments in D. melanogaster. Science (New York, NY) *314*, 1747–1751.

Wang, X., Zhang, M.W., Kim, J.H., Macara, A.M., Sterne, G., Yang, T., and Ye, B. (2015). The Kruppel-Like Factor Dar1 Determines Multipolar Neuron Morphology. J Neurosci *35*, 14251–14259.

Wassarman, D.A., Therrien, M., and Rubin, G.M. (1995). The Ras signaling pathway in Drosophila. Curr Opin Genet Dev *5*, 44–50.

Watson-Scales, S., Kalmar, B., Lana-Elola, E., Gibbins, D., La Russa, F., Wiseman, F., Williamson, M., Saccon, R., Slender, A., Olerinyova, A., *et al.* (2018). Analysis of motor dysfunction in Down Syndrome reveals motor neuron degeneration. PLoS Genet *14*, e1007383.

Wegiel, J., Dowjat, K., Kaczmarski, W., Kuchna, I., Nowicki, K., Frackowiak, J., Mazur Kolecka, B., Wegiel, J., Silverman, W.P., Reisberg, B., *et al.* (2008). The role of overexpressed DYRK1A protein in the early onset of neurofibrillary degeneration in Down syndrome. Acta Neuropathol *116*, 391–407.

Weisberg, E., Manley, P.W., Breitenstein, W., Bruggen, J., Cowan-Jacob, S.W., Ray, A., Huntly, B., Fabbro, D., Fendrich, G., Hall-Meyers, E., *et al.* (2005). Characterization of AMN107, a selective inhibitor of native and mutant Bcr-Abl. Cancer cell *7*, 129–141.

Wisniewski, K.E., Dalton, A.J., McLachlan, C., Wen, G.Y., and Wisniewski, H.M. (1985). Alzheimer's disease in Down's syndrome: clinicopathologic studies. Neurology *35*, 957–961.

Wright, T.C., Orkin, R.W., Destrempes, M., and Kurnit, D.M. (1984). Increased adhesiveness of Down syndrome fetal fibroblasts in vitro. Proceedings of the National Academy of Sciences of the United States of America *81*, 2426–2430.

Wu, Y., Deng, Y., Zhang, S., Luo, Y., Cai, F., Zhang, Z., Zhou, W., Li, T., and Song, W. (2015). Amyloid-beta precursor protein facilitates the regulator of calcineurin 1-mediated apoptosis by downregulating proteasome subunit alpha type-5 and proteasome subunit beta type-7. Neurobiol Aging *36*, 169–177.

Yabut, O., Domogauer, J., and D'Arcangelo, G. (2010). Dyrk1A overexpression inhibits proliferation and induces premature neuronal differentiation of neural progenitor cells. J Neurosci *30*, 4004–4014.

Yamakawa, K., Huo, Y.K., Haendel, M.A., Hubert, R., Chen, X.N., Lyons, G.E., and Korenberg, J.R. (1998). DSCAM: a novel member of the immunoglobulin superfamily maps in a Down syndrome region and is involved in the development of the nervous system. Human Molecular Genetics *7*, 227–237.

Zhang, B., Freeman, M.R., and Waddell, S. (2010). Drosophila neurobiology: a laboratory manual (Cold Spring Harbor, N.Y.: Cold Spring Harbor Laboratory Press).

Zhang, Y.Q., Bailey, A.M., Matthies, H.J., Renden, R.B., Smith, M.A., Speese, S.D., Rubin, G.M., and Broadie, K. (2001). Drosophila fragile X-related gene regulates the MAP1B homolog Futsch to control synaptic structure and function. Cell *107*, 591–603.

Zhu, B., Parsons, T., Stensen, W., Mjoen Svendsen, J.S., Fugelli, A., and Hodge, J.J.L. (2022). DYRK1a Inhibitor Mediated Rescue of Drosophila Models of Alzheimer's Disease-Down Syndrome Phenotypes. Front Pharmacol *13*, 881385.

Zhu, K., Xu, Y., Liu, J., Xu, Q., and Ye, H. (2011). Down syndrome cell adhesion molecule and its functions in neural development. Neurosci Bull *27*, 45–52.

Zipursky, S.L., and Sanes, J.R. (2010). Chemoaffinity revisited: dscams, protocadherins, and neural circuit assembly. Cell *143*, 343–353.

Using Zebrafish to Study the Mechanisms That Underlie Down Syndrome

Anna J. Moyer and Summer B. Thyme

Abstract Zebrafish represent a promising vertebrate model system for the study of triplicated genes that contribute to the pathophysiology of Down syndrome. Their optical transparency, external fertilization, fecundity, and genetic accessibility make them ideal for determining how increased expression of orthologs of Hsa21 genes perturbs embryonic development. Trisomy of human chromosome 21 (Hsa21) alters the levels of hundreds of genes that can have critical developmental roles, leading to pleiotropic effects on the body. In this chapter, we describe the conservation of many organ systems affected in individuals with Down syndrome (e.g., skeletal, cardiovascular, immune, and more) and the tools available in zebrafish to assess their morphology and function. Although a limitation of zebrafish is that aneuploidy cannot be achieved with current genome-engineering technology, studying many individual Hsa21 orthologs and their combinations is immediately accessible. Looking to the future, multiple compounds found using zebrafish have entered the clinic or clinical trials. Establishing zebrafish models of relevant phenotypes will lay the necessary groundwork for large-scale drug screens to discover treatments for aspects of Down syndrome.

Keywords Zebrafish · Down syndrome · Hsa21 · Synteny · Overexpression · Neurodevelopment · Optical transparency · Transgenic line · Drug discovery · Learning · Whole-brain activity · Single-cell RNA-seq · Multi-organ disorder

1 Introduction

Down syndrome is a complex genetic disorder with pleiotropic phenotypes, many of which stem from abnormal early development. The underlying molecular mechanisms responsible for Down syndrome remain unclear, as trisomy of human

---

A. J. Moyer · S. B. Thyme (✉)
Department of Biochemistry and Molecular Biotechnology, The University of Massachusetts Chan Medical School, Worcester, MA, USA
e-mail: Summer.Thyme@umassmed.edu

B. Ye, R. H. Reeves (eds.), *Genetic Models of Down Syndrome*,
https://doi.org/10.1007/978-3-031-78611-2_6

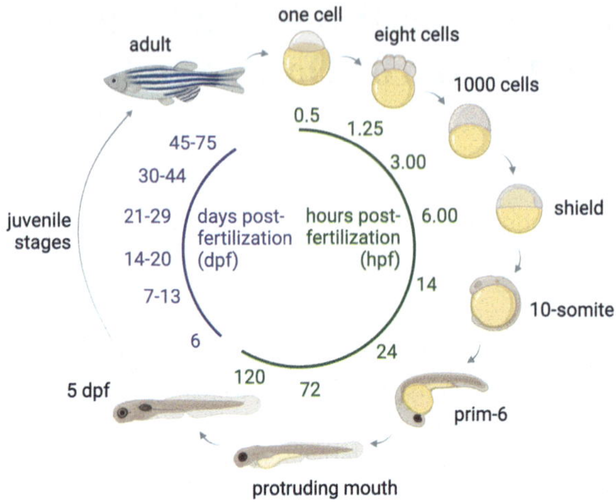

Fig. 1 Zebrafish life cycle. (Figure created with BioRender.com)

chromosome 21 (Hsa21) has direct and indirect effects on the levels of hundreds of factors that can have critical developmental roles. The zebrafish model offers a unique combination of throughput, external development, optical transparency, and conserved genetics that supports characterizing the effects of overexpressing many genes and conducting drug screens. Zebrafish develop rapidly from a single cell to a behaving animal with organs in 1 week (Fig. 1). Here, we introduce the advantages and challenges of using this model for the study of Down syndrome. The many methods and resources available to the zebrafish researcher described in this chapter afford new opportunities to decipher the molecular, cellular, and developmental underpinnings of Down syndrome and its co-occurring conditions.

2 Genetic Approaches to Study Down Syndrome in Zebrafish

Zebrafish represent a promising vertebrate model system for understanding which triplicated genes contribute to Down syndrome-associated phenotypes and for high-throughput drug screening. However, mouse models of Down syndrome have dominated biomedical research for the last 30 years, and zebrafish models need further development to reach their full translational potential. Creating useful zebrafish models will require identifying Hsa21 orthologs (through conservation of genomic position and amino acid identity/similarity), assessing functional conservation of these orthologs (including isoforms and tissue-specific/temporal gene expression patterns), and creating new models that express Hsa21 genes or their zebrafish orthologs.

2.1 Zebrafish Orthologs of Hsa21 Genes

The zebrafish genome encodes an estimated 26,000 protein-coding genes divided between 25 pairs of chromosomes (Howe et al. 2013; Lawson et al. 2020). Genome wide, 71% of human genes have zebrafish orthologs (Howe et al. 2013), and 144 of 164 Hsa21 classical protein-coding genes (excluding the KRTAP gene cluster encoding keratin-associated proteins) have at least one zebrafish ortholog (Hu et al. 2011; Gupta et al. 2016). Although conservation of synteny is lower in zebrafish than in murine models, the orthologs of Hsa21 genes are not randomly scattered among the zebrafish chromosomes. Chromosome 9 encodes 54 Hsa21 orthologs, and an additional 43, 29, and 24 orthologs are located on chromosomes 10, 1, and 15, respectively (Fig. 2a). The largest contiguous regions of Hsa21 orthologs contain up to 13 genes; for example, a 1.45 Mb region of zebrafish chromosome 1 encodes 13 Hsa21 orthologs spanning *mrpl39* to *slc5a3a* (Fig. 2b–d).

However, unequivocal assignment of human/zebrafish ortholog pairs is complicated by an ancient teleost-specific whole-genome duplication event, such that 15% of human genes have more than one zebrafish ortholog (Meyer and Schartl 1999; Howe et al. 2013). Extra copies of genes created by this duplication event, called ohnologs, are present for zebrafish orthologs of at least 35 Hsa21 genes, including *DYRK1A* (*dyrk1aa/dyrk1ab*), *APP* (*appa/appb*), and *RCAN1* (*rcan1a/rcan1b*). Orthologs of these genes are positioned in "double-conserved synteny blocks," which are regions of genes on Hsa21 that have orthologs on two different chromosomes in zebrafish (Howe et al. 2013). The largest double-conserved synteny blocks containing Hsa21 orthologs are located on zebrafish chromosomes 1/9 and 10/15 (Fig. 2c, d). For example, chromosome 1 encodes *jam2a*, *appa*, *slc5a3a*, *tmprss3b*, and *cbsa*, and chromosome 9 encodes *jam2b*, *appb*, *slc5a3b*, *tmprss3a*, and *cbsb*.

Animal models of Down syndrome typically aim to reproduce the overexpression caused by trisomy 21, and for zebrafish genes with four copies (e.g., two copies of *dyrk1aa* and two copies of *dyrk1ab*), it is not obvious whether adding one copy of a gene (e.g., one additional copy of *dyrk1aa*) or two copies of a gene (e.g., additional copies of both *dyrk1aa* and *dyrk1ab*) would more accurately reflect the pathogenesis of human trisomy. Importantly, there are multiple possible fates for duplicated genes, including retention, adaptation, and loss (Lynch and Conery 2000). The most common outcome of duplication is loss of one copy through deletion, and 80% of duplicated genes were lost within an estimated 60 million years of the teleost-specific duplication event (Inoue et al. 2015; Pasquier et al. 2017). Among duplicate genes that are retained in zebrafish, 20% show conservation of both copies with similar expression patterns (Pasquier et al. 2017). One copy can also gain new functions through neofunctionalization, or the functions of the ancestral gene can be divided between the two copies through subfunctionalization. Changes in gene expression between the two ohnologs or between the ohnologs and an unduplicated sister taxon can provide evidence of subfunctionalization and neofunctionalization (Braasch et al. 2016; Pasquier et al. 2017).

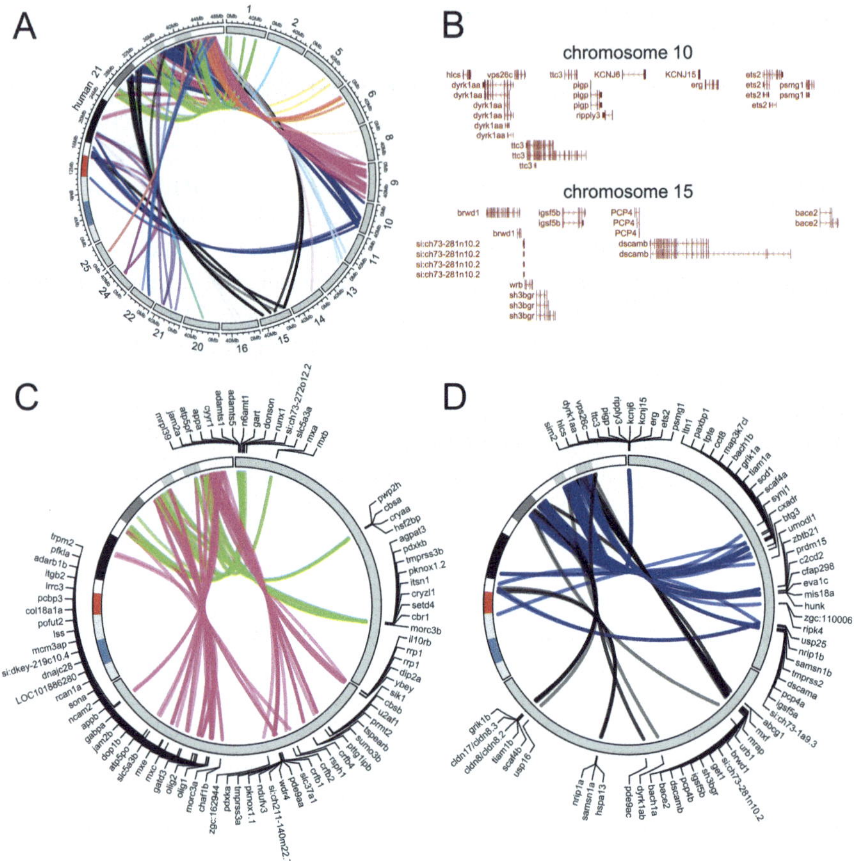

Fig. 2 Chromosomal positions of Hsa21 orthologs in the zebrafish genome. (**a**) Circos plot linking Hsa21 genes (labeled human 21) to their zebrafish orthologs. Each pair of orthologs is connected by one line, and orthologs are colored by zebrafish chromosome. Ortholog predictions were obtained using the Drosophila RNAi Screening Center Integrative Ortholog Prediction Tool, DIOPT (Hu et al. 2011). Circos plots were created using the R package *circlize* (Gu et al. 2014). (**b**) Ensembl gene predictions for examples of Hsa21 orthologs that are located in contiguous regions on zebrafish chromosomes 10 and 15. Gene models were visualized using the UCSC Genome Browser (GRCz11). (**c**) Circos plot linking Hsa21 genes to their orthologs on zebrafish chromosomes 1 (green) and 9 (pink). (**d**) Circos plot linking Hsa21 genes to their orthologs on zebrafish chromosomes 10 (blue) and 15 (black)

We recently profiled single-cell expression patterns in the developing zebrafish forebrain (Pandey et al. 2023), which is a tissue relevant to the hippocampal and cortical phenotypes observed in people with Down syndrome. Expression patterns of zebrafish orthologs of Hsa21 genes indicate both retention and subfunctionalization/neofunctionalization. For example, both *HMGN1* orthologs *si:ch73-281n10.2* and *si:ch73-1a9.3* show ubiquitous expression throughout the telencephalon (Fig. 3a). In contrast, we observed high expression of *pcp4a* and low expression of

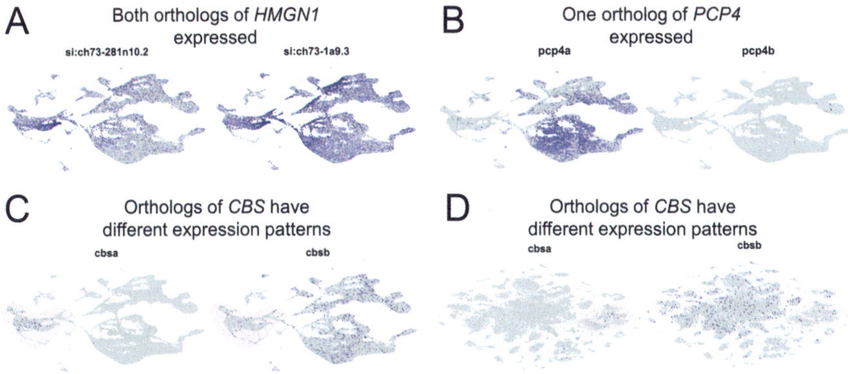

Fig. 3 Expression of orthologs of *HMGN1*, *PCP4*, and *CBS* in zebrafish brain. (**a**–**c**) Uniform Manifold Approximation and Projection (UMAP) representation of 6-day post-fertilization (dpf), 15 dpf, and adult zebrafish telencephalon single-cell RNA-seq data. Expression of orthologs of *HMGN1* (**a**), *PCP4* (**b**), and *CBS* (**c**) is indicated in purple. (**d**) *t*-distributed Stochastic Neighbor Embedding (*t*-SNE) representation of 5 dpf zebrafish brain and eye single-cell RNA-seq data (Raj et al. 2020). Expression of orthologs of *CBS* is indicated in purple. Pink highlights represent populations of neural progenitors

pcp4b (Fig. 3b). In both forebrain (Fig. 3c) and larval whole brain (Fig. 3d), *cbsa* is expressed in a small subset of neural progenitors, while *cbsb* is more broadly expressed in neural progenitors and differentiated neurons (Raj et al. 2020). The variety of expression patterns among duplicated genes highlights the importance of considering individual gene expression when using zebrafish to study Down syndrome. Orthologs of *HMGN1* may need to be expressed at a higher level to reflect the trisomic expression observed in Down syndrome, whereas overexpressing *pcp4a* alone is an appropriate choice for interrogating forebrain phenotypes. Comparison to tissue-specific gene expression in mouse or human is also a first step for determining whether orthologous proteins play functionally equivalent roles in the cell.

3 Methods to Model Down Syndrome

After identifying zebrafish orthologs of Hsa21 genes and interrogating their native expression patterns, overexpression approaches can help to characterize their functions and possible contributions to Down syndrome-associated phenotypes. Unlike murine models, the low conservation of synteny between human and zebrafish prevents simultaneous overexpression of many Hsa21 orthologs in the context of trisomy. However, for single-gene studies, the zebrafish model offers straightforward techniques for transient overexpression through embryonic microinjection of protein or mRNA and for stable overexpression with rapid and efficient transgenesis.

3.1 *Protein or mRNA Injection*

Zebrafish embryos are fertilized and develop externally, and mRNA injected into the yolk or blastomeres is translated into protein, inducing ectopic expression at the earliest stages of embryonic development (Fig. 4a) (Hyatt and Ekker 1999). By contrast, microinjection of DNA is highly mosaic and results in overexpression only after zygotic genome activation. Several factors influence the distribution and stability of injected mRNA. Embryonic stage affected the dispersion of fluorescence

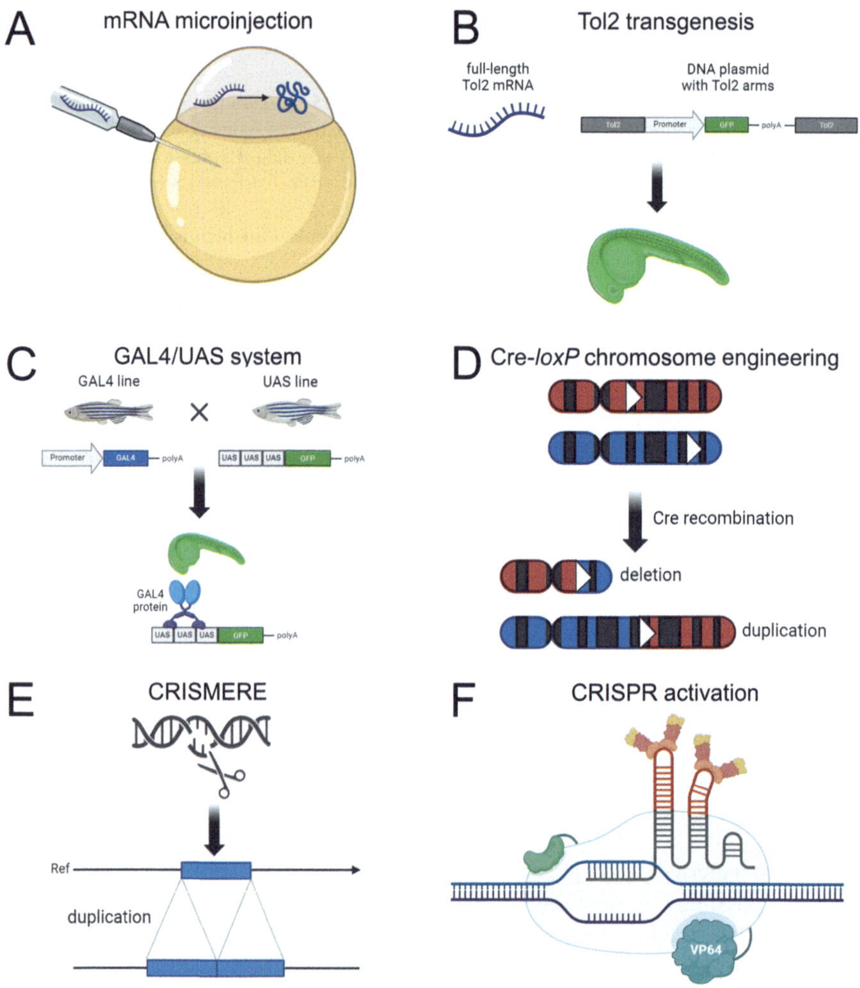

Fig. 4 Methods for creating overexpression models in zebrafish. Techniques discussed in the text include mRNA injection (**a**), single-gene transgenesis using *Tol2* (**b**), the GAL4/UAS system (**c**), Cre-*loxP* recombination (**d**), CRISMERE Cas9-induced rearrangements (**e**), and CRISPR activation (**f**). (Figure created with BioRender.com)

following microinjection of mRNA encoding GFP, and the most uniform distribution of fluorescence was observed in embryos injected at the one-cell stage (Hyatt and Ekker 1999). Although mRNA stability depends on artificial untranslated regions (UTRs), such as those encoding the SV40 poly(A) signal, injected mRNA and endogenous mRNA have similar half-lives in vivo (Fink et al. 2006).

mRNA microinjection offers a rapid and high-throughput method for overexpressing cDNAs at the earliest stages of embryonic development, but it has significant disadvantages compared to other overexpression methods. For most genes, the injected mRNA will be mis-expressed at an inappropriate developmental stage and in cell types where it would normally not be present. This method is particularly problematic for studying the brain because the overexpressed mRNA typically does not persist past day 3 of development (Kardash 2021), and many behavioral and brain imaging studies are conducted at 5 dpf or later. Additionally, the expression level will often be far higher than the 1.5× desired for the study of Down syndrome, and quantitative RT-PCR would be required to optimize the dose. Toxicity from too much mRNA regardless of the gene can produce non-specific phenotypes, which commonly include bent tails and gastrulation defects (Hyatt and Ekker 1999). Ensuring that phenotypes are dose dependent and injecting mRNAs with frameshift mutations as negative controls are two methods for distinguishing between mRNA toxicity and protein-specific overexpression phenotypes.

One of the only zebrafish studies designed to investigate dosage effects in Down syndrome exploited the economy of embryonic microinjection systematically to overexpress mRNAs encoding 163 Hsa21 genes (Edie et al. 2018). Overexpression of eleven mRNAs produced consistent morphological phenotypes, including U-shaped somites (*SOD1*, *RWDD2B*, *RRP1*, *PCBP3*, *YBEY*, and *LINC00313*), cyclopia (LINC00313 and *POFUT2*), craniofacial dysmorphology (*CBR3*), pigment abnormalities (*CCT8*), pericardial edema (*JAM2*), and lethality (*ERG*). The involvement of *JAM2* in heart development was also recapitulated in mice (Li et al. 2016). Of 24 phenotypes uncovered in an initial round, only 11 were similar in a second injection, highlighting the challenges of the mRNA overexpression approach. A second study used the same Hsa21 cDNA library to identify *ATP5PO* as a likely modulator of Hirschsprung disease, which has an increased prevalence in people with Down syndrome (Kuil et al. 2023). Injection of *ATP5PO* mRNA reduced the number of enteric neurons in 4 dpf zebrafish, possibly through impaired ATP production from a paradoxically reduced amount of ATP5PO protein, whereas injection of *RCAN1* mRNA increased the number of enteric neurons. Phenotypes resulting from mRNA injection of other single Hsa21 orthologs have also been reported, including impaired development and possibly motor axon growth from *chodl* (Zhong et al. 2012), excess primary motor neurons and oligodendrocytes from *olig2* (Park et al. 2002), and a posteriorized (smaller head and prominent tail) phenotype from *APP* (Joshi et al. 2009).

Microinjection of purified protein is an alternative strategy for overexpressing Hsa21-encoded proteins or their zebrafish orthologs. For example, injection of recombinant HMGN1 or HMGN2 into *Xenopus laevis* embryos caused body axis malformations, cyclopia, and microcephaly (Körner et al. 2003). Protein injection is not limited to embryos, and one study reported that injection of the APP cleavage product P3 into the hindbrain ventricle of 5 dpf zebrafish larvae affected the length of nighttime sleep bouts (Özcan et al. 2022). Similar to microinjection of mRNA, injection of proteins lacks tissue specificity and is constrained by their half-life. Because injected molecules are degraded over time and diluted by cell division, injection results in low and mosaic overexpression later in larval development.

3.2 *Single-Gene Transgenesis*

In contrast to mRNA injection, stable transgenesis provides the opportunity to monitor the effects of overexpression across the lifetime. Traditional transgenesis by injecting DNA plasmids results in highly mosaic expression, and only ~5% of injected animals transmit the integrated DNA through the germline (Kawakami 2007). The *Tol2* transposase system was first identified in medaka (Kawakami and Shima 1999) and is now the most commonly used method for increasing the frequency of random insertion of plasmid DNA into the zebrafish genome. *Tol2* encodes an autonomous transposase that catalyzes the transposition of non-autonomous *Tol2* elements without creating rearrangements or deletions. To generate fish with germline transmission using the *Tol2* system, a plasmid containing the insertion of interest flanked by *Tol2* terminal inverted repeats is co-injected with *Tol2* mRNA encoding the full-length transposase (Fig. 4b) (Kawakami et al. 2004). When outcrossed to wild-type fish, up to 70% of injected fish produce transgenic F1 offspring (Kawakami 2007). The discovery of *Tol2* transposition represented a radical technological leap over traditional transgenesis and permitted large-scale gene trapping and screening of candidate enhancer elements (Fisher et al. 2006; Kawakami et al. 2010). However, random insertion of plasmid DNA has several technical drawbacks. Multiple *Tol2*-mediated insertions in single F1 offspring necessitate repeated outcrossing to isolate a single insertion, positional effects may silence the transgene, and random insertion can interrupt genes required for normal zebrafish development. In response to these limitations, recent efforts have identified alternative strategies for targeted integration into safe-harbor sites (Lalonde et al. 2024).

For overexpression studies, *Tol2* plasmids typically contain a promoter, a cDNA, UTR elements required for expression, and an optional fluorescent marker for tracking transgenesis. The level of overexpression and how closely the desired 1.5× expression is achieved are dependent on the selected promoter and UTR elements. The total length of these elements affects the efficiency of both cloning and *Tol2* transposition, but insertions of 11 kb are feasible (Urasaki et al. 2006). The selected promoter can drive ubiquitous (e.g., *ubb*), inducible (e.g., *hsp70l*), tissue specific, or endogenous expression of the target transgene. For example, to understand how overexpression of *DYRK1A* affects neurodevelopment, Buchberger et al. used *Tol2* transgenesis to knock in human *DYRK1A* under the control of a Purkinje cell enhancer and the minimal E1b promoter (Buchberger et al. 2021). Overexpression of *DYRK1A* affected cerebellar development and swimming behavior in larval zebrafish. *Tol2* transgenesis can also be used to overexpress mutant transcripts. A new Alzheimer's disease model was created by using *Tol2* transgenesis to knock-in human *APP* with the Swedish mutation under control of the zebrafish *appb* promoter with behavioral and histological consequences (Pu et al. 2017).

3.3 GAL4/UAS

Creating single gene transgenics requires making a new zebrafish line for each promoter/cDNA pair. The GAL4/UAS system, which was pioneered in *Drosophila* (see *Drosophila* chapter) and adapted for use in zebrafish, is a modular approach for cell-type-specific cDNA overexpression without generating a new transgenic line for each combination of cDNA and tissue (Fischer et al. 1988; Scheer and Campos-Ortega 1999). GAL4 is a yeast transcription factor that binds to upstream activation sequence (UAS) repeats to activate transcription (Fig. 4c). GAL4 "driver" lines are generated by placing the GAL4 transcription factor sequence downstream of a tissue-specific promoter, such as the pan-neuronal HuC promoter (Kim et al. 2014). UAS lines are generated by placing the cDNA of interest downstream of a variable number of UAS repeats. When GAL4 and UAS transgenic lines are crossed together, double transgenic progeny overexpress the cDNA only in the tissues where GAL4 is expressed. Technical challenges with employing the GAL4/UAS system, including toxicity of GAL4 and methylation of UAS repeats, have prompted adaptation of both GAL4 and UAS sequences. For example, a modified transcriptional activator, Gal4ff, is less cytotoxic than GAL4 (Asakawa and Kawakami 2009; Zhang et al. 2019). Repetitive UAS elements undergo transgenerational epigenetic silencing, resulting in variegated expression, but a non-repetitive 5× UAS sequence is less likely to undergo silencing than the standard 5× UAS sequence (Akitake et al. 2011; Zhang et al. 2019). The Q system (QF/QUAS) is a similar alternative that is also less susceptible to silencing (Ghosh and Halpern 2016). The GAL4/UAS system was used to express a mutant version of the Hsa21 gene *SOD1* in zebrafish oligodendrocytes, which caused degeneration of both oligodendrocytes and spinal motor neurons (Kim et al. 2019). This method can also be used to ablate populations of cells by overexpressing botulinum toxin or the enzyme nitroreductase (Curado et al. 2008; Sternberg et al. 2016; Sharrock et al. 2022). One study used this approach to identify the subtypes of cerebellar neurons involved in fear conditioning (Koyama et al. 2021).

3.4 BAC Transgenesis

Traditional plasmid transgenesis requires knowledge of the regulatory sequences that drive endogenous expression and selection of a single isoform to overexpress. For some genes and tissues, the zebrafish promoters and enhancers that activate native expression are unknown. Other genes possess tissue-specific alternative splicing or encode cDNAs that approach the size limit of traditional cloning and transgenesis. Bacterial artificial chromosome (BAC) transgenesis offers an alternative approach for knocking in very large genomic fragments. Like plasmid transgenesis, inclusion of *Tol2* ends and co-injection with full-length *Tol2* transposase mRNA significantly increases the frequency of germline transmission, and this method has been used to knock in single copies of BACs up to hundreds of kilobases long (Suster et al. 2011). BAC transgenesis was used to knock in a 21-kb region containing the zebrafish Hsa21 ortholog *sod1* and its regulatory sequences (Ramesh et al. 2010). This method is advantageous for genes without annotated enhancer and promoter sequences, and a single-copy insertion can achieve the desired 1.5× upregulation. The regulatory elements that drive expression of the zebrafish Hsa21 ortholog *olig2* are still unknown, and BAC transgenesis was used to create an *olig2:*GFP reporter line (Shin et al. 2003). However, Quillien et al. recently reported that tissue-specific ATAC-seq data can be used to classify enhancer and promoter elements, and this method's ability to identify short regulatory regions may supplant BAC transgenesis for many applications (Quillien et al. 2017).

3.5 Alternative Approaches

Cre-*loxP* chromosome engineering is the method of choice for creating defined segmental duplications in murine models, and this approach is possible in zebrafish. Zebrafish embryonic stem (ES) cells are not used for genetic manipulation, so knocking in *loxP* sites and inducing recombination must occur in vivo. Generating segmental duplications in zebrafish would require knocking in two *loxP* sites in *trans* upstream and downstream of the region of interest, then expressing Cre to induce reciprocal translocations (Fig. 4d). Although the Cre-*loxP* system has not been used to create duplication lines in zebrafish, it has been used in conjunction with Cas12a to induce conditional deletions (Shin et al. 2023). The number of Hsa21 orthologs in a segmental duplication is limited by the low conservation of synteny between human and zebrafish. Models created using this method could contain up to ~13 triplicated Hsa21 orthologs (Fig. 2a).

Alternative approaches for overexpressing Hsa21 orthologs are technically possible in zebrafish but have not yet met widespread use. CRISMERE is another approach that has been used to generate large duplications in murine models, where Cas9 cleavage upstream and downstream of the region of interest introduces genomic rearrangements (Fig. 4e) (Birling et al. 2017). In zebrafish, CRISPR/Cas9 editing unintentionally produces rare large-scale insertions, indicating that the CRISMERE strategy could be used to generate duplications in zebrafish (Höijer et al. 2022). A final emerging technology for overexpressing genes in zebrafish is CRISPR activation (CRISPRa). A nuclease-dead Cas9 is fused to a transcriptional activator, such as VP64 or VPR, and directed to the promoter of a target gene (Fig. 4f). CRISPR activation has been used to upregulate several target genes in zebrafish (Long et al. 2015; Weuring et al. 2021), and a transgenic dCas9-VP64 zebrafish line has been established (Pelea et al. 2023). Light-modulated control of activation in zebrafish larvae has also been reported (Putri and Chen 2018).

4 Zebrafish Phenotyping Approaches

Trisomy of Hsa21 affects multiple organs, indicating disruptions to early processes in development. As zebrafish are vertebrate organisms, they have counterparts to many human physiological and anatomical systems (Fig. 5). Their optical transparency and external fertilization make zebrafish an ideal model for monitoring development and function in health and under conditions of genetic perturbation such as Down syndrome. The following sections will summarize organ system conservation and phenotyping methods for many conditions occurring in Down syndrome that are typically assessed in mammalian models.

4.1 *Skeletal System*

Craniofacial dysmorphology, lower bone density, and short stature are distinctive features of Down syndrome (Fink et al. 1975; Carfì et al. 2017; Thomas and Roper 2021). Zebrafish share many bone types, developmental processes, and cell types with mammals. These similarities include a vertebral column, complex cranial skeleton, appendicular skeleton, and endochondral ossification, where cartilage is replaced by bone during development (Mork and Crump 2015; Tonelli et al. 2020; Le Pabic et al. 2022). Although zebrafish lack long limb bones, aspects of the growth zones of radial fin bones and underlying signaling pathways are conserved, consistent with both bone types arising from an ancestral homologous structure (Le Pabic et al. 2022). There is a rich literature on craniofacial development in zebrafish in particular; several forward genetics screens uncovered numerous genetic loci that impacted patterning of cartilage and bone in larvae, many of which mapped to genes with conserved functions in mammals [reviewed in (Mork and Crump 2015)]. The anatomy of the craniofacial skeleton has been comprehensively detailed across the lifespan, and there is even an online interactive atlas known as Fish Face (Eames et al. 2013). Although other areas of skeletal research are less prolific than

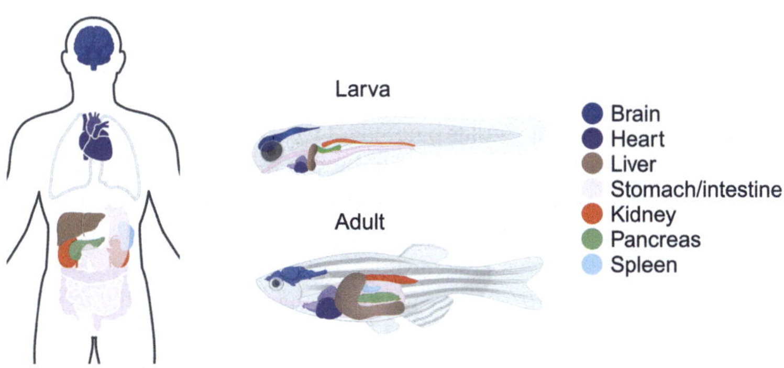

Fig. 5 Anatomical comparison of zebrafish and human organ systems. (Created with BioRender.com)

craniofacial developmental biology, zebrafish has also been proposed as a viable model for osteoporosis (Bergen et al. 2019; Lin et al. 2022; Kague et al. 2023), which is more prevalent in people with Down syndrome than in the general population (Carfì et al. 2017).

Both newer and more classical techniques have yielded insight into skeletal morphogenesis in zebrafish. In larvae, live imaging of development and related signaling processes in real time is possible with transgenic reporters [reviewed in (Mork and Crump 2015)]. This type of approach is complemented by modern methods such as single-cell genomics. A recent study profiled single-cell transcriptomes and chromatin accessibility in cranial neural crest-derived cells across the zebrafish lifespan, nominating lineage-specific transcription factors that may participate in remodeling of chromatin accessibility as diverse cell types emerge and differentiate (Fabian et al. 2022). In addition to basic visual examination of larval size and shape, high-throughput assessment of craniofacial phenotypes can be accomplished with Alcian blue staining of cartilage, as was used in the large forward genetic screens that kickstarted the field of zebrafish developmental biology (Neuhauss et al. 1996; Piotrowski et al. 1996). As bones become mineralized, beginning 3–4 dpf, Alizarin red staining complements cartilage monitoring, and this stain is even compatible with live imaging (Bensimon-Brito et al. 2016). At adult stages, micro computed tomography (microCT, micro-CT, or μCT) uses X-ray technology for whole-body phenotyping [reviewed in (Tonelli et al. 2020), example in (Gistelinck et al. 2018)], and calibration microCT can provide absolute measures of hydroxyapatite to quantify bone density (Liao et al. 2023). A semiautomated quantification of adult craniofacial structure has also been recently developed using image registration to process microCT data (Diamond et al. 2022).

While literature describing zebrafish models of Hsa21 ortholog overexpression are limited, these methods for characterizing skeletal morphology have been extensively used to describe craniofacial phenotypes in similar neurodevelopmental disorders. A recent zebrafish study of Fragile X syndrome (FXS) revealed facial cartilage changes in zebrafish *fmr1* mutants analogous to facial elongation observed in people with FXS (Zhu et al. 2023). Careful quantification based on multiple angles and lengths across developmental stages was necessary to uncover this phenotype, and this work is a good example to follow for similar larval assessments of Hsa21 genes. Although zebrafish Hsa21 orthologs have been implicated in craniofacial development when expression is reduced, such as deformity of cartilage structures observed in *urb1* mutants (Shan et al. 2023), it remains unclear whether overexpression would have a similar outcome. Overexpression of *Dyrk1a* has been clearly implicated in craniofacial dysmorphology in mice (Redhead et al. 2023) and may be one of the key drivers of this phenotype in humans, but existing zebrafish overexpression and mutant lines have not been screened for this phenotype. However, a mutant with lower jaw cartilage defects found by a forward genetic screen was mapped to the *wdr68* gene, which produces a protein that directly interacts with Dyrk1a (Nissen et al. 2006). This protein complex is conserved across eukaryotes and disruption of its associated pathway likely explains some of the characteristic skeletal features of Down syndrome.

4.2 Hematopoiesis and Leukemia

Trisomy 21 perturbs fetal hematopoiesis, causing changes in stem cell proliferation and expansion of megakaryocyte-erythroid progenitor cells (Roy et al. 2012). These developmental shifts result in an at least 40-fold increase in leukemia in people with Down syndrome. In zebrafish, the anatomical sites of hematopoiesis differ from mammals, but the underlying molecular mechanisms are conserved [reviewed in (Paik and Zon 2010; Gore et al. 2018)]. During mammalian development, the first "primitive" wave of hematopoiesis occurs in the yolk sac primarily to yield erythrocytes (red blood cells) that support the developing embryo. In zebrafish, this primitive wave occurs in the intermediate cell mass, with erythrocytes entering circulation at approximately 24 h post-fertilization (hpf). Subsequently, the definitive second wave occurs in the aorta-gonad-mesonephric region in mammals and the ventral wall of the dorsal aorta in zebrafish. The hematopoietic stem cells (HSCs) produced in these areas seed the bone marrow, thymus, and kidney marrow of zebrafish. In mammals, HSCs migrate to the liver, proliferate, and seed the bone marrow. The corresponding niche in zebrafish is in the kidney marrow, which produces mammalian-equivalent cell types, including erythroid, myeloid, lymphoid, and thromboid lineages. Across all developmental hematopoiesis waves, genetic programs and transcription factors are almost entirely shared between zebrafish and mammals.

The main technical advantage of using zebrafish to study hematopoiesis is optical transparency, as the field has generated numerous transgenic lines that fluorescently label HSC-derived lineages for live imaging [list available in (Gore et al. 2018)]. In 2010, back-to-back studies in Nature used zebrafish to image the birth of HSCs and resolve a longstanding question in the field regarding their origin (Bertrand et al. 2010; Kissa and Herbomel 2010). Beyond larvae, the development of the transparent *casper* zebrafish line, lacking all iridophores and melanocytes, has enabled imaging studies in adulthood (White et al. 2008). In vivo monitoring of the growth and engraftment of HSCs and tumor cells at single-cell resolution is possible with this model. Transplantation and engraftment of hematopoietic donor cells in zebrafish following irradiation to ablate the host niche is an established method in the field (Traver et al. 2004). Monitoring the development and clonal diversity of the hematopoietic lineage is possible with imaging using a fluorescent, multicolor labeling method (Bornhorst et al. 2024) or genetically, using CRISPR-based lineage tracing methods (Alemany et al. 2018). Zebrafish models of leukemia and pre-leukemia have been induced in a myriad of ways [reviewed in (Baeten and de Jong 2018; Yi et al. 2022)]. Thus, it is feasible to study this condition in zebrafish and determine Hsa21 orthologs that influence its development and progression.

Among Hsa21 genes, *RUNX1* and *ERG* have roles in hematopoiesis that are conserved in zebrafish. Both have been strongly implicated in disrupted hematopoiesis and promotion of leukemia in Down syndrome (Stankiewicz and Crispino 2009; Roy et al. 2012; Gialesaki et al. 2023). In zebrafish, the first attempt at a model of acute myelogenous leukemia (AML) involved overexpression of the oncogenic human *RUNX1-CBF2T1* (*AML1-ETO*) fusion protein (Kalev-Zylinska et al. 2002). Altered hematopoiesis, as also seen in a similar mammalian model, was observed in this line as well as a later model that expressed the oncogene in a stable transgenic line under control of the inducible heat shock promoter (Yeh et al. 2008). During typical development, *runx1* is essential across species for the formation of HSCs in the definitive wave of hematopoiesis (Gore et al. 2018). In zebrafish, *erg* seems to be mainly involved in endothelial cell development and angiogenesis (Liu and Patient 2008); however, no mutant or overexpression lines have been characterized and the prior mammalian research indicates that ERG has a more complex role in HSC maintenance (Taoudi et al. 2011). A suspected role of *ERG* in the human condition is supported by a cellular study indicating that its increase induces megakaryocyte expansion, which may be related to the increased megakaryoblasts in Down syndrome (Stankiewicz and Crispino 2009). Overall, although more work needs to be done to study individual candidate Hsa21 orthologs in zebrafish, numerous tools make zebrafish an ideal model to discover changes to the development and diversity of HSCs across the lifespan.

4.3 Immune System

Heightened susceptibility to severe infections, including sepsis, and an increased incidence of autoimmune disorders have been reported in individuals with Down syndrome. Persistent cytokine elevation, hundreds of autoantibodies, and hyperactivated T cells are commonly observed in Down syndrome, likely driven by increased JAK-STAT signaling (Malle et al. 2023). The immune system of zebrafish is functionally like mammals, as they have both innate and adaptive branches and similar cell types (e.g., neutrophils, T cells, B cells, and so on) [reviewed in (Meeker and Trede 2008; Renshaw and Trede 2012)]. One difference is that the adaptive immune system only reaches full competence between 4 and 6 weeks of age. Anatomically, zebrafish have a thymus and, as in hematopoiesis, the kidney marrow is the functional equivalent of mammalian bone marrow. Zebrafish lack lymph nodes, but they do have a lymphatic vasculature and a thoracic duct at larval stages (Jung et al. 2017). There is some molecular divergence as well, such as different immunoglobulin subtypes (Danilova et al. 2005; Ji et al. 2021). Practical disadvantages of zebrafish for the study of the immune system include a lack of protocols for establishing cell culture as well as few commercially available antibodies.

Hsa21 encodes four of the six human interferon genes, and they are generally thought to drive immune system hyperactivation in Down syndrome (Malle et al. 2022; Waugh et al. 2023). However, as with all Down syndrome phenotypes, it is likely that multiple genes contribute. For example, the Hsa21 gene *AIRE* encodes a transcription factor critical for removal of self-reactive T-cells; it is paradoxically downregulated in Hsa21 trisomy and may be responsible for thyroid dysfunction (Lima et al. 2011; Giménez-Barcons et al. 2014). The triplicated interferon genes are known to stimulate JAK-STAT signaling (Schindler et al. 2007; Shemesh et al.

2021), a signaling cascade that is also present in teleosts (Robertsen 2006). Zebrafish have an interferon response and orthologs of the trisomic genes (Levraud et al. 2019). While Jak-Stat signaling is also conserved, most of the research involving this pathway has focused on regeneration: following optic nerve resection, the Jak-Stat pathway is required for retinal ganglion cell survival and optic nerve regeneration (Elsaeidi et al. 2014; Chen et al. 2022b). Zebrafish can regenerate many tissues after injury, unlike mammals, and differences in their immune systems likely contribute to this ability.

Both infection response and autoimmune conditions can be studied in zebrafish. However, although there is a conservation of the systems and genes, the specific autoimmune conditions present in Down syndrome do not yet have comparable zebrafish models. Such conditions include alopecia, celiac disease, type I diabetes, and thyroiditis. Although there are zebrafish models of type 1 diabetes, they are typically generated by β-cell ablation through either chemical or genetic means [reviewed in (Zang et al. 2018)]. Demonstrating the power of zebrafish for large-scale chemical screens, molecules that increase pancreatic β-cell mass were discovered by imaging over 500,000 zebrafish larvae (Wang et al. 2015). Thus, even without modeling the specific autoimmune aspects of conditions such as diabetes, treatment options can still be accelerated through zebrafish studies. Similarly, although zebrafish are not infected with the same pathogens as mammals, they are an exceptionally powerful model for characterizing host–pathogen interactions in real-time with live imaging [reviewed in (Torraca et al. 2014)].

4.4 Heart

Nearly half of individuals with Down syndrome are born with a structural heart defect (Freeman et al. 2008). Common types of congenital heart disease observed in infants with trisomy 21 are atrial septal defects (ASDs), ventricular septal defects (VSDs), and atrioventricular septal defects (AVSDs). Zebrafish stands out as an exceptionally useful model for the study of genes that impact early heart development, as embryos can obtain oxygen through passive diffusion and survive with no circulatory system for several days (Stainier et al. 1996). The main difference between the zebrafish and mammalian heart is anatomical, as zebrafish have only one atrium and one ventricle connected to a single circulatory system. The heart connects to circulation through the bulbus arteriosus, which maintains continuous flow to the gills where blood becomes oxygenated. Despite the difference in chamber number, the cellular and molecular mechanisms of heart development are conserved [reviewed in (Staudt and Stainier 2012)]. The stages of development are similar, including the formation of a heart tube from mesodermal cells, the looping of the heart tube, and the formation of valves and chambers.

Studies in humans and animal models have nominated several genes that may contribute to heart defects in Down syndrome. Individuals with partial trisomy suggest contributions of an 840 kb region of Hsa21 near the distal tip (Barlow et al.

2001). This region includes *DSCAM*, an Hsa21 gene implicated in heart defects by a study using *Drosophila* and mice, which also highlighted a synergistic interaction with the Hsa21 gene *COLA2* (Grossman et al. 2011). Most recently, an increased dose of *Dyrk1a* was identified as necessary, although insufficient, to cause heart septation defects in a mouse DS model containing an engineered duplication of a large Hsa21 syntenic region (Lana-Elola et al. 2024). Based on a mouse mapping panel, this study postulates that one of six genes (*Mx2*, *Tmprss2*, *Ripk4*, *Prdm15*, *C2cd2*, and *Zbtb21*) is likely the second causative gene. Thus far, none of these genes have been studied in zebrafish in the context of heart development. It is also possible that a regulatory element controlling the expression of a gene adjacent to these six could be impacted. The necessity of genes in the areas defined by the mapping also does not discount the possibility of additional heart development modifiers across the entirety of Hsa21 and synergistic interactions with genes on other chromosomes. A zebrafish overexpression screen of Hsa21 genes uncovered pericardial edema with increased *JAM2* levels (Edie et al. 2018). *Jam2* was found to potentiate the effects of a heterozygous null *Creld1* allele (Li et al. 2016), a gene associated with AVSD in the normal population that increases heart defect frequency in Down syndrome mouse models (Robinson et al. 2003).

Outside of Down syndrome, zebrafish have been used extensively to characterize the genetic underpinnings of congenital heart defects [examples reviewed in (Brown et al. 2016; Martin and Waxman 2021)]. Mutations in the *NR2F2* transcription factor can cause AVSDs, and loss of the functionally equivalent zebrafish gene (*nr2f1a*) results in reduced numbers of atrial cardiomyocytes and smaller atria, also observed in mouse *Nr2f2* knockouts (Al Turki et al. 2014; Duong et al. 2018). Across other diverse genes, zebrafish have been used to corroborate human genetics studies of septal defects and discover disrupted developmental mechanisms (Smith et al. 2009; Glenn et al. 2012; Wang et al. 2018). Endocardial cushion malformation was shared by two of these mutants, indicating that it may be a common phenotype observed in zebrafish models of AVSDs.

4.5 Muscle

Motor deficits, including hypotonia and postural control, in children with Down syndrome manifest in delays in gross and fine motor skills (Malak et al. 2015). Low oral tone can also impact feeding and speech intelligibility. Although muscular differences are near ubiquitous in humans and prevalent in mouse models (Glass and Connor 2016; Glass et al. 2024), the involved genes are relatively understudied. Transcriptional profiling of gastrocnemius muscles of the Ts3Yah mouse model, which contains three copies of a 9.4 Mb region from *Hspa13* to *App*, identified dysregulated energy metabolism as a possible contributor to motor phenotypes (Brault et al. 2015). While genes *Gabpa* and *Nrip1* were suggested as candidate drivers of the phenotype, no additional mechanistic characterization has been published. It is reasonable to postulate these genes impact metabolism and energy balance, as they are involved in mitochondrial biogenesis. However, motor neuron function and muscle innervation are important contributors to muscle tone. Work in zebrafish revealed that *chodl*, whose ortholog is duplicated in the Ts3Yah mouse model, is involved in development and possibly motor axon growth (Zhong et al. 2012). In zebrafish larvae, the muscle can be assessed visually using birefringence, in addition to typical imaging with transgenic lines and histology, and functionally with behavioral assays [reviewed in (Karuppasamy et al. 2024)]. Beyond understanding muscle dysfunction in Down syndrome, motor impairments are also an important consideration when interpreting behavioral data from assays assessing cognitive phenotypes.

4.6 Enteric Nervous System and Hirschsprung Disease

While Hirschsprung disease is relatively rare, occurring in 2–3% of people with Down syndrome, the risk of this condition is increased >40-fold compared to the general population (Friedmacher and Puri 2013). The development of the neural crest-derived enteric neurons, absent in the distal intestine in Hirschsprung disease, has been an active area of zebrafish research for over 20 years [reviewed in (Uribe 2024)]. A core gene regulatory network driving the formation of the enteric nervous system (ENS) is conserved across vertebrates, including zebrafish. Coding mutations in the gene *RET*, encoding a tyrosine kinase central to the ENS development, cause 50% of familial Hirschsprung disease. In zebrafish, *ret* homozygous mutants lack all intestinal neurons and heterozygosity leads to neuron loss in the distal colon due to reduced progenitor migration rates (Heanue et al. 2016). Intestinal motility and gut peristalsis were also impaired in the heterozygous mutants. The transparency of zebrafish facilitates multiple methods for detailed monitoring of motility and gut transit time in live animals [reviewed in (Kuil et al. 2020)]. A zebrafish study found fewer enteric neurons when *ATP5PO* was overexpressed, and the phenotype significantly worsened in the *ret* heterozygous background (Kuil et al. 2023). An epistatic interaction of these Hsa21 genes and underlying genetic vulnerabilities that impact levels of other critical regulators such as *RET* could explain why only a subset of people with Down syndrome develop Hirschsprung disease. New regulators of ENS development have also recently been discovered through G0 CRISPANT screening—analysis of embryos injected with CRISPR/Cas9 and guides—in zebrafish and scRNA-seq (Davidson et al. 2021; Moreno-Campos et al. 2023).

4.7 Hearing

Hearing loss impacts 30–80% of children with Down syndrome (Shott et al. 2001; Raut et al. 2011). Fluid accumulation (otitis media) is common and related to narrow ear canals (Austeng et al. 2013; Chin et al. 2014). In addition to the middle ear, where fluid buildup and infection occur, inner ear structural abnormalities (e.g., malformed semicircular canals) have been reported (Intrapiromkul et al. 2012). Hearing loss has been described in Down syndrome mouse models (Chen et al. 2022a), as has inflammation of the middle ear (Han et al. 2009). Zebrafish are a particularly attractive model for studying the inner ear because of its accessibility and conservation (Abbas and Whitfield 2010; Shi et al. 2023; Baeza-Loya and Raible 2023). As in mammals, zebrafish have three semicircular canals, an utricle, and saccule. The main difference between the species is the absence of a cochlea, a specific sound sensing organ, and that sound is detected by multiple structures in the zebrafish ear including one not in mammals known as the lagena (Lau and Vasconcelos 2023). While zebrafish lack an obvious middle ear, there is strong evidence for homology between bones in the fish jaw with the mammalian middle ear ossicles (Mork and Crump 2015; Teng et al. 2017). From a functional perspective, simple behavioral tests can reveal vestibular abnormalities through altered swimming patterns or reduced auditory responsivity to stimulation. Auditory processing in the central nervous system could also be impacted in Down syndrome, and the transparency of the zebrafish brain makes it an ideal model for understanding the circuitry that transforms sensory input into motor output (Privat et al. 2019; Favre-Bulle et al. 2020).

4.8 Vision

Multiple ophthalmic conditions are more prevalent in individuals with Down syndrome than in the general population (Sun and Kraus 2023). Eye alignment issues (e.g., strabismus or crossed eyes) and refractive errors are among the most common. The visual system of zebrafish is well characterized and has similar morphology and development to the human eye (Gestri et al. 2012; Chhetri et al. 2014; Richardson et al. 2017). Unlike mice, which have rod-dominated vision, zebrafish have cone-dominated vision similar to humans. Using zebrafish, a genetic model of a congenital form of strabismus revealed disrupted development of the neuromuscular connections of the abducens motor neurons (Asakawa and Kawakami 2018). Abnormal eye movements in genetic models can be uncovered by inducing gaze stabilization reflexes with visual stimulation; a zebrafish mutant for *dscam1l*, an ortholog of the Hsa21 *DSCAM* gene, displayed such oculomotor phenotypes as well as retinal patterning defects (Ma et al. 2020). Eye measurements can be collected across the lifetime of the zebrafish using spectral domain-optical coherence tomography (SD-OCT) to uncover refractive errors (Collery et al. 2014; Quint et al. 2021). While zebrafish is a powerful model for the discovery of visual phenotypes related to Down syndrome, one difference between the human and zebrafish eye is that the zebrafish lens does not accommodate and instead, its focus depends on a gradient of refractive indices (Easter and Nicola 1996; Wang et al. 2020). Accommodation defects that affect the eye's focusing ability are also common in Down syndrome and could not be studied in zebrafish. Within the central nervous system there are also major circuitry differences, mainly the absence of a pathway similar to the mammalian extrageniculate, although the geniculate pathway to optic tectum, analogous to the superior colliculus, is conserved across species (Mueller 2012; Isa et al. 2021).

4.9 Brain Structure and Function

Intellectual disability is a near universal attribute among individuals with Down syndrome [reviewed in (Chapman and Hesketh 2000)]. Anatomical phenotypes observed in people with Down syndrome include a smaller hippocampus and cerebellum [reviewed in (Haydar and Reeves 2012)]. Changes to brain structure could result from abnormal proliferation or survival or subtler effects such as misspecification of neuronal types. Neurogenesis is reduced in people with Down syndrome (Guidi et al. 2008). Mouse models have revealed an underproduction of cortical excitatory neurons, cerebellar granule cells, and mature myelinating oligodendrocytes, and an overproduction of inhibitory interneurons (Haydar and Reeves 2012; Olmos-Serrano et al. 2016).

Zebrafish share major structural brain organization with mammals and other vertebrates [reviewed in (Panula et al. 2010; Kozol et al. 2016)] (Fig. 6a). Gene expression patterns and neurochemical identities in most regions are similar, including the cerebellum, thalamus, optic tectum, and hypothalamus. The most substantial structural differences between zebrafish and mammals occur in the telencephalon (Folgueira et al. 2012), which undergoes eversion to yield two lobes separated by a ventricle instead of evagination to produce internal ventricles. This process leads to an inversion of the medial and lateral regions; for example, the hippocampal equivalent is lateral in zebrafish versus medial and deep inside the mammalian brain. Although structurally different, the pallium contains regions that are functionally analogous to the mammalian hippocampus and amygdala (Vargas et al. 2009; Aoki

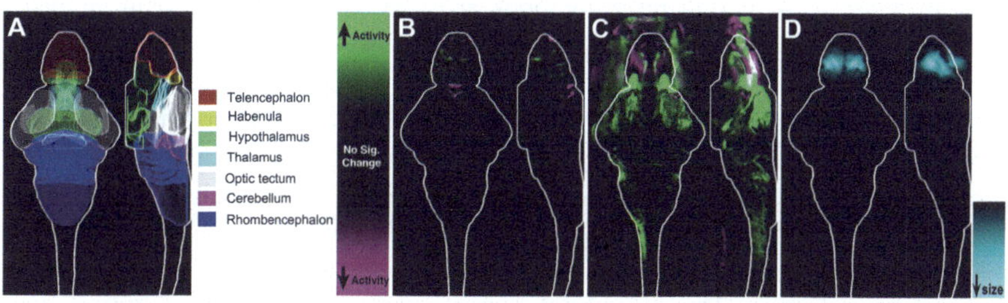

Fig. 6 Brain activity mapping in larval zebrafish. (**a**) Major regions of the 6 dpf brain from the Z-Brain atlas. Created with ImageJ. (**b**) Example of brain activity mapping with phospho-Erk, comparing a mutant to a control population. Green indicates increased activity and magenta indicates decreased. This image represents a mild, but specific phenotype. Sum-of-slices intensity projection is shown. (**c**) A severe brain activity phenotype example. (**d**) Example of a structural size reduction in the forebrain identified by the deformation-based morphometry that is part of the phospho-Erk mapping method

et al. 2013; Porter and Mueller 2020). The telencephalic subpallium contains regions homologous to the mammalian basal ganglia (Ganz et al. 2012; Cheng et al. 2014; Aguda et al. 2021). Many transcriptionally defined neuron types in both the pallium and subpallium also have zebrafish counterparts (Raj et al. 2020; Pandey et al. 2023), including the types implicated in Down syndrome. Several atlases have been built to explore the anatomy of the zebrafish brain at larval and adult stages (Randlett et al. 2015; Kunst et al. 2019; Kenney et al. 2021), and gene expression data is beginning to be included as spatial transcriptomic methods improve (Shainer et al. 2023). Regions orthologous to those with altered size in Down syndrome, such as the hippocampus and cerebellum (Bae et al. 2009; Porter and Mueller 2020), could be impacted in zebrafish models.

Whole-Brain Activity Mapping

Zebrafish larvae are an exceptionally powerful model for the functional dissection of neural circuitry. As they are transparent, whole-brain single-cell resolution imaging of neuronal activity became possible with the advent of improved genetically encoded calcium indicators (Ahrens et al. 2012, 2013; Akerboom et al. 2012). Typically whole-brain imaging is limited to 10 days post-fertilization, although similar data from 2- to 3-week animals has been reported (Andalman et al. 2019; Kappel et al. 2022) and the dorsal telencephalon remains accessible in adults (Aoki et al. 2013; Huang et al. 2020). Combining imaging with two-photon-mediated laser ablation of individual neurons whose activity correlates with a behavior can affirm their role in a circuit (Lacoste et al. 2015; Naumann et al. 2016). Optogenetic activation can be similarly useful, although interpretation of both types of manipulations can sometimes lead to unanticipated indirect effects on other circuitry (Otchy et al. 2015). Recent advances in this field have focused on new microscopy technology for tracking neuronal activity in freely swimming animals (Cong et al. 2017; Kim et al. 2017; Zhang et al. 2021); previous studies required the head of the larva to be fixed in agarose. Although these methods are very powerful for circuit analysis, their main constraint is throughput, limiting their use for studying many genes or drug molecules.

An alternative to calcium imaging is phospho-Erk activity mapping, which is a high-resolution and large-scale approach for comparing the neural activity differences in freely swimming larvae (Randlett et al. 2015). This method rests on the observation that the extracellular signal-related kinase (Erk) is phosphorylated (pErk) as a consequence of calcium signaling in active neurons (Rosen et al. 1994; Dolmetsch et al. 2001; Dai et al. 2002; Greer and Greenberg 2008; Randlett et al. 2015). The approach is analogous to c-Fos activity mapping (Curran and Morgan 1995; Renier et al. 2016) but is more direct, as Erk phosphorylation is upstream of immediate early gene expression and responds within 10 s of calcium influx. Antibodies for Erk and pErk are used to identify active neurons. By registering groups of fish labeled with both antibodies into a common reference brain and quantifying statistically significant differential staining between mutant and control siblings, regions with altered activity can be identified (Fig. 6b, c) using the Z-Brain atlas (Rohlfing and Maurer 2003; Jefferis et al. 2007; Randlett et al. 2015; Thyme et al. 2019). Rather than showing wild type and mutant side-by-side, these images directly compare the groups, with green indicating increased brain activity in the treatment. Activity measurements can be performed in the presence or absence of sensory challenges, and in wild type compared to genetic or chemical perturbation (i.e., stimuli, genes, and drugs). Additionally, a general statistical quantification of brain morphology can be derived using deformation-based morphometry metrics calculated as part of the image registration (Cachero et al. 2010; Thyme et al. 2019) (Fig. 6d). Using these quantification measurements, over 100 genes associated with schizophrenia were screened and over half were found to have changes to brain activity (Thyme et al. 2019).

Imaging and Manipulation of Markers and Pathways

The optical transparency of zebrafish coupled with genetically encoded tools for visualizing, activating, and ablating subtypes of neurons allows detailed investigations into the mechanisms of disease pathogenesis. While mouse studies have implicated cell types involved in Down syndrome, the external development of zebrafish offers an opportunity to further dissect the origins of shared cellular phenotypes. Further, the model's throughput supports the rapid and simultaneous testing of many hypotheses regarding the underlying disrupted genetic circuits and signaling pathways. Proliferation of zebrafish neural progenitor cells can be quantified by staining for phosphorylation of histone H3 (pHH3) or proliferating cell nuclear antigen (PCNA) (Escamilla et al. 2017; Male et al. 2020). Live imaging of cell cycle state is also possible using fluorescent ubiquitin-mediated cell cycle indicator (FUCCI) lines (Sugiyama et al. 2009; Bouldin et al. 2014) or cyclin-dependent kinase (CDK) biosensors (Adikes et al. 2020). Transgenic reporter lines are available for key developmental pathways that may be dysregulated in Down syndrome (Roper et al. 2009; Moon and Lawrence 2022; Martinez et al. 2024), including Notch (e.g., Tp1bglob:eGFP), Sonic hedgehog (e.g., *ptch2*:Kaede), and Wnt (e.g., 6xTCF/LEF-miniP:dGFP) (Parsons et al. 2009; Huang et al. 2012; Shimizu et al. 2012). Once a brain phenotype is established, the neuronal subtypes responsible can be identified using the large library of cell-type-specific GAL4 drivers (Table 1). Hundreds of additional GAL4 lines created through gene trapping have brain-specific expression patterns and are cataloged in databases such as zTrap (Kawakami et al. 2010). Subsets of neurons can be visualized by expressing a fluorescent protein (e.g., UAS-eGFP) or calcium indicator (e.g., UAS:GCaMP6s) and activated or ablated. Optogenetic activation or suppression of neurons is accomplished by expressing channelrhodopsins and exposing fish to light (Mu et al. 2019; Antinucci et al. 2020), and chemogenetic activation is also possible by expressing the rat TRPV1 channel and exposing fish to capsaicin (Lee et al. 2017). Chemogenetic ablation of neuronal subtypes is typically achieved by overexpressing the enzyme nitroreductase, which sensitizes cells to the prodrug metronidazole (Sharrock et al.

Table 1 Representative GAL4/UAS transgenic lines for interrogating neuronal phenotypes

GAL4 driver lines with expression in brain		
Line	Neuronal subtype	PMID
Tg(elavl3:KALTA4)	Pan-neuronal	25505384
TgBAC(gad1b:GAL4-VP16,myl7:mCherry)	GABAergic neurons	28701772
TgBAC(slc17a6b:GAL4FF)	Glutamatergic neurons	23946442
TgBAC(slc6a5:GAL4FF)	Glycinergic neurons	23946442
Tg(Pet1:KalTA4)	Serotonergic raphe neurons	Steve Wilson lab
Tg(tph2:Gal4ff)	Serotonergic raphe neurons	23100441
Tg(dbh:KALTA4)	Noradrenergic neurons	31230713
Tg(th-P2A-Gal4)	Dopaminergic/noradrenergic neurons	25849248
TgBAC(chata:GAL4-VP16)	Cholinergic neurons	28701772
Tg(cbln12:Gal4FF)	Cerebellar granule cells	33952613
Tg(aldoca:GAL4FF)	Purkinje cells	25300581
Tg(olig1:KALTA4)	Oligodendrocyte precursor cells (OPCs)	32066987
Tg(sox10:KALTA4,myl7:EGFP)	Subset of spinal cord neurons, OPCs, oligodendrocytes	26485616
TgBAC(foxb1a:GAL4-VP16,myl7:mCherry)	Neural progenitor cells, OPCs	28701772
TgBAC(avp:KALTA4)	Hypothalamic avp+ neurons	23404108
Tg(npvf:KALTA4)	Hypothalamic npvf+ neurons	29106375
TgBAC(gal:GAL4-VP16,myl7:mCherry)	Hypothalamic galanin+ neurons	28701772
TgBAC(pomca:KALTA4)	Hypothalamic pomca+ neurons	29742429
Tg(npy:kalta4)	Neuropeptide Y+ neurons	29225025
TgBAC(sst7:GAL4-VP16,myl7:mCherry)	Cortistatin+ neurons	28701772
Tg(mpeg1.1:KALTA4,he1.1:BFP2)	Macrophages, microglia, B-lymphocytes	32116293
TgBAC(lhx2b:GAL4-VP16,myl7:mCherry)	Subset of CNS neurons including forebrain	28701772
TgBAC(lhx9:GAL4-VP16,myl7:mCherry)	Subset of CNS neurons including forebrain, hypocretin+ neurons	28701772
TgBAC(otx1:GAL4-VP16,myl7:mCherry)	Subset of CNS neurons	28701772
TgBAC(slit1a:GAL4-VP16,myl7:mCherry)	Broad CNS expression	28701772
TgBAC(neurod1:KALTA4)	Subset of neurons, pancreas	24179142
Tg(her3:KALTA4)	Mesencephalon/rhombencephalon and spinal cord	20735570
Tg(pvalb6:KALTA4)	Subset of neurons including pretectal neurons	31591961
Tg(nefma-2A-KALTA4-nefma)	Neurons with large axon caliber	31571409

(continued)

Table 1 (continued)

GAL4 driver lines with expression in brain		
Line	Neuronal subtype	PMID
Tg(dlx5a/6a:kalTA4)	Telencephalon, optic tectum, ventral diencephalon, cerebellum	30057306
Tg(isl2b:GAL4)	Rohon-Beard neurons, dorsal root ganglia, retinal ganglion cells	20702722
TgBAC(vsx2:GAL4FF)	Hindbrain and spinal V2a neurons, retina	23623549
Tg(eng1b-hs:Gal4)	Subset of CNS neurons including midbrain–hindbrain boundary	25293390
Tg(evx2-hs:Gal4)	Spinal cord V0v neurons	25293390
Tg(ptf1a:Gal4-VP16)	Retina, hindbrain, pancreas	19378102
TgBAC(cxcr4b:GAL4-VP16)	Sensory neurons, motoneurons, cerebellum	28701772
TgBAC(dmbx1b:GAL4-VP16)	Optic tectum, hindbrain	28701772
TgBAC(dbx1a:GAL4-VP16,myl7:mCherry)	Subset of CNS neurons	28701772
TgBAC(drd2a:GAL4-VP16,myl7:mCherry)	Subset of CNS neurons	28701772
UAS lines for activating or ablating neuronal subtypes		
Line	**Purpose**	**PMID**
Tg(UAS:CoChR-eGFP)	Cation channelrhodopsin activated by blue light	31230713
Tg(UAS:ChrimsonR-tdTomato)	Cation channelrhodopsin activated by red light	32216873
Tg(UAS:GtACR1-tdTomato)	Anion channelrhodopsin activated by blue light	32216873
Tg(UAS:TRPV1-TagRFP-T)	Neurons expressing TRPV1 are stimulated with the agonist capsaicin, ablation occurs with prolonged exposure	29106375
Tg(5xUAS:GAP-mCherry-P2A-NfsB_Vv F70A/F108Y,he:tagBFP2)	Chemogenetic ablation of cells expressing NTR and treated with MTZ	35132245
Tg(UAS:BoTxBLC-GFP)	Overexpression of botulinum neurotoxin blocks presynaptic release	27524486
UAS lines for visualizing neuronal activity and connectivity		
Line	**Purpose**	**PMID**
Tg(UAS-kaede)	Photoconvertible fluorescent protein with green and red fluorescence	17406330
Tg(UAS:GCaMP6s)	Calcium indicator	28425439

2022). Neuronal silencing can also be induced by overexpressing the botulinum neurotoxin, which blocks presynaptic release (Sternberg et al. 2016).

Larval Behavior

Locomotion during day and night conditions is an established approach for studying the behavior of zebrafish mutants (Rihel et al. 2010; Kokel et al. 2010; Hoffman et al. 2016; Thyme et al. 2019). The method involves analyzing the position of larvae and monitoring the number, length, and intensity of activity bouts (Joo et al. 2020). Transgenic overexpression of *DYRK1A* in zebrafish cerebellar Purkinje neurons resulted in reduced number of bouts and distance traveled per bout (Buchberger et al. 2021). While a difference in these simple behaviors provides little insight into whether an overexpressed gene contributes to intellectual disability, the method can highlight dosage-sensitive genes involved in zebrafish brain development or function. Both short- and long-term habituation learning have been demonstrated in zebrafish larvae (Wolman et al. 2011; Randlett et al. 2019), and it is possible that a reduction in this primitive form of learning could correlate with more complex learning and memory tasks. With larval behavioral assays, comorbidities of abnormal sleep patterns, seizures, and reduced muscle tone may also be detected (Lagan et al. 2020). Seizures are increased in people with Down syndrome compared to the typically developing population (Rahman and Fatema 2019); they can be measured behaviorally in larvae (Baraban et al. 2005) and confirmed with electrophysiological recording (Robens et al. 2022). Finally, as described previously, visual and acoustic stimulation may also uncover hearing and vision deficits, and a possible dysfunction of these sensory systems must be considered when interpreting neural phenotypes and designing control experiments.

Learning and Memory in Juveniles and Adults

Four learning and memory tests often used to assess neural phenotypes in Down syndrome mouse models include novel object recognition, spontaneous alternation, Morris water maze, and fear conditioning [reviewed in (Das and Reeves 2011)]. It is possible to test similar behaviors in zebrafish, but not at the larval stages (Valente et al. 2012; Gaspary et al. 2018; Bruzzone et al. 2020). Such experiments require growing zebrafish until at least juvenile stages, reducing assay throughput.

A lack of discrimination between a novel and familiar object following a retention period is considered a test of recognition memory. This assay involves the following steps: (1) habituation to the open area without objects, (2) two objects provided for exploration, and (3) one familiar and one novel object after a time delay (typically 24 h for mouse experiments). In this test, wild-type mice prefer to explore the novel object more than the familiar one. Adult zebrafish can behave as expected in this assay (Gaspary et al. 2018), with exploration increasing for the new object compared to the familiar one after a 1-h retention interval between the training and test sessions. However, there are many considerations when designing these tests for zebrafish, as the protocol is less well established than for mammalian models. For example, the complexity, color, size, and buoyancy of the object can all impact the animal's preference (May et al. 2016; Faillace et al. 2017; Magyary 2019). Neophobia, avoidance of the novel object, can also complicate interpretation. Performance at 1, 2, and 3 weeks of age (larval and juvenile stages) indicated that recognition memory may function as early as 2 weeks (Bruzzone et al. 2020), which would enable far higher throughput than if adults are required.

Short-term working memory can be assessed by measuring spontaneous alternation, typically in a Y-shaped maze. Alternations refer to the number of times an animal enters the new arm compared to the one it recently visited, and this number divided by all possible choices is the common performance measure. Mouse models of Down syndrome show lower rates of spontaneous alternation compared to wild-type mice. Analyzing the movement pattern of adult zebrafish in the Y-maze indicates that it is a robust behavioral test in this species, and developmental exposure to ethanol impaired later adult performance (Cleal and Parker 2018; Cleal et al. 2021). In addition to the Y-maze, several other types of mazes have also been used to test various behavioral paradigms in zebrafish, including in combination with reward [reviewed in (Benvenutti et al. 2021)]. Difficulty in drawing conclusions from this assay could arise if animals have changes to locomotion rates or anxiety-related avoidance of novelty. Considering results from multiple tests of memory, anxiety, and locomotion can aid in clarifying interpretations. Secondary tests of anxiety-like behavior are the light–dark preference test and the novel tank diving assay (Fontana et al. 2022). While memory was not assessed in the transgenic lines that overexpressed DYRK1A in zebrafish cerebellar Purkinje neurons (Buchberger et al. 2021), a novel tank diving assay revealed anxiety-like behavior, as measured by fewer transitions to the upper tank zone and reduced dwell time in the upper part.

Pairing a conditioned stimulus with an unconditioned stimulus such as an electric shock is a commonly used method for testing associative memory. In mice, the conditioned stimulus, such a tone or visual context, can induce a freezing response in the absence of the shock. To evaluate contextual fear conditioning, the freezing response is measured in both the original chamber and a novel chamber. The contextual fear conditioning assay for adult zebrafish functions very similarly to rodents (Kenney et al. 2017). The requirement to change tanks in this method and only use adult animals substantially reduces its throughput. Associative learning emerges at approximately 4 weeks of age (Valente et al. 2012), as demonstrated with an operant conditioning paradigm that paired electric shock to entering a section of the arena with specific visual cues (e.g., colored or patterned floor). This protocol is robust and can be conducted with reasonable throughput (Palumbo et al. 2021). Combining aversive conditioning with calcium imaging of the adult telencephalon linked a pallial region to long-term memory recall (Aoki et al. 2013), consistent with the known role of the hippocampus in learning and memory.

Reduced performance in the Morris water maze spatial navigation task is well established in multiple Down syndrome mouse models (Das and Reeves 2011). The mice must learn to navigate opaque water to locate a submerged transparent platform, initially marked by a visible cue that is removed in the testing trial. For obvious reasons, the specifics of this behavioral paradigm are not replicable in an aquatic species. Thus, other spatial learning tasks must be employed in its place (Karnik and Gerlai 2012; Buatois and Gerlai 2020). Recently, a spatial navigation task akin to the Morris water maze was developed for the related species *Danionella cerebrum* (Lee and Briggman 2023). This species prefers to swim into dark regions, which could be used as motivating stimuli. Distant, partially occluded visual cues were sufficient for navigation, indicating a memory-based strategy. While successful performance of zebrafish in this paradigm has not yet been confirmed, it offers the possibility of a relatively high-throughput approach to assess spatial memory.

Establishing learning and memory phenotypes in zebrafish models of Down syndrome is a first step. Once these models are identified, the many tools available in zebrafish can be leveraged to define disrupted circuits and cell types that are critical to the development and execution of these complex behaviors. Shortcomings of mouse models that can be overcome using zebrafish include their highly inbred nature and limited possible sample sizes. The genetic variation of commonly used zebrafish strains is more reflective of the diversity of the human population. The large number of zebrafish available for experiments could enable studies infeasible in mice, such as the discovery of genetic modifiers with forward or reverse genetic screens. While the described behavioral assays do not have as high throughput because of the requirement for the more mature 3–4-week stage, many fish can still be tested simultaneously. Establishing models with phenotypes in these assays would put small-scale drug screening for modulators of learning and memory deficits associated with Down syndrome within reach.

4.10 Alzheimer's Disease

Almost every person with Down syndrome is at risk for developing early-onset Alzheimer's disease (Iulita et al. 2022). This co-occurring condition is one of the driving factors of reduced life expectancy in this population. Triplication of *APP*, located on Hsa21, is the cause, as the extra copy in the absence of Hsa21 trisomy results in early-onset Alzheimer's disease without Down syndrome or intellectual disability (Rovelet-Lecrux et al. 2006; Cabrejo et al. 2006). APP is cleaved by secretases to produce amyloid-β (Aβ) peptides that can aggregate and deposit as plaques. A zebrafish model of *appa* (Fig. 3b) triplication has not been generated, but several approaches have been used to study Alzheimer's disease in this species, including other genetic models of early-onset familial Alzheimer's disease mutations (Barthelson et al. 2022) and injection of Aβ peptides into brain ventricles (Bhattarai et al. 2016; Özcan et al. 2020). While all previously described neural phenotyping strategies are relevant to characterizing zebrafish models of Alzheimer's, the importance of this condition and body of work merits a separate section in this chapter.

A series of papers by Michael Lardelli's group culminated in the transcriptomic analysis of the adult brains of 11 zebrafish lines generated with early-onset Alzheimer's mutations (Barthelson et al. 2022; Lardelli et al. 2023). These lines, with mutations in zebrafish orthologs of *PSEN1*, *PSEN2*, and *SORL1*, mimicked the human genetics as closely as possible by testing the impact of the single heterozygous state on young adult brains. Although protein-coding alleles in *APP* have been found to cause early-onset Alzheimer's (TCW and Goate 2017), this gene was avoided because of the complications of gene duplication. The main takeaway from the transcriptomic analysis of these zebrafish lines was a shared disruption of oxidative phosphorylation, with most mutations leading to downregulation of the pathway. Reduced expression of the associated gene set was present in one model characterized as early as the larval 7 dpf stage, indicating that it may be an early cellular stress response to the mutations. Importantly, they also found that disrupted oxidative phosphorylation was conserved in a mouse model with an *APOE* risk allele.

Despite gene duplication, the zebrafish orthologs of *APP* have been studied by other groups. Tol2 transgenesis was used to knock-in *APP* with the Swedish mutation under control of the zebrafish *appb* promoter (Pu et al. 2017). This overexpression resulted in behavioral impairments in a T-maze-based memory test, differences in the novel tank assay, and pathological changes including Aβ deposition, neuron loss, mitochondrial damage, and disrupted microvasculature. While not a model of Alzheimer's disease, knock-out of either ortholog or the combination of the two resulted in changes to early development in zebrafish, including smaller brain ventricles (Banote et al. 2020; Chebli et al. 2021). In another study, movement and sleep behavior were found to be disrupted in larvae with mutations for either ortholog (Özcan et al. 2022). Although most research on *APP* focuses on the aged population, the zebrafish is an ideal model for exploring how increased *APP* levels on impact neurodevelopment and cognitive function prior to an Alzheimer's diagnosis.

The most widely used model for Aβ toxicity in zebrafish is based on microinjection of peptides into the brain ventricles or heart (Bhattarai et al. 2016; Özcan et al. 2020). In one of the first of these studies, Aβ42 peptides were coupled to cell-penetrating tags and injected in the adult brain ventricle (Bhattarai et al. 2016). The peptide deposited in neurons and led to disease-relevant phenotypes including microglial activity and loss of synapses and neurons. Subsequently, the same approach was found to have different outcomes when old (1.5 years of age) and young (6 months of age) zebrafish were injected (Bhattarai et al. 2017). More neurodegeneration was observed in older animals, and neurogenesis was stimulated at the younger age. Single-cell RNA sequencing of this model has revealed numerous transcriptional changes to both neuronal and glial clusters (Cosacak et al. 2019, 2022). In larvae, a similar peptide-based strategy was used by a different group, injecting either into the heart or brain ventricles (Özcan et al. 2020, 2022). In both cases, larval sleep was disrupted, in different ways depending on the length of the injected peptide. Given the presence of larval phenotypes, it is possible that large-scale drug screening could be conducted in a zebrafish model of Alzheimer's disease. While no such screen has been reported for a genetic model, predictive pharmacology has been attempted based on zebrafish behavioral data (Kroll et al. 2023), and a recent drug-repurposing screen for calcineurin inhibitors yielded several potentially neuroprotective hits (Del Rosario et al. 2024).

5 Drug Discovery for Down Syndrome

More than ten compounds discovered using zebrafish have entered the clinic or clinical trials (Patton et al. 2021). While the number is small compared to molecules discovered using mice, it is far more impressive when considering that only around 2000 manuscripts had been published using zebrafish before the twenty-first century (Kinth et al. 2013). Molecules in clinical trials include potential treatments for several kinds of cancer, anemia, and hearing loss. In one exciting bench-to-bedside story, a child with a rare lymphatic disorder was sequenced and a zebrafish model was used to identify a treatment with clinical success (Li et al. 2019). While most molecules were discovered using fluorescent-based reporter assays [reviewed in (Patton et al. 2021)], behavioral-based screening (Griffin et al. 2020) was used to find a treatment for a severe seizure disorder that is currently in phase II trials. Zebrafish can also be used in conjunction with high-content cell-based screens to test hits in an in vivo setting. Although not in clinical trials, chemical screens for phenotypes relevant to co-occurring conditions such as pancreatic β-cell number and hematopoietic stem cell transplantation have been reported (North et al. 2007; Wang et al. 2015). Among Hsa21 orthologs, a zebrafish-based screen identified compounds that rescued reduced length of motor neuron axons in *chodl* mutants (Oprişoreanu et al. 2021), which might also ameliorate phenotypes in the overexpression condition (Zhong et al. 2012).

In addition to genetic screens, zebrafish could be used to test rationally designed drug molecules. A highly selective Dyrk1a inhibitor was developed starting from structural models, and this molecule rescued cerebellar disorganization in the *DYRK1A*-overexpressing zebrafish line (Buchberger et al. 2021). Computationally screening chemical libraries for binding to a protein of interest has been gaining momentum in recent years with the advent of new technologies (Stein et al. 2020; Sadybekov et al. 2022). Over one billion small molecules are available for purchase, a number that is inconceivable to test experimentally. Given that Down syndrome is the result of overexpressing genes on Hsa21, it is logical to suggest that inhibitors of the encoded proteins could offer possible treatments. Although the increased level is not pathogenic for every Hsa21 gene, a subset contributes to Down syndrome clinical phenotypes.

However, not every class of protein is a viable target for rational drug targeting, as many proteins have been classically considered "undruggable" because of a lack of defined pockets (Henley and Koehler 2021). Here, we have collated a list of Hsa21 genes that would be the most straightforward candidates for computational discovery and zebrafish testing, based on protein class and conservation in zebrafish (Table 2). As new biology is revealed for each candidate, additional proteins that

Table 2 Zebrafish orthologs of human chromosome 21 genes in commonly drug-targeted classes of proteins. The genes on this list code for G-protein-coupled receptors, ion channels, and enzymes, including kinases, and have >50% identity to the human protein sequence. Active ligands, approved drugs, and lowest number of known off-targets for approved drugs were retrieved from the Pharos interface to the Illuminating the Druggable Genome program (Sharma et al. 2023)

Human gene name	Zebrafish gene name	Percent identity to human (%)	Percent similarity to human (%)	Protein class	Active ligands	Approved drugs	Lowest number of known off-targets
HSPA13	*hspa13*	64	82	Enzyme	0	0	NA
USP25	*usp25*	72	84	Enzyme	111	0	NA
ADAMTS1	*adamts1*	56	80	Enzyme	4	0	NA
ADAMTS5	*adamts5*	58	74	Enzyme	252	0	NA
N6AMT1	*n6amt1*	62	86	Enzyme	0	0	NA
LTN1	*ltn1*	52	80	Enzyme	0	0	NA
GRIK1	*grik1a*	78	90	Ion channel	160	3	7 (kainic acid)
	grik1b	83	93				
TIAM1	*tiam1a*	61	80	Enzyme	0	0	NA
	tiam1b	59	76				
SOD1	*sod1*	70	87	Enzyme	8	0	NA
SYNJ1	*synj1*	70	86	Enzyme	0	1	1 (pyrvinium)
ITSN1	*itsn1*	75	86	Enzyme	0	0	NA
CLIC6	*si:ch73-272o12.2*	70	85	Ion channel	0	0	NA
CBR1	*cbr1*	66	87	Enzyme	12	1	1 (triclosan)
	cbr1l	62	82				
PIGP	*pigp*	64	91	Enzyme	0	0	NA
DYRK1A	*dyrk1aa*	76	86	Enzyme	673	0	NA
	dyrk1ab	80	90				
KCNJ6	*kcnj6*	82	93	Ion channel	3	11	8 (desipramine)
KCNJ15	*kcnj15*	60	85	Ion channel	0	0	NA
BACE2	*bace2*	55	77	Enzyme	933	0	NA
MX1	*mxa*	53	79	Enzyme	0	0	NA
	mxb	52	80				
RIPK4	*ripk4*	59	81	Enzyme	1	0	NA
PDE9A	*pde9aa*	68	83	Enzyme	443	1	4 (vardenafil)
	pde9ab	64	82				
	pde9ac	65	83				
	si:dkey-219c10.4	38	66				
CBS	*cbsa*	73	89	Enzyme	2	0	NA
	cbsb	70	85				
SIK1	*sik1*	55	70	Enzyme	11	2	50 (dasatinib)

(continued)

Table 2 (continued)

Human gene name	Zebrafish gene name	Percent identity to human (%)	Percent similarity to human (%)	Protein class	Active ligands	Approved drugs	Lowest number of known off-targets
PDXK	*pdxka*	73	90	Enzyme	0	0	NA
	pdxkb	72	90				
AGPAT3	*agpat3*	75	92	Enzyme	0	0	NA
GATD3	*gatd3*	75	89	Enzyme	0	0	NA
	si:ch211-153b23.5	59	81				
	zgc:162944	64	82				
	gatd3l	48	75				
PFKL	*pfkla*	78	94	Enzyme	0	0	NA
	pfklb	80	95				
UBE2G2	*ube2g2*	94	98	Enzyme	0	0	NA
FTCD	*ftcd*	66	89	Enzyme	0	0	NA
LSS	*lss*	75	91	Enzyme	94	0	NA
YBEY	*ybey*	59	88	Enzyme	0	0	NA
PRMT2	*prmt2*	53	78	Enzyme	0	0	NA

function in the same pathways will be discovered (Yousefelahiyeh et al. 2018), which may ultimately be better or more specific targets.

6 Discussion

The zebrafish offers a powerful compromise between the widely accepted rodent models and more high-throughput fly, worm, and cellular systems. A main strength of zebrafish is that detailed phenotypic characterization is possible at the earliest stages of life (Table 3), enabled by optical transparency and external development. The pleiotropic effects on the brain and body from Hsa21 trisomy suggest disruptions in early embryogenesis, making zebrafish an ideal choice to unravel the basis of this disorder. They are small, cost-effective organisms, enabling the study of many individual Hsa21 orthologs and their combinations, as well as potential drug treatments. A main limitation of zebrafish is that aneuploidy cannot be achieved with current genome-engineering technology. Even duplication models like those established in rodents would be challenging to generate because of the low conservation of synteny. Although it is not expected that zebrafish carrying extra copies of one or several Hsa21 orthologs will fully recapitulate the human condition, it is well established that zebrafish genetic studies can uncover the basic molecular, cellular, and developmental functions of human disease genes. Thus far, only a small number

Table 3 Strengths and weaknesses of zebrafish models

	Strengths	Weaknesses
Genetics	88% of classical protein-coding genes on Hsa21 have at least one zebrafish ortholog Some Hsa21 orthologs are highly conserved—For example, human and zebrafish *UBE2G2* share 94% identity across 165 amino acids Vertebrate model with conservation of major developmental signaling pathways Transparent embryos and high fecundity permit rapid microinjection and screening of thousands of embryos Genetic diversity of lab strains is more analogous to the human population than inbred mouse lines Low cost allows for the study of many genetic mutants and transgenic lines	Compared to mouse and rat models, low conservation of synteny between human and zebrafish Low sequence conservation of non-coding RNAs (Ulitsky et al. 2011) and regulatory sequences Genome duplication complicates interpretation of experiments assessing dosage sensitivity Potential for compensation between ohnologs—for example, *appa* is upregulated in *appb* mutants (Banote et al. 2020) Single genes are not overexpressed in the context of trisomy Most single transgenics are limited to one isoform—No alternative splicing Large site-specific knock-in alleles are historically challenging to generate in zebrafish compared to mouse, although GeneWeld method has improved efficiency (Welker et al. 2021) Antibiotic resistance genes are not used for selection, so rare genetic events such as recombination over large distances cannot be selected for
Phenotyping	External development enables easy visualization at early stages Transparent tissues for imaging of individual cells across the lifespan Whole-brain imaging at single-cell resolution up to early juvenile stages Larvae survive under conditions lethal to mice, enabling access to important developmental processes Scale allows for analysis of multiple effects, such as environmental interactions with a genetic mutation	Some anatomical, cellular, and molecular differences between zebrafish and mammals Lack of a layered cortex Robust tissue regeneration may limit study of some diseased and degenerative processes Life-long brain neurogenesis
Drug discovery	High fecundity with individual zebrafish producing hundreds of embryos weekly for screens	Not all Hsa21 orthologs are sufficiently conserved to be targeted by the same small molecule Gene-therapy- and ASO-based approaches are not translatable between species, limiting zebrafish to small molecule or peptide therapeutics

of studies have overexpressed Hsa21 genes in zebrafish (Pu et al. 2017; Edie et al. 2018; Buchberger et al. 2021; Kuil et al. 2023), with only *DYRK1A* and *APP* undergoing detailed phenotyping. The new, highly sensitive phenotyping technologies available for monitoring whole-brain activity in zebrafish could be used to nominate genes that contribute to brain development, function, and, ultimately, intellectual disability. Its unique combination of throughput, transparency, available tools, and conservation makes zebrafish an exciting option to consider for the future of Down syndrome research.

Acknowledgments We thank Bing Ye, Cecilia Moens, Randall Roper, and Eduardo Torres for helpful comments on this manuscript. We also thank the Jerome Lejeune Foundation for their support of Anna J. Moyer.

Bibliography

Abbas L, Whitfield TT (2010) The zebrafish inner ear. Zebrafish. Elsevier, pp 123–171

Adikes RC, Kohrman AQ, Martinez MAQ, et al (2020) Visualizing the metazoan proliferation-quiescence decision in vivo. Elife. doi: https://doi.org/10.7554/eLife.63265

Aguda V, Chasiotis H, Riadi I, Thiele T (2021) Identification of Subpallial Neuronal Populations Across Zebrafish Larval Stages that Express Molecular Markers for the Striatum. BioRxiv. doi: https://doi.org/10.1101/2021.08.11.455880

Ahrens MB, Li JM, Orger MB, et al (2012) Brain-wide neuronal dynamics during motor adaptation in zebrafish. Nature 485:471–477. doi: https://doi.org/10.1038/nature11057

Ahrens MB, Orger MB, Robson DN, et al (2013) Whole-brain functional imaging at cellular resolution using light-sheet microscopy. Nat Methods 10:413–420. doi: https://doi.org/10.1038/nmeth.2434

Akerboom J, Chen T-W, Wardill TJ, et al (2012) Optimization of a GCaMP calcium indicator for neural activity imaging. J Neurosci 32:13819–13840. doi: https://doi.org/10.1523/JNEUROSCI.2601-12.2012

Akitake CM, Macurak M, Halpern ME, Goll MG (2011) Transgenerational analysis of transcriptional silencing in zebrafish. Dev Biol 352:191–201. doi: https://doi.org/10.1016/j.ydbio.2011.01.002

Al Turki S, Manickaraj AK, Mercer CL, et al (2014) Rare variants in NR2F2 cause congenital heart defects in humans. Am J Hum Genet 94:574–585. doi: https://doi.org/10.1016/j.ajhg.2014.03.007

Alemany A, Florescu M, Baron CS, et al (2018) Whole-organism clone tracing using single-cell sequencing. Nature 556:108–112. doi: https://doi.org/10.1038/nature25969

Andalman AS, Burns VM, Lovett-Barron M, et al (2019) Neuronal dynamics regulating brain and behavioral state transitions. Cell 177:970–985.e20. doi: https://doi.org/10.1016/j.cell.2019.02.037

Antinucci P, Dumitrescu A, Deleuze C, et al (2020) A calibrated optogenetic toolbox of stable zebrafish opsin lines. Elife. doi: https://doi.org/10.7554/eLife.54937

Aoki T, Kinoshita M, Aoki R, et al (2013) Imaging of neural ensemble for the retrieval of a learned behavioral program. Neuron 78:881–894. doi: https://doi.org/10.1016/j.neuron.2013.04.009

Asakawa K, Kawakami K (2009) The Tol2-mediated Gal4-UAS method for gene and enhancer trapping in zebrafish. Methods 49:275–281. doi: https://doi.org/10.1016/j.ymeth.2009.01.004

Asakawa K, Kawakami K (2018) Protocadherin-Mediated Cell Repulsion Controls the Central Topography and Efferent Projections of the Abducens Nucleus. Cell Rep 24:1562–1572. doi: https://doi.org/10.1016/j.celrep.2018.07.024

Austeng ME, Akre H, Øverland B, et al (2013) Otitis media with effusion in children with in Down syndrome. Int J Pediatr Otorhinolaryngol 77:1329–1332. doi: https://doi.org/10.1016/j.ijporl.2013.05.027

Bae Y-K, Kani S, Shimizu T, et al (2009) Anatomy of zebrafish cerebellum and screen for mutations affecting its development. Dev Biol 330:406–426. doi: https://doi.org/10.1016/j.ydbio.2009.04.013

Baeten JT, de Jong JLO (2018) Genetic models of leukemia in zebrafish. Front Cell Dev Biol 6:115. doi: https://doi.org/10.3389/fcell.2018.00115

Baeza-Loya S, Raible DW (2023) Vestibular physiology and function in zebrafish. Front Cell Dev Biol 11:1172933. doi: https://doi.org/10.3389/fcell.2023.1172933

Banote RK, Chebli J, Şatır TM, et al (2020) Amyloid precursor protein-b facilitates cell adhesion during early development in zebrafish. Sci Rep 10:10127. doi: https://doi.org/10.1038/s41598-020-66584-8

Baraban SC, Taylor MR, Castro PA, Baier H (2005) Pentylenetetrazole induced changes in zebrafish behavior, neural activity and c-fos expression. Neuroscience 131:759–768. doi: https://doi.org/10.1016/j.neuroscience.2004.11.031

Barlow GM, Chen XN, Shi ZY, et al (2001) Down syndrome congenital heart disease: a narrowed region and a candidate gene. Genet Med 3:91–101. doi: https://doi.org/10.1097/00125817-200103000-00002

Barthelson K, Newman M, Lardelli M (2022) Brain transcriptomes of zebrafish and mouse Alzheimer's disease knock-in models imply early disrupted energy metabolism. Dis Model Mech. doi: https://doi.org/10.1242/dmm.049187

Bensimon-Brito A, Cardeira J, Dionísio G, et al (2016) Revisiting in vivo staining with alizarin red S--a valuable approach to analyse zebrafish skeletal mineralization during development and regeneration. BMC Dev Biol 16:2. doi: https://doi.org/10.1186/s12861-016-0102-4

Benvenutti R, Marcon M, Gallas-Lopes M, et al (2021) Swimming in the maze: An overview of maze apparatuses and protocols to assess zebrafish behavior. Neurosci Biobehav Rev 127:761–778. doi: https://doi.org/10.1016/j.neubiorev.2021.05.027

Bergen DJM, Kague E, Hammond CL (2019) Zebrafish as an Emerging Model for Osteoporosis: A Primary Testing Platform for Screening New Osteo-Active Compounds. Front Endocrinol (Lausanne) 10:6. doi: https://doi.org/10.3389/fendo.2019.00006

Bertrand JY, Chi NC, Santoso B, et al (2010) Haematopoietic stem cells derive directly from aortic endothelium during development. Nature 464:108–111. doi: https://doi.org/10.1038/nature08738

Bhattarai P, Thomas AK, Cosacak MI, et al (2016) IL4/STAT6 Signaling Activates Neural Stem Cell Proliferation and Neurogenesis upon Amyloid-β42 Aggregation in Adult Zebrafish Brain. Cell Rep 17:941–948. doi: https://doi.org/10.1016/j.celrep.2016.09.075

Bhattarai P, Thomas AK, Zhang Y, Kizil C (2017) The effects of aging on Amyloid-β42-induced neurodegeneration and regeneration in adult zebrafish brain. Neurogenesis (Austin) 4:e1322666. doi: https://doi.org/10.1080/23262133.2017.1322666

Birling M-C, Schaeffer L, André P, et al (2017) Efficient and rapid generation of large genomic variants in rats and mice using CRISMERE. Sci Rep 7:43331. doi: https://doi.org/10.1038/srep43331

Bornhorst D, Gheller B, Zon LI (2024) Colorimetric barcoding to track, isolate, and analyze hematopoietic stem cell clones. Methods Mol Biol 2707:265–277. doi: https://doi.org/10.1007/978-1-0716-3401-1_18

Bouldin CM, Snelson CD, Farr GH, Kimelman D (2014) Restricted expression of cdc25a in the tailbud is essential for formation of the zebrafish posterior body. Genes Dev 28:384–395. doi: https://doi.org/10.1101/gad.233577.113

Braasch I, Gehrke AR, Smith JJ, et al (2016) The spotted gar genome illuminates vertebrate evolution and facilitates human-teleost comparisons. Nat Genet 48:427–437. doi: https://doi.org/10.1038/ng.3526

Brault V, Duchon A, Romestaing C, et al (2015) Opposite phenotypes of muscle strength and locomotor function in mouse models of partial trisomy and monosomy 21 for the proximal Hspa13-App region. PLoS Genet 11:e1005062. doi: https://doi.org/10.1371/journal.pgen.1005062
Brown DR, Samsa LA, Qian L, Liu J (2016) Advances in the study of heart development and disease using zebrafish. J Cardiovasc Dev Dis. doi: https://doi.org/10.3390/jcdd3020013
Bruzzone M, Gatto E, Lucon Xiccato T, et al (2020) Measuring recognition memory in zebrafish larvae: issues and limitations. PeerJ 8:e8890. doi: https://doi.org/10.7717/peerj.8890
Buatois A, Gerlai R (2020) Elemental and configural associative learning in spatial tasks: could zebrafish be used to advance our knowledge? Front Behav Neurosci 14:570704. doi: https://doi.org/10.3389/fnbeh.2020.570704
Buchberger A, Schepergerdes L, Flaßhoff M, et al (2021) A novel inhibitor rescues cerebellar defects in a zebrafish model of Down syndrome-associated kinase Dyrk1A overexpression. J Biol Chem 297:100853. doi: https://doi.org/10.1016/j.jbc.2021.100853
Cabrejo L, Guyant-Maréchal L, Laquerrière A, et al (2006) Phenotype associated with APP duplication in five families. Brain 129:2966–2976. doi: https://doi.org/10.1093/brain/awl237
Cachero S, Ostrovsky AD, Yu JY, et al (2010) Sexual dimorphism in the fly brain. Curr Biol 20:1589–1601. doi: https://doi.org/10.1016/j.cub.2010.07.045
Carfì A, Liperoti R, Fusco D, et al (2017) Bone mineral density in adults with Down syndrome. Osteoporos Int 28:2929–2934. doi: https://doi.org/10.1007/s00198-017-4133-x
Chapman RS, Hesketh LJ (2000) Behavioral phenotype of individuals with Down syndrome. Ment Retard Dev Disabil Res Rev 6:84–95. doi: https://doi.org/10.1002/1098-2779(2000)6:2<84::AID-MRDD2>3.0.CO;2-P
Chebli J, Rahmati M, Lashley T, et al (2021) The localization of amyloid precursor protein to ependymal cilia in vertebrates and its role in ciliogenesis and brain development in zebrafish. Sci Rep 11:19115. doi: https://doi.org/10.1038/s41598-021-98487-7
Chen G-D, Li L, McCall A, et al (2022a) Hearing impairment in murine model of Down syndrome. Front Genet 13:936128. doi: https://doi.org/10.3389/fgene.2022.936128
Chen S, Lathrop KL, Kuwajima T, Gross JM (2022b) Retinal ganglion cell survival after severe optic nerve injury is modulated by crosstalk between Jak/Stat signaling and innate immune responses in the zebrafish retina. Development. doi: https://doi.org/10.1242/dev.199694
Cheng R-K, Jesuthasan SJ, Penney TB (2014) Zebrafish forebrain and temporal conditioning. Philos Trans R Soc Lond B, Biol Sci 369:20120462. doi: https://doi.org/10.1098/rstb.2012.0462
Chhetri J, Jacobson G, Gueven N (2014) Zebrafish--on the move towards ophthalmological research. Eye (Lond) 28:367–380. doi: https://doi.org/10.1038/eye.2014.19
Chin CJ, Khami MM, Husein M (2014) A general review of the otolaryngologic manifestations of Down Syndrome. Int J Pediatr Otorhinolaryngol 78:899–904. doi: https://doi.org/10.1016/j.ijporl.2014.03.012
Cleal M, Fontana BD, Ranson DC, et al (2021) The Free-movement pattern Y-maze: A cross-species measure of working memory and executive function. Behav Res Methods 53:536–557. doi: https://doi.org/10.3758/s13428-020-01452-x
Cleal M, Parker MO (2018) Moderate developmental alcohol exposure reduces repetitive alternation in a zebrafish model of fetal alcohol spectrum disorders. Neurotoxicol Teratol 70:1–9. doi: https://doi.org/10.1016/j.ntt.2018.09.001
Collery RF, Veth KN, Dubis AM, et al (2014) Rapid, accurate, and non-invasive measurement of zebrafish axial length and other eye dimensions using SD-OCT allows longitudinal analysis of myopia and emmetropization. PLoS One 9:e110699. doi: https://doi.org/10.1371/journal.pone.0110699
Cong L, Wang Z, Chai Y, et al (2017) Rapid whole brain imaging of neural activity in freely behaving larval zebrafish (Danio rerio). Elife. doi: https://doi.org/10.7554/eLife.28158
Cosacak MI, Bhattarai P, De Jager PL, et al (2022) Single cell/nucleus transcriptomics comparison in zebrafish and humans reveals common and distinct molecular responses to alzheimer's disease. Cells. doi: https://doi.org/10.3390/cells11111807

Cosacak MI, Bhattarai P, Reinhardt S, et al (2019) Single-Cell Transcriptomics Analyses of Neural Stem Cell Heterogeneity and Contextual Plasticity in a Zebrafish Brain Model of Amyloid Toxicity. Cell Rep 27:1307–1318.e3. doi: https://doi.org/10.1016/j.celrep.2019.03.090

Curado S, Stainier DYR, Anderson RM (2008) Nitroreductase-mediated cell/tissue ablation in zebrafish: a spatially and temporally controlled ablation method with applications in developmental and regeneration studies. Nat Protoc 3:948–954. doi: https://doi.org/10.1038/nprot.2008.58

Curran T, Morgan JI (1995) Fos: an immediate-early transcription factor in neurons. J Neurobiol 26:403–412. doi: https://doi.org/10.1002/neu.480260312

Dai Y, Iwata K, Fukuoka T, et al (2002) Phosphorylation of extracellular signal-regulated kinase in primary afferent neurons by noxious stimuli and its involvement in peripheral sensitization. J Neurosci 22:7737–7745. doi: https://doi.org/10.1523/JNEUROSCI.22-17-07737.2002

Danilova N, Bussmann J, Jekosch K, Steiner LA (2005) The immunoglobulin heavy-chain locus in zebrafish: identification and expression of a previously unknown isotype, immunoglobulin Z. Nat Immunol 6:295–302. doi: https://doi.org/10.1038/ni1166

Das I, Reeves RH (2011) The use of mouse models to understand and improve cognitive deficits in Down syndrome. Dis Model Mech 4:596–606. doi: https://doi.org/10.1242/dmm.007716

Davidson AE, Straquadine NR, Cook SA, et al (2021) A rapid F0 CRISPR screen in zebrafish to identify regulators of neuronal development in the enteric nervous system. BioRxiv. doi: https://doi.org/10.1101/2021.07.17.452230

Del Rosario Hernández T, Gore SV, Kreiling JA, Creton R (2024) Drug repurposing for neurodegenerative diseases using Zebrafish behavioral profiles. Biomed Pharmacother 171:116096. doi: https://doi.org/10.1016/j.biopha.2023.116096

Diamond KM, Rolfe SM, Kwon RY, Maga AM (2022) Computational anatomy and geometric shape analysis enables analysis of complex craniofacial phenotypes in zebrafish. Biol Open. doi: https://doi.org/10.1242/bio.058948

Dolmetsch RE, Pajvani U, Fife K, et al (2001) Signaling to the nucleus by an L-type calcium channel-calmodulin complex through the MAP kinase pathway. Science 294:333–339. doi: https://doi.org/10.1126/science.1063395

Duong TB, Ravisankar P, Song YC, et al (2018) Nr2f1a balances atrial chamber and atrioventricular canal size via BMP signaling-independent and -dependent mechanisms. Dev Biol 434:7–14. doi: https://doi.org/10.1016/j.ydbio.2017.11.010

Eames BF, DeLaurier A, Ullmann B, et al (2013) FishFace: interactive atlas of zebrafish craniofacial development at cellular resolution. BMC Dev Biol 13:23. doi: https://doi.org/10.1186/1471-213X-13-23

Easter SS, Nicola GN (1996) The development of vision in the zebrafish (Danio rerio). Dev Biol 180:646–663. doi: https://doi.org/10.1006/dbio.1996.0335

Edie S, Zaghloul NA, Leitch CC, et al (2018) Survey of Human Chromosome 21 Gene Expression Effects on Early Development in Danio rerio. G3 (Bethesda) 8:2215–2223. doi: https://doi.org/10.1534/g3.118.200144

Elsaeidi F, Bemben MA, Zhao X-F, Goldman D (2014) Jak/Stat signaling stimulates zebrafish optic nerve regeneration and overcomes the inhibitory actions of Socs3 and Sfpq. J Neurosci 34:2632–2644. doi: https://doi.org/10.1523/JNEUROSCI.3898-13.2014

Escamilla CO, Filonova I, Walker AK, et al (2017) Kctd13 deletion reduces synaptic transmission via increased RhoA. Nature 551:227–231. doi: https://doi.org/10.1038/nature24470

Fabian P, Tseng K-C, Thiruppathy M, et al (2022) Lifelong single-cell profiling of cranial neural crest diversification in zebrafish. Nat Commun 13:13. doi: https://doi.org/10.1038/s41467-021-27594-w

Faillace MP, Pisera-Fuster A, Medrano MP, et al (2017) Short- and long-term effects of nicotine and the histone deacetylase inhibitor phenylbutyrate on novel object recognition in zebrafish. Psychopharmacology 234:943–955. doi: https://doi.org/10.1007/s00213-017-4532-x

Favre-Bulle IA, Taylor MA, Marquez-Legorreta E, et al (2020) Sound generation in zebrafish with Bio-Opto-Acoustics. Nat Commun 11:6120. doi: https://doi.org/10.1038/s41467-020-19982-5

Fink GB, Madaus WK, Walker GF (1975) A quantitative study of the face in Down's syndrome. Am J Orthod 67:540–553.

Fink M, Flekna G, Ludwig A, et al (2006) Improved translation efficiency of injected mRNA during early embryonic development. Dev Dyn 235:3370–3378. doi: https://doi.org/10.1002/dvdy.20995

Fischer JA, Giniger E, Maniatis T, Ptashne M (1988) GAL4 activates transcription in Drosophila. Nature 332:853–856. doi: https://doi.org/10.1038/332853a0

Fisher S, Grice EA, Vinton RM, et al (2006) Evaluating the biological relevance of putative enhancers using Tol2 transposon-mediated transgenesis in zebrafish. Nat Protoc 1:1297–1305. doi: https://doi.org/10.1038/nprot.2006.230

Folgueira M, Bayley P, Navratilova P, et al (2012) Morphogenesis underlying the development of the everted teleost telencephalon. Neural Dev 7:32. doi: https://doi.org/10.1186/1749-8104-7-32

Fontana BD, Alnassar N, Parker MO (2022) The zebrafish (Danio rerio) anxiety test battery: comparison of behavioral responses in the novel tank diving and light-dark tasks following exposure to anxiogenic and anxiolytic compounds. Psychopharmacology 239:287–296. doi: https://doi.org/10.1007/s00213-021-05990-w

Freeman SB, Bean LH, Allen EG, et al (2008) Ethnicity, sex, and the incidence of congenital heart defects: a report from the National Down Syndrome Project. Genet Med 10:173–180. doi: https://doi.org/10.1097/GIM.0b013e3181634867

Friedmacher F, Puri P (2013) Hirschsprung's disease associated with Down syndrome: a meta-analysis of incidence, functional outcomes and mortality. Pediatr Surg Int 29:937–946. doi: https://doi.org/10.1007/s00383-013-3361-1

Ganz J, Kaslin J, Freudenreich D, et al (2012) Subdivisions of the adult zebrafish subpallium by molecular marker analysis. J Comp Neurol 520:633–655. doi: https://doi.org/10.1002/cne.22757

Gaspary KV, Reolon GK, Gusso D, Bonan CD (2018) Novel object recognition and object location tasks in zebrafish: Influence of habituation and NMDA receptor antagonism. Neurobiol Learn Mem 155:249–260. doi: https://doi.org/10.1016/j.nlm.2018.08.005

Gestri G, Link BA, Neuhauss SCF (2012) The visual system of zebrafish and its use to model human ocular diseases. Dev Neurobiol 72:302–327. doi: https://doi.org/10.1002/dneu.20919

Ghosh A, Halpern ME (2016) Transcriptional regulation using the Q system in transgenic zebrafish. Methods Cell Biol 135:205–218. doi: https://doi.org/10.1016/bs.mcb.2016.05.001

Gialesaki S, Bräuer-Hartmann D, Issa H, et al (2023) RUNX1 isoform disequilibrium promotes the development of trisomy 21-associated myeloid leukemia. Blood 141:1105–1118. doi: https://doi.org/10.1182/blood.2022017619

Giménez-Barcons M, Casteràs A, Armengol M del P, et al (2014) Autoimmune predisposition in Down syndrome may result from a partial central tolerance failure due to insufficient intrathymic expression of AIRE and peripheral antigens. J Immunol 193:3872–3879. doi: https://doi.org/10.4049/jimmunol.1400223

Gistelinck C, Kwon RY, Malfait F, et al (2018) Zebrafish type I collagen mutants faithfully recapitulate human type I collagenopathies. Proc Natl Acad Sci USA 115:E8037–E8046. doi: https://doi.org/10.1073/pnas.1722200115

Glass TJ, Connor NP (2016) Digastric muscle phenotypes of the ts65dn mouse model of down syndrome. PLoS One 11:e0158008. doi: https://doi.org/10.1371/journal.pone.0158008

Glass TJ, Russell JA, Fisher EH, et al (2024) Altered tongue muscle contractile properties coincide with altered swallow function in the adult Ts65Dn mouse model of down syndrome. Front Neurol 15:1384572. doi: https://doi.org/10.3389/fneur.2024.1384572

Glenn NO, McKane M, Kohli V, et al (2012) The W-loop of alpha-cardiac actin is critical for heart function and endocardial cushion morphogenesis in zebrafish. Mol Cell Biol 32:3527–3540. doi: https://doi.org/10.1128/MCB.00486-12

Gore AV, Pillay LM, Venero Galanternik M, Weinstein BM (2018) The zebrafish: A fintastic model for hematopoietic development and disease. Wiley Interdiscip Rev Dev Biol 7:e312. doi: https://doi.org/10.1002/wdev.312

Greer PL, Greenberg ME (2008) From synapse to nucleus: calcium-dependent gene transcription in the control of synapse development and function. Neuron 59:846–860. doi: https://doi.org/10.1016/j.neuron.2008.09.002

Griffin A, Anvar M, Hamling K, Baraban SC (2020) Phenotype-Based Screening of Synthetic Cannabinoids in a Dravet Syndrome Zebrafish Model. Front Pharmacol 11:464. doi: https://doi.org/10.3389/fphar.2020.00464

Grossman TR, Gamliel A, Wessells RJ, et al (2011) Over-expression of DSCAM and COL6A2 cooperatively generates congenital heart defects. PLoS Genet 7:e1002344. doi: https://doi.org/10.1371/journal.pgen.1002344

Gu Z, Gu L, Eils R, et al (2014) circlize Implements and enhances circular visualization in R. Bioinformatics 30:2811–2812. doi: https://doi.org/10.1093/bioinformatics/btu393

Guidi S, Bonasoni P, Ceccarelli C, et al (2008) Neurogenesis impairment and increased cell death reduce total neuron number in the hippocampal region of fetuses with Down syndrome. Brain Pathol 18:180–197. doi: https://doi.org/10.1111/j.1750-3639.2007.00113.x

Gupta M, Dhanasekaran AR, Gardiner KJ (2016) Mouse models of Down syndrome: gene content and consequences. Mamm Genome 27:538–555. doi: https://doi.org/10.1007/s00335-016-9661-8

Han F, Yu H, Zhang J, et al (2009) Otitis media in a mouse model for Down syndrome. Int J Exp Pathol 90:480–488. doi: https://doi.org/10.1111/j.1365-2613.2009.00677.x

Haydar TF, Reeves RH (2012) Trisomy 21 and early brain development. Trends Neurosci 35:81–91. doi: https://doi.org/10.1016/j.tins.2011.11.001

Heanue TA, Boesmans W, Bell DM, et al (2016) A Novel Zebrafish ret Heterozygous Model of Hirschsprung Disease Identifies a Functional Role for mapk10 as a Modifier of Enteric Nervous System Phenotype Severity. PLoS Genet 12:e1006439. doi: https://doi.org/10.1371/journal.pgen.1006439

Henley MJ, Koehler AN (2021) Advances in targeting "undruggable" transcription factors with small molecules. Nat Rev Drug Discov 20:669–688. doi: https://doi.org/10.1038/s41573-021-00199-0

Hoffman EJ, Turner KJ, Fernandez JM, et al (2016) Estrogens suppress a behavioral phenotype in zebrafish mutants of the autism risk gene, CNTNAP2. Neuron 89:725–733. doi: https://doi.org/10.1016/j.neuron.2015.12.039

Höijer I, Emmanouilidou A, Östlund R, et al (2022) CRISPR-Cas9 induces large structural variants at on-target and off-target sites in vivo that segregate across generations. Nat Commun 13:627. doi: https://doi.org/10.1038/s41467-022-28244-5

Howe K, Clark MD, Torroja CF, et al (2013) The zebrafish reference genome sequence and its relationship to the human genome. Nature 496:498–503. doi: https://doi.org/10.1038/nature12111

Hu Y, Flockhart I, Vinayagam A, et al (2011) An integrative approach to ortholog prediction for disease-focused and other functional studies. BMC Bioinformatics 12:357. doi: https://doi.org/10.1186/1471-2105-12-357

Huang K-H, Rupprecht P, Frank T, et al (2020) A virtual reality system to analyze neural activity and behavior in adult zebrafish. Nat Methods 17:343–351. doi: https://doi.org/10.1038/s41592-020-0759-2

Huang P, Xiong F, Megason SG, Schier AF (2012) Attenuation of Notch and Hedgehog signaling is required for fate specification in the spinal cord. PLoS Genet 8:e1002762. doi: https://doi.org/10.1371/journal.pgen.1002762

Hyatt TM, Ekker SC (1999) Vectors and techniques for ectopic gene expression in zebrafish. Methods Cell Biol 59:117–126.

Inoue J, Sato Y, Sinclair R, et al (2015) Rapid genome reshaping by multiple-gene loss after whole-genome duplication in teleost fish suggested by mathematical modeling. Proc Natl Acad Sci USA 112:14918–14923. doi: https://doi.org/10.1073/pnas.1507669112

Intrapiromkul J, Aygun N, Tunkel DE, et al (2012) Inner ear anomalies seen on CT images in people with Down syndrome. Pediatr Radiol 42:1449–1455. doi: https://doi.org/10.1007/s00247-012-2490-3

Isa T, Marquez-Legorreta E, Grillner S, Scott EK (2021) The tectum/superior colliculus as the vertebrate solution for spatial sensory integration and action. Curr Biol 31:R741–R762. doi: https://doi.org/10.1016/j.cub.2021.04.001

Iulita MF, Garzón Chavez D, Klitgaard Christensen M, et al (2022) Association of alzheimer disease with life expectancy in people with down syndrome. JAMA Netw Open 5:e2212910. doi: https://doi.org/10.1001/jamanetworkopen.2022.12910

Jefferis GSXE, Potter CJ, Chan AM, et al (2007) Comprehensive maps of Drosophila higher olfactory centers: spatially segregated fruit and pheromone representation. Cell 128:1187–1203. doi: https://doi.org/10.1016/j.cell.2007.01.040

Ji J-F, Hu C-B, Shao T, et al (2021) Differential immune responses of immunoglobulin Z subclass members in antibacterial immunity in a zebrafish model. Immunology 162:105–120. doi: https://doi.org/10.1111/imm.13269

Joo W, Vivian MD, Graham BJ, et al (2020) A Customizable Low-Cost System for Massively Parallel Zebrafish Behavioral Phenotyping. Front Behav Neurosci 14:606900. doi: https://doi.org/10.3389/fnbeh.2020.606900

Joshi P, Liang JO, DiMonte K, et al (2009) Amyloid precursor protein is required for convergent-extension movements during Zebrafish development. Dev Biol 335:1–11. doi: https://doi.org/10.1016/j.ydbio.2009.07.041

Jung HM, Castranova D, Swift MR, et al (2017) Development of the larval lymphatic system in zebrafish. Development 144:2070–2081. doi: https://doi.org/10.1242/dev.145755

Kague E, Kwon RY, Winkler C (2023) Editorial: Fish as model organism for skeletal diseases. Front Endocrinol (Lausanne) 14:1331690. doi: https://doi.org/10.3389/fendo.2023.1331690

Kalev-Zylinska ML, Horsfield JA, Flores MVC, et al (2002) Runx1 is required for zebrafish blood and vessel development and expression of a human RUNX1-CBF2T1 transgene advances a model for studies of leukemogenesis. Development 129:2015–2030. doi: https://doi.org/10.1242/dev.129.8.2015

Kappel JM, Förster D, Slangewal K, et al (2022) Visual recognition of social signals by a tectothalamic neural circuit. Nature 608:146–152. doi: https://doi.org/10.1038/s41586-022-04925-5

Kardash E (2021) Zebrafish Research Methods. Materials and Methods

Karnik I, Gerlai R (2012) Can zebrafish learn spatial tasks? An empirical analysis of place and single CS-US associative learning. Behav Brain Res 233:415–421. doi: https://doi.org/10.1016/j.bbr.2012.05.024

Karuppasamy M, English KG, Henry CA, et al (2024) Standardization of zebrafish drug testing parameters for muscle diseases. Dis Model Mech. doi: https://doi.org/10.1242/dmm.050339

Kawakami K (2007) Tol2: a versatile gene transfer vector in vertebrates. Genome Biol 8 Suppl 1:S7. doi: https://doi.org/10.1186/gb-2007-8-s1-s7

Kawakami K, Abe G, Asada T, et al (2010) zTrap: zebrafish gene trap and enhancer trap database. BMC Dev Biol 10:105. doi: https://doi.org/10.1186/1471-213X-10-105

Kawakami K, Shima A (1999) Identification of the Tol2 transposase of the medaka fish Oryzias latipes that catalyzes excision of a nonautonomous Tol2 element in zebrafish Danio rerio. Gene 240:239–244. doi: https://doi.org/10.1016/S0378-1119(99)00444-8

Kawakami K, Takeda H, Kawakami N, et al (2004) A transposon-mediated gene trap approach identifies developmentally regulated genes in zebrafish. Dev Cell 7:133–144. doi: https://doi.org/10.1016/j.devcel.2004.06.005

Kenney JW, Scott IC, Josselyn SA, Frankland PW (2017) Contextual fear conditioning in zebrafish. Learn Mem 24:516–523. doi: https://doi.org/10.1101/lm.045690.117

Kenney JW, Steadman PE, Young O, et al (2021) A 3D adult zebrafish brain atlas (AZBA) for the digital age. Elife. doi: https://doi.org/10.7554/eLife.69988

Kim CK, Miri A, Leung LC, et al (2014) Prolonged, brain-wide expression of nuclear-localized GCaMP3 for functional circuit mapping. Front Neural Circuits 8:138. doi: https://doi.org/10.3389/fncir.2014.00138

Kim DH, Kim J, Marques JC, et al (2017) Pan-neuronal calcium imaging with cellular resolution in freely swimming zebrafish. Nat Methods 14:1107–1114. doi: https://doi.org/10.1038/nmeth.4429

Kim S, Chung A-Y, Na JE, et al (2019) Myelin degeneration induced by mutant superoxide dismutase 1 accumulation promotes amyotrophic lateral sclerosis. Glia 67:1910–1921. doi: https://doi.org/10.1002/glia.23669

Kinth P, Mahesh G, Panwar Y (2013) Mapping of zebrafish research: a global outlook. Zebrafish 10:510–517. doi: https://doi.org/10.1089/zeb.2012.0854

Kissa K, Herbomel P (2010) Blood stem cells emerge from aortic endothelium by a novel type of cell transition. Nature 464:112–115. doi: https://doi.org/10.1038/nature08761

Kokel D, Bryan J, Laggner C, et al (2010) Rapid behavior-based identification of neuroactive small molecules in the zebrafish. Nat Chem Biol 6:231–237. doi: https://doi.org/10.1038/nchembio.307

Körner U, Bustin M, Scheer U, Hock R (2003) Developmental role of HMGN proteins in Xenopus laevis. Mech Dev 120:1177–1192. doi: https://doi.org/10.1016/j.mod.2003.07.001

Koyama W, Hosomi R, Matsuda K, et al (2021) Involvement of cerebellar neural circuits in active avoidance conditioning in zebrafish. eNeuro. doi: https://doi.org/10.1523/ENEURO.0507-20.2021

Kozol RA, Abrams AJ, James DM, et al (2016) Function over form: modeling groups of inherited neurological conditions in zebrafish. Front Mol Neurosci 9:55. doi: https://doi.org/10.3389/fnmol.2016.00055

Kroll F, Donnelly J, Özcan GG, et al (2023) Behavioural pharmacology predicts disrupted signalling pathways and candidate therapeutics from zebrafish mutants of Alzheimer's disease risk genes. BioRxiv. doi: https://doi.org/10.1101/2023.11.28.568940

Kuil LE, Chauhan RK, Cheng WW, et al (2020) Zebrafish: A model organism for studying enteric nervous system development and disease. Front Cell Dev Biol 8:629073. doi: https://doi.org/10.3389/fcell.2020.629073

Kuil LE, Chauhan RK, de Graaf BM, et al (2023) ATP5PO levels regulate enteric nervous system development in zebrafish, linking Hirschsprung disease to Down Syndrome. Biochim Biophys Acta Mol Basis Dis 1870:166991. doi: https://doi.org/10.1016/j.bbadis.2023.166991

Kunst M, Laurell E, Mokayes N, et al (2019) A Cellular-Resolution Atlas of the Larval Zebrafish Brain. Neuron 103:21–38.e5. doi: https://doi.org/10.1016/j.neuron.2019.04.034

Lacoste AMB, Schoppik D, Robson DN, et al (2015) A convergent and essential interneuron pathway for Mauthner-cell-mediated escapes. Curr Biol 25:1526–1534. doi: https://doi.org/10.1016/j.cub.2015.04.025

Lagan N, Huggard D, Mc Grane F, et al (2020) Multiorgan involvement and management in children with Down syndrome. Acta Paediatr 109:1096–1111. doi: https://doi.org/10.1111/apa.15153

Lalonde RL, Wells HH, Kemmler CL, et al (2024) pIGLET: Safe harbor landing sites for reproducible and efficient transgenesis in zebrafish. Sci Adv 10:eadn6603. doi: https://doi.org/10.1126/sciadv.adn6603

Lana-Elola E, Aoidi R, Llorian M, et al (2024) Increased dosage of DYRK1A leads to congenital heart defects in a mouse model of Down syndrome. Sci Transl Med 16:eadd6883. doi: https://doi.org/10.1126/scitranslmed.add6883

Lardelli M, Baer L, Hin N, et al (2023) The use of zebrafish in transcriptome analysis of the early effects of mutations causing early onset familial alzheimer's disease and other inherited neurodegenerative conditions. J Alzheimers Dis. doi: https://doi.org/10.3233/JAD-230522

Lau IH, Vasconcelos RO (2023) Noise-induced damage in the zebrafish inner ear endorgans: evidence for higher acoustic sensitivity of saccular and lagenar hair cells. J Exp Biol. doi: https://doi.org/10.1242/jeb.245992

Lawson ND, Li R, Shin M, et al (2020) An improved zebrafish transcriptome annotation for sensitive and comprehensive detection of cell type-specific genes. Elife. doi: https://doi.org/10.7554/eLife.55792
Le Pabic P, Dranow DB, Hoyle DJ, Schilling TF (2022) Zebrafish endochondral growth zones as they relate to human bone size, shape and disease. Front Endocrinol (Lausanne) 13:1060187. doi: https://doi.org/10.3389/fendo.2022.1060187
Lee DA, Andreev A, Truong TV, et al (2017) Genetic and neuronal regulation of sleep by neuropeptide VF. Elife. doi: https://doi.org/10.7554/eLife.25727
Lee TJ, Briggman KL (2023) Visually guided and context-dependent spatial navigation in the translucent fish Danionella cerebrum. Curr Biol 33:5467–5477.e4. doi: https://doi.org/10.1016/j.cub.2023.11.030
Levraud J-P, Jouneau L, Briolat V, et al (2019) IFN-Stimulated Genes in Zebrafish and Humans Define an Ancient Arsenal of Antiviral Immunity. J Immunol 203:3361–3373. doi: https://doi.org/10.4049/jimmunol.1900804
Li D, March ME, Gutierrez-Uzquiza A, et al (2019) ARAF recurrent mutation causes central conducting lymphatic anomaly treatable with a MEK inhibitor. Nat Med 25:1116–1122. doi: https://doi.org/10.1038/s41591-019-0479-2
Li H, Edie S, Klinedinst D, et al (2016) Penetrance of congenital heart disease in a mouse model of down syndrome depends on a trisomic potentiator of a disomic modifier. Genetics 203:763–770. doi: https://doi.org/10.1534/genetics.116.188045
Liao W-N, You M-S, Ulhaq ZS, et al (2023) Micro-CT analysis reveals the changes in bone mineral density in zebrafish craniofacial skeleton with age. J Anat 242:544–551. doi: https://doi.org/10.1111/joa.13780
Lima FA, Moreira-Filho CA, Ramos PL, et al (2011) Decreased AIRE expression and global thymic hypofunction in Down syndrome. J Immunol 187:3422–3430. doi: https://doi.org/10.4049/jimmunol.1003053
Lin W-Y, Dharini KK, Peng C-H, et al (2022) Zebrafish models for glucocorticoid-induced osteoporosis. Tzu Chi Medical Journal 34:373–380. doi: https://doi.org/10.4103/tcmj.tcmj_80_22
Liu F, Patient R (2008) Genome-wide analysis of the zebrafish ETS family identifies three genes required for hemangioblast differentiation or angiogenesis. Circ Res 103:1147–1154. doi: https://doi.org/10.1161/CIRCRESAHA.108.179713
Long L, Guo H, Yao D, et al (2015) Regulation of transcriptionally active genes via the catalytically inactive Cas9 in C. elegans and D. rerio. Cell Res 25:638–641. doi: https://doi.org/10.1038/cr.2015.35
Lynch M, Conery JS (2000) The evolutionary fate and consequences of duplicate genes. Science 290:1151–1155. doi: https://doi.org/10.1126/science.290.5494.1151
Ma M, Ramirez AD, Wang T, et al (2020) Zebrafish dscaml1 Deficiency Impairs Retinal Patterning and Oculomotor Function. J Neurosci 40:143–158. doi: https://doi.org/10.1523/JNEUROSCI.1783-19.2019
Magyary I (2019) Floating novel object recognition in adult zebrafish: a pilot study. Cogn Process 20:359–362. doi: https://doi.org/10.1007/s10339-019-00910-5
Malak R, Kostiukow A, Krawczyk-Wasielewska A, et al (2015) Delays in Motor Development in Children with Down Syndrome. Med Sci Monit 21:1904–1910. doi: https://doi.org/10.12659/MSM.893377
Male I, Ozacar AT, Fagan RR, et al (2020) Hedgehog signaling regulates neurogenesis in the larval and adult zebrafish hypothalamus. eNeuro. doi: https://doi.org/10.1523/ENEURO.0226-20.2020
Malle L, Martin-Fernandez M, Buta S, et al (2022) Excessive negative regulation of type I interferon disrupts viral control in individuals with Down syndrome. Immunity 55:2074–2084.e5. doi: https://doi.org/10.1016/j.immuni.2022.09.007
Malle L, Patel RS, Martin-Fernandez M, et al (2023) Autoimmunity in Down's syndrome via cytokines, CD4 T cells and CD11c+ B cells. Nature 615:305–314. doi: https://doi.org/10.1038/s41586-023-05736-y

Martin KE, Waxman JS (2021) Atrial and sinoatrial node development in the zebrafish heart. J Cardiovasc Dev Dis. doi: https://doi.org/10.3390/jcdd8020015

Martinez JL, Piciw JG, Crockett M, et al (2024) Transcriptional consequences of trisomy 21 on neural induction. Front Cell Neurosci 18:1341141. doi: https://doi.org/10.3389/fncel.2024.1341141

May Z, Morrill A, Holcombe A, et al (2016) Object recognition memory in zebrafish. Behav Brain Res 296:199–210. doi: https://doi.org/10.1016/j.bbr.2015.09.016

Meeker ND, Trede NS (2008) Immunology and zebrafish: spawning new models of human disease. Dev Comp Immunol 32:745–757. doi: https://doi.org/10.1016/j.dci.2007.11.011

Meyer A, Schartl M (1999) Gene and genome duplications in vertebrates: the one-to-four (-to-eight in fish) rule and the evolution of novel gene functions. Curr Opin Cell Biol 11:699–704. doi: https://doi.org/10.1016/s0955-0674(99)00039-3

Moon JE, Lawrence JB (2022) Chromosome silencing in vitro reveals trisomy 21 causes cell-autonomous deficits in angiogenesis and early dysregulation in Notch signaling. Cell Rep 40:111174. doi: https://doi.org/10.1016/j.celrep.2022.111174

Moreno-Campos R, Singleton EW, Uribe RA (2023) A targeted CRISPR-Cas9 mediated F0 screen identifies genes involved in establishment of the enteric nervous system. BioRxiv. doi: https://doi.org/10.1101/2023.12.28.573581

Mork L, Crump G (2015) Zebrafish Craniofacial Development: A Window into Early Patterning. Curr Top Dev Biol 115:235–269. doi: https://doi.org/10.1016/bs.ctdb.2015.07.001

Mu Y, Bennett DV, Rubinov M, et al (2019) Glia Accumulate Evidence that Actions Are Futile and Suppress Unsuccessful Behavior. Cell 178:27–43.e19. doi: https://doi.org/10.1016/j.cell.2019.05.050

Mueller T (2012) What is the Thalamus in Zebrafish? Front Neurosci 6:64. doi: https://doi.org/10.3389/fnins.2012.00064

Naumann EA, Fitzgerald JE, Dunn TW, et al (2016) From Whole-Brain Data to Functional Circuit Models: The Zebrafish Optomotor Response. Cell 167:947–960.e20. doi: https://doi.org/10.1016/j.cell.2016.10.019

Neuhauss SC, Solnica-Krezel L, Schier AF, et al (1996) Mutations affecting craniofacial development in zebrafish. Development 123:357–367. doi: https://doi.org/10.1242/dev.123.1.357

Nissen RM, Amsterdam A, Hopkins N (2006) A zebrafish screen for craniofacial mutants identifies wdr68 as a highly conserved gene required for endothelin-1 expression. BMC Dev Biol 6:28. doi: https://doi.org/10.1186/1471-213X-6-28

North TE, Goessling W, Walkley CR, et al (2007) Prostaglandin E2 regulates vertebrate haematopoietic stem cell homeostasis. Nature 447:1007–1011. doi: https://doi.org/10.1038/nature05883

Olmos-Serrano JL, Kang HJ, Tyler WA, et al (2016) Down syndrome developmental brain transcriptome reveals defective oligodendrocyte differentiation and myelination. Neuron 89:1208–1222. doi: https://doi.org/10.1016/j.neuron.2016.01.042

Oprişoreanu A-M, Smith HL, Krix S, et al (2021) Automated in vivo drug screen in zebrafish identifies synapse-stabilising drugs with relevance to spinal muscular atrophy. Dis Model Mech. doi: https://doi.org/10.1242/dmm.047761

Otchy TM, Wolff SBE, Rhee JY, et al (2015) Acute off-target effects of neural circuit manipulations. Nature 528:358–363. doi: https://doi.org/10.1038/nature16442

Özcan GG, Lim S, Leighton PL, et al (2020) Sleep is bi-directionally modified by amyloid beta oligomers. Elife. doi: https://doi.org/10.7554/eLife.53995

Özcan GG, Lim S, Rihel J (2022) Genetic and chemical disruption of Amyloid Precursor Protein processing impairs zebrafish sleep maintenance. BioRxiv. doi: https://doi.org/10.1101/2022.06.08.495312

Paik EJ, Zon LI (2010) Hematopoietic development in the zebrafish. Int J Dev Biol 54:1127–1137. doi: https://doi.org/10.1387/ijdb.093042ep

Palumbo F, Serneels B, Yaksi E (2021) Optimized protocol for conditioned place avoidance learning in juvenile zebrafish. STAR Protocols 2:100465. doi: https://doi.org/10.1016/j.xpro.2021.100465

Pandey S, Moyer AJ, Thyme SB (2023) A single-cell transcriptome atlas of the maturing zebrafish telencephalon. Genome Res 33:658–671. doi: https://doi.org/10.1101/gr.277278.122

Panula P, Chen YC, Priyadarshini M, et al (2010) The comparative neuroanatomy and neurochemistry of zebrafish CNS systems of relevance to human neuropsychiatric diseases. Neurobiol Dis 40:46–57. doi: https://doi.org/10.1016/j.nbd.2010.05.010

Park H-C, Mehta A, Richardson JS, Appel B (2002) olig2 is required for zebrafish primary motor neuron and oligodendrocyte development. Dev Biol 248:356–368. doi: https://doi.org/10.1006/dbio.2002.0738

Parsons MJ, Pisharath H, Yusuff S, et al (2009) Notch-responsive cells initiate the secondary transition in larval zebrafish pancreas. Mech Dev 126:898–912. doi: https://doi.org/10.1016/j.mod.2009.07.002

Pasquier J, Braasch I, Batzel P, et al (2017) Evolution of gene expression after whole-genome duplication: New insights from the spotted gar genome. J Exp Zool B Mol Dev Evol 328:709–721. doi: https://doi.org/10.1002/jez.b.22770

Patton EE, Zon LI, Langenau DM (2021) Zebrafish disease models in drug discovery: from preclinical modelling to clinical trials. Nat Rev Drug Discov 20:611–628. doi: https://doi.org/10.1038/s41573-021-00210-8

Pelea O, Mayes S, Ferry QR, et al (2023) Specific Modulation of CRISPR Transcriptional Activators through RNA-Sensing Guide RNAs in Mammalian Cells and Zebrafish Embryos. doi: https://doi.org/10.7554/eLife.87722.1

Piotrowski T, Schilling TF, Brand M, et al (1996) Jaw and branchial arch mutants in zebrafish II: anterior arches and cartilage differentiation. Development 123:345–356. doi: https://doi.org/10.1242/dev.123.1.345

Porter BA, Mueller T (2020) The Zebrafish Amygdaloid Complex - Functional Ground Plan, Molecular Delineation, and Everted Topology. Front Neurosci 14:608. doi: https://doi.org/10.3389/fnins.2020.00608

Privat M, Romano SA, Pietri T, et al (2019) Sensorimotor transformations in the zebrafish auditory system. Curr Biol 29:4010–4023.e4. doi: https://doi.org/10.1016/j.cub.2019.10.020

Pu Y-Z, Liang L, Fu A-L, et al (2017) Generation of Alzheimer's Disease Transgenic Zebrafish Expressing Human APP Mutation Under Control of Zebrafish appb Promotor. Curr Alzheimer Res 14:668–679. doi: https://doi.org/10.2174/1567205013666161201202000

Putri RR, Chen L (2018) Spatiotemporal control of zebrafish (Danio rerio) gene expression using a light-activated CRISPR activation system. Gene 677:273–279. doi: https://doi.org/10.1016/j.gene.2018.07.077

Quillien A, Abdalla M, Yu J, et al (2017) Robust Identification of Developmentally Active Endothelial Enhancers in Zebrafish Using FANS-Assisted ATAC-Seq. Cell Rep 20:709–720. doi: https://doi.org/10.1016/j.celrep.2017.06.070

Quint WH, Tadema KCD, de Vrieze E, et al (2021) Loss of Gap Junction Delta-2 (GJD2) gene orthologs leads to refractive error in zebrafish. Commun Biol 4:676. doi: https://doi.org/10.1038/s42003-021-02185-z

Rahman MM, Fatema K (2019) Seizures in down syndrome: an update. Mymensingh Med J 28:712–715.

Raj B, Farrell JA, Liu J, et al (2020) Emergence of Neuronal Diversity during Vertebrate Brain Development. Neuron 108:1058–1074.e6. doi: https://doi.org/10.1016/j.neuron.2020.09.023

Ramesh T, Lyon AN, Pineda RH, et al (2010) A genetic model of amyotrophic lateral sclerosis in zebrafish displays phenotypic hallmarks of motoneuron disease. Dis Model Mech 3:652–662. doi: https://doi.org/10.1242/dmm.005538

Randlett O, Haesemeyer M, Forkin G, et al (2019) Distributed plasticity drives visual habituation learning in larval zebrafish. Curr Biol 29:1337–1345.e4. doi: https://doi.org/10.1016/j.cub.2019.02.039

Randlett O, Wee CL, Naumann EA, et al (2015) Whole-brain activity mapping onto a zebrafish brain atlas. Nat Methods 12:1039–1046. doi: https://doi.org/10.1038/nmeth.3581

Raut P, Sriram B, Yeoh A, et al (2011) High prevalence of hearing loss in Down syndrome at first year of life. Ann Acad Med Singap 40:493–498.

Redhead Y, Gibbins D, Lana-Elola E, et al (2023) Craniofacial dysmorphology in Down syndrome is caused by increased dosage of Dyrk1a and at least three other genes. Development. doi: https://doi.org/10.1242/dev.201077

Renier N, Adams EL, Kirst C, et al (2016) Mapping of brain activity by automated volume analysis of immediate early genes. Cell 165:1789–1802. doi: https://doi.org/10.1016/j.cell.2016.05.007

Renshaw SA, Trede NS (2012) A model 450 million years in the making: zebrafish and vertebrate immunity. Dis Model Mech 5:38–47. doi: https://doi.org/10.1242/dmm.007138

Richardson R, Tracey-White D, Webster A, Moosajee M (2017) The zebrafish eye-a paradigm for investigating human ocular genetics. Eye (Lond) 31:68–86. doi: https://doi.org/10.1038/eye.2016.198

Rihel J, Prober DA, Arvanites A, et al (2010) Zebrafish behavioral profiling links drugs to biological targets and rest/wake regulation. Science 327:348–351. doi: https://doi.org/10.1126/science.1183090

Robens BK, Yang X, McGraw CM, et al (2022) Mosaic and non-mosaic protocadherin 19 mutation leads to neuronal hyperexcitability in zebrafish. Neurobiol Dis 105738. doi: https://doi.org/10.1016/j.nbd.2022.105738

Robertsen B (2006) The interferon system of teleost fish. Fish Shellfish Immunol 20:172–191. doi: https://doi.org/10.1016/j.fsi.2005.01.010

Robinson SW, Morris CD, Goldmuntz E, et al (2003) Missense mutations in CRELD1 are associated with cardiac atrioventricular septal defects. Am J Hum Genet 72:1047–1052. doi: https://doi.org/10.1086/374319

Rohlfing T, Maurer CR (2003) Nonrigid image registration in shared-memory multiprocessor environments with application to brains, breasts, and bees. IEEE Trans Inf Technol Biomed 7:16–25. doi: https://doi.org/10.1109/titb.2003.808506

Roper RJ, VanHorn JF, Cain CC, Reeves RH (2009) A neural crest deficit in Down syndrome mice is associated with deficient mitotic response to Sonic hedgehog. Mech Dev 126:212–219. doi: https://doi.org/10.1016/j.mod.2008.11.002

Rosen LB, Ginty DD, Weber MJ, Greenberg ME (1994) Membrane depolarization and calcium influx stimulate MEK and MAP kinase via activation of Ras. Neuron 12:1207–1221. doi: https://doi.org/10.1016/0896-6273(94)90438-3

Rovelet-Lecrux A, Hannequin D, Raux G, et al (2006) APP locus duplication causes autosomal dominant early-onset Alzheimer disease with cerebral amyloid angiopathy. Nat Genet 38:24–26. doi: https://doi.org/10.1038/ng1718

Roy A, Cowan G, Mead AJ, et al (2012) Perturbation of fetal liver hematopoietic stem and progenitor cell development by trisomy 21. Proc Natl Acad Sci USA 109:17579–17584. doi: https://doi.org/10.1073/pnas.1211405109

Sadybekov AA, Sadybekov AV, Liu Y, et al (2022) Synthon-based ligand discovery in virtual libraries of over 11 billion compounds. Nature 601:452–459. doi: https://doi.org/10.1038/s41586-021-04220-9

Scheer N, Campos-Ortega JA (1999) Use of the Gal4-UAS technique for targeted gene expression in the zebrafish. Mech Dev 80:153–158. doi: https://doi.org/10.1016/S0925-4773(98)00209-3

Schindler C, Levy DE, Decker T (2007) JAK-STAT signaling: from interferons to cytokines. J Biol Chem 282:20059–20063. doi: https://doi.org/10.1074/jbc.R700016200

Shainer I, Kuehn E, Laurell E, et al (2023) A single-cell resolution gene expression atlas of the larval zebrafish brain. Sci Adv 9:eade9909. doi: https://doi.org/10.1126/sciadv.ade9909

Shan L, Xu G, Yao R-W, et al (2023) Nucleolar URB1 ensures 3' ETS rRNA removal to prevent exosome surveillance. Nature 615:526–534. doi: https://doi.org/10.1038/s41586-023-05767-5

Sharma KR, Colvis CM, Rodgers GP, Sheeley DM (2023) Illuminating the druggable genome: Pathways to progress. Drug Discov Today 29:103805. doi: https://doi.org/10.1016/j.drudis.2023.103805

Sharrock AV, Mulligan TS, Hall KR, et al (2022) NTR 2.0: a rationally engineered prodrug-converting enzyme with substantially enhanced efficacy for targeted cell ablation. Nat Methods 19:205–215. doi: https://doi.org/10.1038/s41592-021-01364-4

Shemesh M, Lochte S, Piehler J, Schreiber G (2021) IFNAR1 and IFNAR2 play distinct roles in initiating type I interferon-induced JAK-STAT signaling and activating STATs. Sci Signal 14:eabe4627. doi: https://doi.org/10.1126/scisignal.abe4627

Shi T, Beaulieu MO, Saunders LM, et al (2023) Single-cell transcriptomic profiling of the zebrafish inner ear reveals molecularly distinct hair cell and supporting cell subtypes. Elife. doi: https://doi.org/10.7554/eLife.82978

Shimizu N, Kawakami K, Ishitani T (2012) Visualization and exploration of Tcf/Lef function using a highly responsive Wnt/β-catenin signaling-reporter transgenic zebrafish. Dev Biol 370:71–85. doi: https://doi.org/10.1016/j.ydbio.2012.07.016

Shin J, Park H-C, Topczewska JM, et al (2003) Neural cell fate analysis in zebrafish using olig2 BAC transgenics. Methods Cell Sci 25:7–14. doi: https://doi.org/10.1023/B:MICS.0000006847.09037.3a

Shin M, Yin H-M, Shih Y-H, et al (2023) Generation and application of endogenously floxed alleles for cell-specific knockout in zebrafish. Dev Cell 58:2614–2626.e7. doi: https://doi.org/10.1016/j.devcel.2023.07.022

Shott SR, Joseph A, Heithaus D (2001) Hearing loss in children with Down syndrome. Int J Pediatr Otorhinolaryngol 61:199–205. doi: https://doi.org/10.1016/s0165-5876(01)00572-9

Smith KA, Joziasse IC, Chocron S, et al (2009) Dominant-negative ALK2 allele associates with congenital heart defects. Circulation 119:3062–3069. doi: https://doi.org/10.1161/CIRCULATIONAHA.108.843714

Stainier DY, Fouquet B, Chen JN, et al (1996) Mutations affecting the formation and function of the cardiovascular system in the zebrafish embryo. Development 123:285–292. doi: https://doi.org/10.5167/uzh-237

Stankiewicz MJ, Crispino JD (2009) ETS2 and ERG promote megakaryopoiesis and synergize with alterations in GATA-1 to immortalize hematopoietic progenitor cells. Blood 113:3337–3347. doi: https://doi.org/10.1182/blood-2008-08-174813

Staudt D, Stainier D (2012) Uncovering the molecular and cellular mechanisms of heart development using the zebrafish. Annu Rev Genet 46:397–418. doi: https://doi.org/10.1146/annurev-genet-110711-155646

Stein RM, Kang HJ, McCorvy JD, et al (2020) Virtual discovery of melatonin receptor ligands to modulate circadian rhythms. Nature 579:609–614. doi: https://doi.org/10.1038/s41586-020-2027-0

Sternberg JR, Severi KE, Fidelin K, et al (2016) Optimization of a neurotoxin to investigate the contribution of excitatory interneurons to speed modulation in vivo. Curr Biol 26:2319–2328. doi: https://doi.org/10.1016/j.cub.2016.06.037

Sugiyama M, Sakaue-Sawano A, Iimura T, et al (2009) Illuminating cell-cycle progression in the developing zebrafish embryo. Proc Natl Acad Sci USA 106:20812–20817. doi: https://doi.org/10.1073/pnas.0906464106

Sun E, Kraus CL (2023) The ophthalmic manifestations of down syndrome. Children (Basel). doi: https://doi.org/10.3390/children10020341

Suster ML, Abe G, Schouw A, Kawakami K (2011) Transposon-mediated BAC transgenesis in zebrafish. Nat Protoc 6:1998–2021. doi: https://doi.org/10.1038/nprot.2011.416

Taoudi S, Bee T, Hilton A, et al (2011) ERG dependence distinguishes developmental control of hematopoietic stem cell maintenance from hematopoietic specification. Genes Dev 25:251–262. doi: https://doi.org/10.1101/gad.2009211

TCW J, Goate AM (2017) Genetics of β-Amyloid Precursor Protein in Alzheimer's Disease. Cold Spring Harb Perspect Med. doi: https://doi.org/10.1101/cshperspect.a024539

Teng CS, Yen H-Y, Barske L, et al (2017) Requirement for Jagged1-Notch2 signaling in patterning the bones of the mouse and human middle ear. Sci Rep 7:2497. doi: https://doi.org/10.1038/s41598-017-02574-7

Thomas JR, Roper RJ (2021) Current analysis of skeletal phenotypes in down syndrome. Curr Osteoporos Rep 19:338–346. doi: https://doi.org/10.1007/s11914-021-00674-y

Thyme SB, Pieper LM, Li EH, et al (2019) Phenotypic Landscape of Schizophrenia-Associated Genes Defines Candidates and Their Shared Functions. Cell 177:478–491.e20. doi: https://doi.org/10.1016/j.cell.2019.01.048

Tonelli F, Bek JW, Besio R, et al (2020) Zebrafish: A resourceful vertebrate model to investigate skeletal disorders. Front Endocrinol (Lausanne) 11:489. doi: https://doi.org/10.3389/fendo.2020.00489

Torraca V, Masud S, Spaink HP, Meijer AH (2014) Macrophage-pathogen interactions in infectious diseases: new therapeutic insights from the zebrafish host model. Dis Model Mech 7:785–797. doi: https://doi.org/10.1242/dmm.015594

Traver D, Winzeler A, Stern HM, et al (2004) Effects of lethal irradiation in zebrafish and rescue by hematopoietic cell transplantation. Blood 104:1298–1305. doi: https://doi.org/10.1182/blood-2004-01-0100

Ulitsky I, Shkumatava A, Jan CH, et al (2011) Conserved function of lincRNAs in vertebrate embryonic development despite rapid sequence evolution. Cell 147:1537–1550. doi: https://doi.org/10.1016/j.cell.2011.11.055

Urasaki A, Morvan G, Kawakami K (2006) Functional dissection of the Tol2 transposable element identified the minimal cis-sequence and a highly repetitive sequence in the subterminal region essential for transposition. Genetics 174:639–649. doi: https://doi.org/10.1534/genetics.106.060244

Uribe RA (2024) Genetic regulation of enteric nervous system development in zebrafish. Biochem Soc Trans. doi: https://doi.org/10.1042/BST20230343

Valente A, Huang K-H, Portugues R, Engert F (2012) Ontogeny of classical and operant learning behaviors in zebrafish. Learn Mem 19:170–177. doi: https://doi.org/10.1101/lm.025668.112

Vargas JP, López JC, Portavella M (2009) What are the functions of fish brain pallium? Brain Res Bull 79:436–440. doi: https://doi.org/10.1016/j.brainresbull.2009.05.008

Wang G, Rajpurohit SK, Delaspre F, et al (2015) First quantitative high-throughput screen in zebrafish identifies novel pathways for increasing pancreatic β-cell mass. Elife. doi: https://doi.org/10.7554/eLife.08261

Wang J, Luo J, Chen Q, et al (2018) Identification of LBX2 as a novel causal gene of atrial septal defect. Int J Cardiol 265:188–194. doi: https://doi.org/10.1016/j.ijcard.2018.04.038

Wang K, Vorontsova I, Hoshino M, et al (2020) Optical development in the zebrafish eye lens. FASEB J 34:5552–5562. doi: https://doi.org/10.1096/fj.201902607R

Waugh KA, Minter R, Baxter J, et al (2023) Triplication of the interferon receptor locus contributes to hallmarks of Down syndrome in a mouse model. Nat Genet 55:1034–1047. doi: https://doi.org/10.1038/s41588-023-01399-7

Welker JM, Wierson WA, Almeida MP, et al (2021) Geneweld: efficient targeted integration directed by short homology in zebrafish. Bio Protoc 11:e4100. doi: https://doi.org/10.21769/BioProtoc.4100

Weuring WJ, Dilevska I, Hoekman J, et al (2021) CRISPRa-Mediated Upregulation of scn11aa During Early Development Causes Epileptiform Activity and dCas9-Associated Toxicity. The CRISPR Journal 4:575–582. doi: https://doi.org/10.1089/crispr.2021.0013

White RM, Sessa A, Burke C, et al (2008) Transparent adult zebrafish as a tool for in vivo transplantation analysis. Cell Stem Cell 2:183–189. doi: https://doi.org/10.1016/j.stem.2007.11.002

Wolman MA, Jain RA, Liss L, Granato M (2011) Chemical modulation of memory formation in larval zebrafish. Proc Natl Acad Sci USA 108:15468–15473. doi: https://doi.org/10.1073/pnas.1107156108

Yeh J-RJ, Munson KM, Chao YL, et al (2008) AML1-ETO reprograms hematopoietic cell fate by downregulating scl expression. Development 135:401–410. doi: https://doi.org/10.1242/dev.008904

Yi Z-N, Chen X-K, Ma AC-H (2022) Modeling leukemia with zebrafish (Danio rerio): Towards precision medicine. Exp Cell Res 421:113401. doi: https://doi.org/10.1016/j.yexcr.2022.113401

Yousefelahiyeh M, Xu J, Alvarado E, et al (2018) DCAF7/WDR68 is required for normal levels of DYRK1A and DYRK1B. PLoS One 13:e0207779. doi: https://doi.org/10.1371/journal.pone.0207779
Zang L, Maddison LA, Chen W (2018) Zebrafish as a model for obesity and diabetes. Front Cell Dev Biol 6:91. doi: https://doi.org/10.3389/fcell.2018.00091
Zhang Y, Ouyang J, Qie J, et al (2019) Optimization of the Gal4/UAS transgenic tools in zebrafish. Appl Microbiol Biotechnol 103:1789–1799. doi: https://doi.org/10.1007/s00253-018-09591-0
Zhang Z, Bai L, Cong L, et al (2021) Imaging volumetric dynamics at high speed in mouse and zebrafish brain with confocal light field microscopy. Nat Biotechnol 39:74–83. doi: https://doi.org/10.1038/s41587-020-0628-7
Zhong Z, Ohnmacht J, Reimer MM, et al (2012) Chondrolectin mediates growth cone interactions of motor axons with an intermediate target. J Neurosci 32:4426–4439. doi: https://doi.org/10.1523/JNEUROSCI.5179-11.2012
Zhu SI, McCullough MH, Pujic Z, et al (2023) fmr1 Mutation Alters the Early Development of Sensory Coding and Hunting and Social Behaviors in Larval Zebrafish. J Neurosci 43:1211–1224. doi: https://doi.org/10.1523/JNEUROSCI.1721-22.2022

Human Models of Down Syndrome

Megan Jandy, Hao Hu, Yan Liu, and Anita Bhattacharyya

Abstract Trisomy 21 results in a group of clinical features known as Down syndrome (DS), the most common form of congenital developmental disability (Presson et al., J Pediatr 163:1163–1168, 2013; Shin, Pediatrics 124:1565–1571, 2009; Parker, Mol Teratol 88:1008–1016, 2010). The fundamental question for the field is how an extra copy of human chromosome 21 (HSA21) translates into the organ-specific pathologies in individuals with DS. Tissue and cells obtained from individuals with DS have been a primary source of analysis since before the discovery that trisomy 21 causes DS in 1959 (Lejeune, Ann Genet 1:41–49, 1959). Examination of primary tissue and cells established the features of DS in different organ systems. More recently, human pluripotent stem cells carrying trisomy 21 enable the generation of affected cell types and, importantly, provide insight into the development and degeneration of affected cell types. Further, exogenous and genetic manipulation of trisomy 21 cells facilitates identification of the contribution of genes and pathways to the pathogenesis of organ systems affected in DS.

Keywords Cells · Tissue · Human · Post-mortem · Stem cells · Neural · Organ

M. Jandy
Waisman Center, University of Wisconsin-Madison, Madison, WI, USA

Cellular and Molecular Biology Graduate Program, University of Wisconsin-Madison, Madison, WI, USA

H. Hu · Y. Liu (✉)
Institute for Stem Cell and Neural Regeneration, School of Pharmacy, State Key Laboratory of Reproductive Medicine, Nanjing Medical University, Nanjing, China
e-mail: yanliu@njmu.edu.cn

A. Bhattacharyya (✉)
Waisman Center, University of Wisconsin-Madison, Madison, WI, USA

Department of Cell and Regenerative Biology, School of Medicine and Public Health, University of Wisconsin-Madison, Madison, WI, USA
e-mail: bhattacharyy@waisman.wisc.edu

B. Ye, R. H. Reeves (eds.), *Genetic Models of Down Syndrome*,
https://doi.org/10.1007/978-3-031-78611-2_7

1 Genetics of Human Chromosome 21 and Trisomy 21

Trisomy 21 (T21) is the only viable human autosomal aneuploidy. Most individuals with DS have full T21 caused by non-disjunction during gametogenesis. A small percentage of individuals have a Robertson translocation (3%) or mosaicism (5–7%) (Devlin and Morrison 2004; Papavassiliou et al. 2009, 2015). Rare individuals have partial trisomy in which only segments of HSA21 are triplicated (Delabar et al. 1993; Korenberg et al. 1990).

HSA21 is the smallest human chromosome, comprising about 48 million base pairs and representing 1.5 to 2 percent of the total DNA in cells (Hattori et al. 2000; Gardiner et al. 2002). The p-arm of HSA21 consists of highly repetitive DNA sequences including one of the five nucleolus organizing regions in the genome. The q-arm contains protein-coding genes and non-protein-coding sequences (Fig. 1). Of the proteins encoded on HSA21, those that are dosage sensitive are thought to drive phenotypes in DS. However, we have an incomplete understanding of which HSA21 encoded genes are sensitive to copy number. This gap in knowledge makes it difficult to attribute DS phenotypes to any one gene or small groups of genes. Further, overexpression of HSA21 genes affects the expression of genes across the genome.

HSA21 genes are encoded across at least three mouse chromosomes, while other model organisms have more segmented HSA21 homologous regions throughout

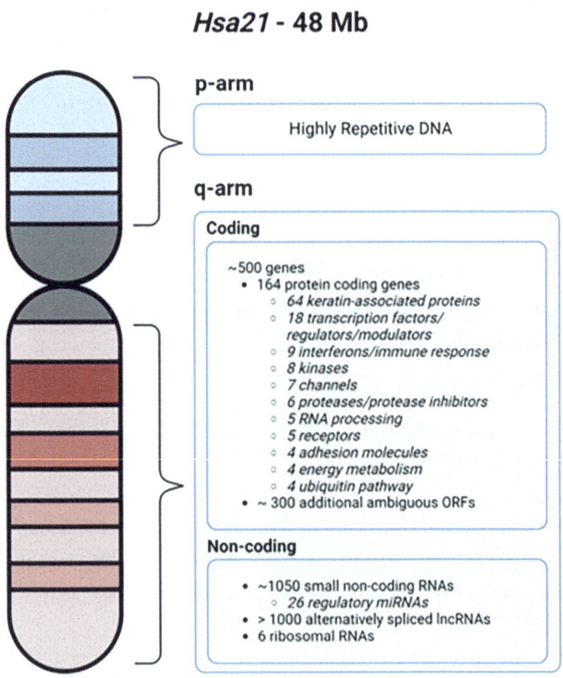

Fig. 1 Human chromosome 21, the smallest human autosome, contains hundreds of genes and thousands of noncoding sequences (Ganguly 2022). (Created with BioRender.com)

their genomes (https://www.ncbi.nlm.nih.gov/projects/homology/maps/) (Moyer et al. 2021). Comparison of DNA between human and animal models reveals subtle sequence differences and human specific genes that likely contribute to gene regulation. For example, comparative genomics indicates different expression patterns of the same genes in human and mouse (Breschi et al. 2017; Denas et al. 2015; Kwan et al. 2012). Thus, differences in the timing, pattern, and level of gene expression between human and non-human models may result in distinct molecular mechanisms and diverse phenotypic consequences. Further, non-protein-coding sequences that are transcribed, including many species of RNA, also modulate gene function and regulation. How these non-protein-coding sequences influence the expression of HSA21 and non-HSA21 genes is largely unknown.

2 Examination of Primary Cells and Tissue from Individuals with DS Reveals Deficits in Brain and Other Organs

Clinical examination of individuals with DS together with analyses of post-mortem tissues established the structural pathologies that typify the syndrome and generally include the presence of smaller organs (Heinke et al. 2021; Antonarakis et al. 2020). Analysis of post-mortem tissue and human cells from individuals with DS provides some, albeit static, mechanistic insight.

2.1 Human Postmortem Tissue

Post-mortem tissues have historically been the major source for analyses to inform our understanding of the anatomical, cellular, and molecular features of DS (Hasina et al. 2022). Tissue and cells are generally available due to the widespread diagnosis of DS. Formalin-fixed tissue enables structural analyses and provides a static representation of organ features. Challenges to the use of this tissue include long post-mortem interval (PMI) and extensive periods of fixation. Frozen tissue is used for molecular studies, including analysis of proteins, gene expression, and other omics. Although available, tissue samples can be limited in number and amount so it can be difficult to age, sex, and race/ethnicity match samples to controls for comparison (Risgaard et al. 2022).

Structural pathologies in DS are obvious in the neurological, cardiovascular (Rehder 1981), and craniofacial systems, although most organ systems are affected (Antonarakis et al. 2020; Catalano 1990; Pueschel 1990). Because the most penetrant feature of DS is intellectual disability (Silverman 2007; Lott 2012), post-mortem brain tissue has been used to define and explore structural defects. Neural correlates of intellectual and language function include cerebral cortical structures, specifically temporal and frontal lobes, which are smaller in DS (Friederici and

Gierhan 2013; Jung and Haier 2007; Van Den Heuvel and Sporns 2013; Menghini et al. 2011; Carducci et al. 2013).

Pathology is present at birth, indicating that developmental defects occur during the prenatal period. Prenatal tissue can be obtained from elective abortions because T21 can be diagnosed through amniocentesis in pregnancy. Availability of prenatal tissue is dependent on local and national policy. Prenatal DS tissue enables analyses of developmental processes and how they differ in DS (Contestabile et al. 2007; Guidi et al. 2008, 2011, 2018; Stagni et al. 2019; Olmos-Serrano et al. 2016). Together, the anatomical features in DS at birth suggest that organs do not develop or grow normally. These data and others advocate for a developmental biology approach to understand the structural features that emerge in DS.

2.2 *Primary Cells*

Primary cells obtained from individuals with DS are used to study cell and molecular biology of T21. Various cell types can be acquired from living individuals and, in some cases, from post-mortem tissue. Lymphocytes from blood and fibroblasts from skin and other tissues can be readily acquired from individuals of any age. T21 amniocytes can be acquired during amniocentesis during a pregnancy in which the fetus has been diagnosed with T21. All these cells can be maintained and expanded for a short time in culture to assess cellular processes that may be affected by T21 including proliferation, cell viability, mitochondrial function, and molecular dysregulation analyzed through proteomics and transcriptomics. Blood-derived cells have been used to identify biomarkers, most recently in adults with DS to assess Alzheimer's disease pathology (Montoliu-Gaya et al. 2021). Despite the accessibility of these cells, primary cells may not represent aspects of biology of all organs and cell types.

3 Human Pluripotent Stem Cells Offer Experimental Models to Study Trisomy 21

Investigation of stages of human development provides insight into how T21 results in smaller organs. Human pluripotent stem cell (hPSC) models offer a dynamic and manipulatable platform to study the effects of T21 on early development of different organ systems. Differentiation of trisomic hPSCs lines into specific mature cell types provides insight into the cellular and molecular processes altered in development, function, and degeneration of organ-specific T21 cells.

3.1 Human Embryonic Stem Cells

In 1998, seminal work describing the first sustainable cultures of human embryonic stem cells (hESCs) revolutionized research in human cell models (Thomson et al. 1998). The inner cell masses harvested from donated in vitro fertilized human blastocysts were used to create five genetically unique hESC lines. These hESCs are proliferative and karyotypically stable in their undifferentiated state and have high levels of telomerase activity necessary to provide a longer proliferative lifespan beyond human somatic noncancerous cell lines. Pluripotency, the ability to differentiate into all cell types, is validated by the fact that hESC lines form teratomas. Thus, hESCs in culture retain hallmarks of stem cells in vivo: self-renewal and pluripotency. The establishment of hESC cultures ushered in a new era of research in which investigation of human development in vitro was possible.

Scientists around the world began creating hESC lines to model human development and disease (Ludwig et al. 2018). To investigate a specific genetic abnormality, either the blastocyst must naturally contain the disease mutation or the stem cells must be genetically edited. Several T21 hESC lines were developed from donated embryos obtained during assisted reproductive technology procedures (Biancotti et al. 2010; Dumevska et al. 2016; Halevy et al. 2016; Salari et al. 2023; Shahbazi et al. 2020). While hESC lines are used to study development, researchers studying complex genetic diseases rely more on human induced pluripotent stem cells due to the ability to generate them from living donors whose features are known.

3.2 Human Induced Pluripotent Stem Cells

Building on the unique characteristics of hESCs, human induced pluripotent stem cells (hiPSCs) were first described in 2007 and enabled the reprogramming of somatic cells into cells with the hallmark features of hESCs through forced expression of transcription factors known to mediate pluripotency (Oct3/4, Sox2, and either Klf4, and Myc-C, or NANOG, and LIN28) (Park et al. 2008b; Takahashi et al. 2007; Yu et al. 2007) (Fig. 2). Reprogramming provides a way to create stem cells from diverse individuals and those with specific diseases (Park et al. 2008a). Further, hiPSC generation avoids the ethical controversy of generating stem cell lines from human embryos. Initially, hiPSCs were generated from fibroblasts using viral transfection. However, use of certain viral vectors leads to genetic integration; therefore, the highly efficient and non-integrating Sendai RNA virus has become the standard for hiPSC reprogramming (Tokusumi et al. 2002; Fusaki et al. 2009).

Importantly, hiPSCs retain genetic mutations of the somatic cells through the reprogramming process. Many hiPSC lines are generated from individuals with complex genetic diseases, including those from individuals with DS. These hiPSC lines can be produced from donated fibroblasts, amniotic fluid, and urinary cells by

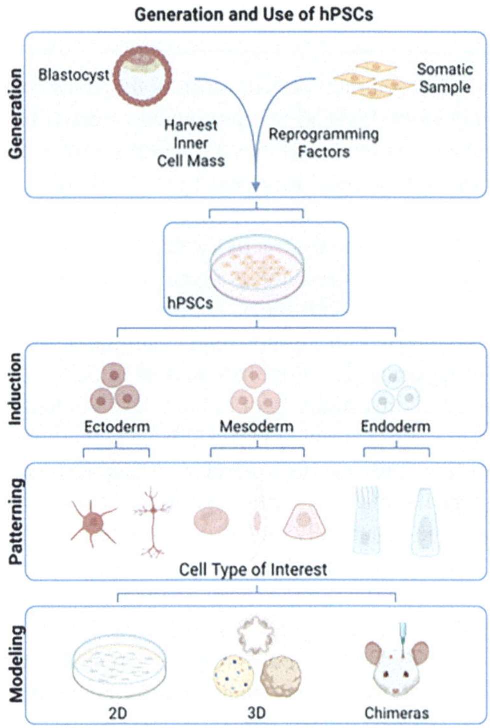

Fig. 2 Schematic depicting the generation and differentiation of hPSCs for modeling development and disease. (Created with BioRender.com)

a variety of reprogramming mechanisms, and with both full and partial triplication of chromosome 21 (Hibaoui and Feki 2015).

hiPSCs generated from age- and sex-matched unaffected individuals are typically used as controls for disease modeling, including for Down syndrome. Yet, for Down syndrome specifically, inherent genetic variability in the cells can either identify phenotypes unrelated to T21 or mask T21 specific phenotypes. Thus, isogenic control iPSCs provide a paradigm to identify only T21 related phenotypes. The advent of gene editing enables the correction of single genes but is not yet useful for correcting or introducing triplication of an entire chromosome. Somatic cells from individuals mosaic for T21 can be reprogrammed to create clonal isogenic T21 and euploid/disomic hiPSCs with the same genetic background. These isogenic cells limit the individual variability and allow direct investigation of the effects of T21 (Papavassiliou et al. 2015; Weick et al. 2013; Murray et al. 2015).

Isogenic disomic cell lines can be generated through chromosomal silencing of HSA21 in T21 hPSCs. Inducible *XIST* transgenes inserted into HSA21 repress transcription through a chromosome wide heterochromatin state, functionally silencing the third copy of HSA21 (Jiang et al. 2013). However, further transcriptomic analysis is warranted to evaluate the similarity of these cells to euploid hPSCs.

hiPSCs generated from individuals with DS provide the opportunity to study how triplication of HSA21 affects these processes through manipulations and real-time observations. To date, T21 hPSC models have been utilized to identify a myriad of cellular and molecular differences in DS development with increasing numbers of publications featuring their use. Analyses of these stem cell models will hopefully identify altered pathway drivers and, potentially, therapeutic targets to alleviate some characteristics of DS.

3.3 Maintenance and Differentiation of hPSC Models

Human PSCs, hESCs and hiPSCs collectively, are maintained on embryonic fibroblast feeders or in feeder-free conditions on matrices (Thomson et al. 1998; Chen et al. 2014b). Regardless of culture conditions, hPSCs rely on basic fibroblast growth factor to maintain their pluripotency. Directed differentiation of hPSCs in vitro mimics in vivo development by inducing specific germ layer precursors followed by patterning to organ-specific progenitors and ultimately to the cell type of interest (Itskovitz-Eldor et al. 2000; Trounson 2006). Mimicking of developmental biological processes in hPSC differentiation protocols enables the generation of many cell types from organ systems affected by T21, including multiple neural types, cardiomyocytes, hematopoietic cells, and skeletal muscle cells (Chambers et al. 2009; Shi et al. 2012a; Zhang et al. 2001; Burridge et al. 2014; Maffioletti et al. 2015; Kennedy et al. 2007).

3.4 Two- and Three-Dimensional Models of hPSC Differentiation

Due to its efficiency, many differentiation protocols use a traditional two-dimensional (2D) culture method that results in an enriched or homogenous monolayer culture of cells due to equal access to oxygen, nutrients, and added signaling molecules (Edmondson et al. 2014). 2D differentiation methods are particularly good for quantitative and developmental fate decision studies as the cells can be easily analyzed. However, 2D cultures do not fully recapitulate some features present in in vivo development such as complex cellular interactions and concentration gradients of morphogens, nutrients, and oxygen. Nonetheless, cell type homogeneity and ease of examination of monolayer cultures promotes high-throughput analyses, which is particularly useful for pharmacological testing.

Three-dimensional (3D) culture techniques including organoids, organs-on-chips, and bioprinting promote complexity and encourage cell–cell interactions that more accurately recapitulate tissue heterogeneity and microenvironments found in vivo. Human PSC-derived 3D models of organs, including brain, kidney, intestine, and lung, are utilized to study both complex biological processes and pharmacological interactions (Kim et al. 2020). Due to their ability to mimic cell diversity and cell–cell interactions of in vivo tissues, it has been argued that applying these 3D hPSC models to investigate the development of organ systems could lead to a reduction in animal in vivo modeling — both saving money and reducing the need for animal sacrifice. 3D models also allow for the direct investigation of pharmacological effects on complex human tissues, closing the gap between bench and human preclinical trials (Bhatia and Ingber 2014). In fact, pharmacological testing in 3D culture systems provides more accuracy in modeling in vivo drug response in comparison to 2D cultures, potentially accelerating the timeline from drug discovery to human application (Fang and Eglen 2017). Analysis of 3D cultures also enables the study of early developmental events that are not accessible in humans. Most organoid models can be cultured for long times in efforts to expand the developmental time window that can be analyzed. Yet, 3D modeling cannot yet be used to study the entire prenatal development timeline. Human brain organoids, for example, cultured in vitro are equivalent to the second trimester of human fetal brain at the transcriptome level and structural level (Paşca et al. 2015; Amiri et al. 2018; Camp et al. 2015).

Additional limitations of 3D hPSC cultures include the lack of control over self-organization, limited cell types, lack of vascularization, and variability. 3D models contain a limited number of cell types because they are formed by induction of a specific germ layer, while cell types within organs commonly arise from different germ layers. For instance, brain organoid models derived from neural induction lack the precursors required to form microglia and endothelial cells, which play important roles in synaptic pruning and signaling, respectively (Zhang et al. 2023). Attempts to overcome these challenges include distinct 3D culture systems that are patterned separately and then combined or co-cultured (Sloan et al. 2018). Lack of vascularization results in size limitations in 3D tissues, as diffusion-limited gradients of oxygen and nutrients lead to necrotic cores in long-term cultures. However, high O2, microfluidics, and methods to expose the inside of 2D cultures to the culture environment are potential solutions. The major challenges in 3D modeling are the heterogeneity of cultures and lack of standardization across the field, resulting in inconsistent reproducibility (Sandoval et al. 2024).

3.5 Chimeras

Human stem cells can be transplanted into another species, predominantly mice, to create chimeras (Levine and Grabel 2017; Mascetti and Pedersen 2016; Masaki and Nakauchi 2017). The gold standard for assessing pluripotency of hPSCs is through the transplantation of hPSCs into immunodeficient mice to generate teratomas. Chimeras also offer an experimental tool for interrogating the fate potential of cells derived from hPSCs and address the inherent limitations of 2D and 3D cell culture models in their ability to recapitulate the complexity of in vivo environments. For example, human brain organoids or human neural and glial progenitor cells derived from hPSCs integrate within the host brain upon transplantation and enable a deeper analysis of mature cell function.

4 Investigating Effects of Trisomy 21 on Development of the Cerebral Cortex with Trisomy 21 hPSCs

Intellectual disability results largely from abnormal formation or function of the cerebral cortex. Consequently, most investigations using hPSC models have focused on the effects of T21 on neural development and neuron function of the cerebral cortex (Fig. 3).

Human PSC modeling of cortical development attempts to mimic neurodevelopmental processes in the embryo. In vivo, neuroectoderm derives from ectoderm followed by formation of a polarized epithelial layer lining the surface of the developing ventricles (Cadwell et al. 2019). The neural stem cells of the ventricular zone first divide to self-renew, and subsequently divide asymmetrically to differentiate into radial glia cells that give rise to progenitors which produce neurons. The newly born neurons migrate radially toward the cortical plate and settle in an inside-out manner so that the first-born neurons occupying the deepest layers of the cortex. In

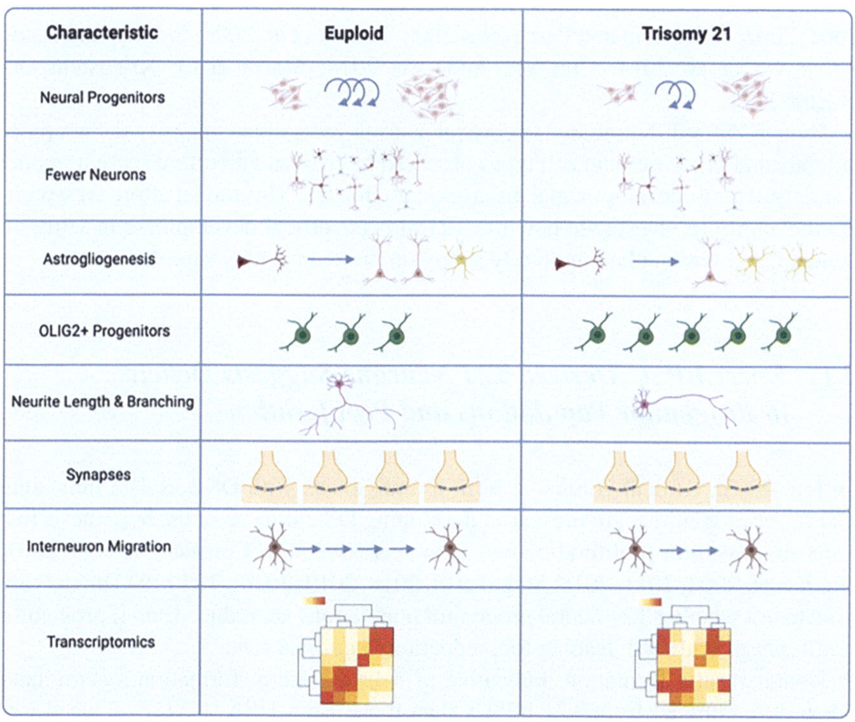

Fig. 3 Phenotypes observed in T21 hPSC-derived cortical differentiations including reduced number and proliferation of neural progenitors, reduced layer specific neurons, astroglial lineage bias, increased OLIG2+ progenitors, reduced neurite length and branching, reduced synaptic markers, reduced migration length, and global transcriptomic changes. (Created with BioRender.com)

humans, outer radial glia (oRG) are also generated that produce the expanded outer cortical layer neuron population unique to higher-level primates (Hansen et al. 2010). After neurogenesis, radial glia become astrocyte precursor cells giving rise to astrocytes (astrogliogenesis), while remaining radial glia become oligodendrocyte precursor cells and eventually oligodendrocytes (oligodendrogenesis) perinatally. Thus, the development of the human cerebral cortex follows a highly synchronized temporal pattern.

Proper functioning of the cerebral cortex relies on the correct number and laminar placement of excitatory and inhibitory neurons. This process is remarkably complex because excitatory and inhibitory neurons have different developmental origins, thus cortical formation relies on the synchronization of these two developmental programs (Kepecs and Fishell 2014, Chu and Anderson 2014, Arber and Li 2013, Gelman et al. 2012, Batista-Brito and Fishell 2009, Wonders and Anderson 2006, Flames and Marin 2005, Kessaris et al. 2014). Excitatory glutamatergic projection neurons arise from progenitor cells in the ventricular zone of the *dorsal* forebrain while inhibitory interneurons arise from progenitors in the *ventral* forebrain. All respond to cues in their environment to migrate to their proper laminar destination to mature and function (Parnavelas et al. 2000; Anderson et al. 1997, 2001, 2002; Kriegstein and Parnavelas 2006; Noctor et al. 2007; Sidman and Rakic 1973; Wu et al. 2014; Chu and Anderson 2014; Marin 2013; Kriegstein and Noctor 2004).

Human PSCs differentiated to cortical neurons recapitulate many of the neurodevelopmental processes and cell types observed in vivo, and therefore serve to model a variety of neurodevelopmental diseases including DS. This model affords researchers the ability to investigate how the T21 affects cortical development in vitro, on demand, and without having to rely solely on limited primary samples.

4.1 Fewer hPSC-Derived T21 Neurons Suggests Deficits in Progenitor Populations and Proliferation

Differences in cortical volume at birth in individuals with DS and data indicating that fewer progenitors are present in developing DS cortex raise the hypothesis that reduced progenitor proliferation results in decreased size (Contestabile et al. 2007; Guidi et al. 2008, 2011, 2018; Stagni et al. 2018, 2019, 2020). T21 hPSC models are used to test whether T21 neural progenitor populations are reduced and if progenitor proliferation is altered, leading to a reduction in cortical size.

Neural rosette formation, indicative of neuroectoderm formation, occurs later when differentiated from T21 hPSCs than those from HSA21 *XIST* silenced isogenic controls (Jiang et al. 2013). However, rosettes form prematurely in T21 hPSC embryoid bodies (EBs) (Prutton et al. 2022). This contradiction could be explained by the differences between 2D and 3D neural induction techniques and/or the unknown effects of *XIST* chromosome silencing. T21 neural progenitor cells (NPCs) show significantly smaller 3D analogs of neural rosettes (Tang et al. 2021). Taken together, these data indicate that the trisomy has an effect on the early stages of NPC proliferation, and investigations using T21 NPCs support reduced proliferative phenotypes as fewer NPCs and proliferative markers were observed in both 2D and 3D cultures (Murray et al. 2015; Hibaoui et al. 2014; Tang et al. 2021; Czermiński and Lawrence 2020).

In these and other studies, T21 hPSC-derived progenitor populations are identified using common NPC markers. It is possible that specific progenitor populations are altered, and that some are more affected than others by T21. Further, single cell sequencing will continue to reveal subpopulations based on gene expression rather than marker expression (Giffin-Rao et al. 2022). Future studies using hPSC models to quantify specific progenitor subtypes, such vRG, oRG, and intermediate progenitors, along with their proliferative capacities will help define the differential effects of T21 on neural progenitor populations.

4.2 Reduction of hPSC-Derived Neurons May Be Due to Increased Apoptosis in T21 Progenitor and Neuron Populations

An increase in apoptosis is also a potential mechanism for the hypo-cellularity in the developing DS cortex. In vivo studies of prenatal DS brains and primary cultures have demonstrated conflicting results as to whether there is increased apoptosis in DS (Rueda et al. 2013). These results may be due in part to the challenges associated with multiple mechanisms of cell death, temporal changes in these processes, and the inability to detect dead cells. Human PSC models of T21 cortical development enable observation of apoptotic changes early in development and in real time. Nuclear morphology and immunocytochemical data suggest that apoptosis is increased in T21 hPSC-derived NPCs (Hibaoui et al. 2014). Manual NPC cell viability counts performed during passaging suggest no difference in cell death; however, this method is less precise than immunocytochemical quantitation (Murray et al. 2015). Multiple studies that utilize the TUNEL assay (terminal deoxynucleotidyl transferase dUTP nick end labeling) that detects the DNA breaks formed when DNA fragmentation occurs in the last phase of apoptosis, found no differences in apoptosis between T21 and control neurons (Briggs et al. 2013; Weick et al. 2013; Tang et al. 2021). However, upon treatment with H_2O_2, T21 neurons show twice the apoptotic activity as their disomic counterparts, suggesting that these neurons are more sensitive to reactive oxygen species (ROS) and consequently cell death (Briggs et al. 2013). More recent sophisticated longitudinal analysis of T21 neurons using a cell death indicator that is activated before the cells are lost from culture supports excess cell death in T21 neurons, which can be corrected by silencing the extra chromosome (Bansal et al. 2024).Taken together, the role of apoptosis in T21 hPSC models and thus in DS cortical development is not completely resolved. It is possible, for example, that some T21 progenitors and/or neurons have increased apoptosis while others do not (Bansal et al. 2024). As apoptosis is a naturally programmed process during late-stage cortical development, further studies are needed to determine whether neurons die aberrantly throughout development in DS.

4.3 Reduced Populations of hPSC-Derived T21 Cortical Neuron Subtypes Suggest Altered Neurogenesis

The presence of fewer neurons in DS primary cortical tissue samples leads to the hypothesis that there are global or subtype-specific neurogenesis defects in the developing DS cortex (Ross et al. 1984; Golden and Hyman 1994). Differentiation of T21 NPCs into post-mitotic cortical neurons revealed that some progenitors have reduced neurogenic capacity that may account for the differences observed in vivo (Bhattacharyya et al. 2009; Giffin-Rao et al. 2022; Lu et al. 2013).

Excitatory neurons in the dorsal cortex are responsible for higher-order cognition that is affected in DS, and analysis of fetal cortex suggests defects in DS (Golden and Hyman 1994). However, conclusions from in vivo studies are limited due to small sample sizes. T21 2D hPSC models demonstrate conflicting results with some describing no difference in hPSC-derived excitatory neurons (Shi et al. 2012b; Murray et al. 2015; Weick et al. 2013). In contrast, results using 3D hPSCs indicate significant reductions in both the lower- and upper-layer cortical neuron populations (Tang et al. 2021; Li et al. 2022). It is interesting to consider how 2D vs. 3D culture techniques may affect neural subtype specification resulting in divergent conclusions. Further subtype analysis in primary tissues is required for confirmation.

DS mouse models have an increased ratio of inhibitory to excitatory neurons compared to euploid mice (Chakrabarti et al. 2010), raising a hypothesis that there is an imbalance in the production of excitatory and inhibitory neurons in DS. The observations in mouse appear to contradict human tissue data that indicate decreases in inhibitory neurons (Ross et al. 1984; Bhattacharyya et al. 2009; Guidi et al. 2018; Giffin-Rao et al. 2022; Golden and Hyman 1994; Džaja et al. 2014). However, it is important to consider differences in neurodevelopmental timelines and relative abundance of specific cell populations in mice versus humans, as well as the specific functional subtypes of interneurons. Whether the population of excitatory neurons in DS is affected has not been well established in humans, precluding ratio calculations.

Differences in inhibitory neuron development in DS can occur in progenitor proliferation, differentiation, and/or migration. T21 hPSC-derived models have been used to specifically assess inhibitory neuron progenitors and show differences in proliferation, differentiation, and migration (Giffin-Rao et al. 2022; Huo et al. 2018; Xu et al. 2019). Expression of PAK1, a kinase important for cell migration in mouse cortical development, was increased in T21 neurons and inhibition of PAK1 partially restored migration deficits, providing potential mechanistic insight (Huo et al. 2018).

Further investigation using additional T21 cell lines to corroborate primary data is required to establish which specific neuronal subtypes are affected in DS and how they might contribute to cognitive phenotypes. Once this data is established, the value of T21 models will be to identify underlying mechanisms.

4.4 Altered Gliogenesis and Glial Function in hPSC Models of DS

Neuroglia, specifically astrocytes and oligodendrocytes, are critical for brain formation and function, and are generated after neurons during brain development (Allen and Lyons 2018). Premature cessation of neurogenesis has been shown and can be rescued by XIST silencing of HSA21 (Kawatani et al. 2021; Czerminski et al. 2022; Reiche et al. 2019; Bansal et al. 2024). Reduced neurogenesis in DS may be accompanied by increased gliogenesis, which is supported by reports that DS fetal-derived NPCs favor glial over neurogenic lineages (Mito and Becker 1993; Zdaniuk et al. 2011; Bahn et al. 2002). An increased proportion of astrocytes concomitant with a reduction in neurons in T21 hPSC neural differentiation supports the idea that astrogliogenesis is affected by T21 (Briggs et al. 2013; Hibaoui et al. 2014). Together, these data suggest that astrogliogenesis and/or the developmental switch from neurogenesis to gliogenesis is altered in DS.

Glial genes *OLIG1*, *OLIG2*, and *S100B* are encoded on HSA21, and their overexpression may contribute to changes in gliogenesis and glial function in DS. T21 hPSC-derived astrocytes have increased cell size due to spreading, differences in cell adhesion, and increased motility compared to controls (Ponroy Bally et al. 2020). When cultured with hiPSC-derived T21 astrocytes or in T21 astrocyte conditioned media, both trisomic and disomic NPCs demonstrate decreased neurogenesis, decreased neurite outgrowth, and increased neural apoptosis (Araujo et al. 2018). Co-culture of hiPSC-derived T21 astrocytes with hiPSC-derived neurons negatively affects the function of neurons; T21 astroglia exhibited more-frequent spontaneous calcium fluctuations, which reduced the excitability of the cocultured neurons (Mizuno et al. 2018). T21-elicited changes in astrocyte function also results in decreased neuron function in the ability to maturation and establish neurites when cocultured (Ponroy Bally et al. 2020). Astrocyte conditioned media from hPSC-derived T21 astrocytes increased ROS production and decreased secreted neuroprotective factors that may underlie the cellular changes to neurons (Chen et al. 2014a). In a chimeric mouse study, transplanted T21 hiPSC-derived astroglia in newborn mice have reduced neurogenesis (Chen et al. 2014a). Together these results suggest that aberrant development and function of astrocytes in DS may contribute to neuronal dysfunction that may ultimately underlie intellectual disability.

Oligodendrocytes myelinate axons in the central nervous system and imaging studies report reduced white matter in DS (Reiche et al. 2019; Olmos-Serrano et al. 2016). This structural defect is bolstered by transcriptomics of primary DS tissue samples that indicates defects in oligodendrogenesis and myelination (Reiche et al. 2019, Olmos-Serrano et al. 2016). Studies using T21 hPSC models to explore oligodendrocyte differentiation and phenotypes are limited, in large part because oligodendrocytes are technically difficult to differentiate in culture. Most existing studies characterize OLIG2+ progenitors as oligodendrocyte progenitor cells (OPCs); however, *OLIG2* expression is not specific to OPCs. Nevertheless, multiple studies describe increased proportions of OLIG2+ T21 NPCs compared to controls (Klein et al. 2022; Xu et al. 2019). In contrast, no differences were observed in mature T21 oligodendrocytes or astrocytes produced in chimeric mouse models (Xu et al. 2019). More work is needed to establish differences, investigate mechanisms for oligodendrocyte development and function in DS, so as to define their contribution to the neuropathology in DS.

4.5 hPSC-Derived T21 Neurons Have Altered Electrophysiological Properties and Synaptic Deficits

Function of the cerebral cortex relies on proper synaptogenesis, synaptic activity, and synaptic transmission, and thus alterations in any of these processes may contribute to the cognitive disability in individuals with DS. Fresh T21 post-mortem tissue that is amenable to electrophysiological recording is exceedingly rare and no studies have thus far assessed synaptic function in DS primary cortical tissue. Human PSC models provide a tractable system for investigating intrinsic neuronal properties and network connectivity. The hypothesis that neuron function is altered in DS comes largely from studies of Golgi staining of neurons and electron microscopy of primary brain tissue that detected differences in the number and morphology of dendritic spines, a proxy for synapses (Marin-Padilla 1972; Purpura 1975; Becker et al. 1986; Petit et al. 1984; Weitzdoerfer et al. 2001). However, these static observations cannot provide insight into the mechanism of the alterations that hPSC-derived models afford. T21 hPSC-derived neurons enable the testing of hypotheses of aberrant synaptogenesis and function caused by T21.

Neurons differentiated from T21 hPSCs have fewer synapses, as identified by presynaptic markers, suggesting reduced synaptogenesis or increased synaptic pruning in DS (Weick et al. 2013; Hibaoui et al. 2014; Cao et al. 2017). Furthermore, both excitatory and inhibitory T21 neurons demonstrate less complex dendritic branching, and fewer synapses in culture (Araujo et al. 2018; Huo et al. 2018; Jin et al. 2022; Foliaki et al. 2021; Chen et al. 2014a).

Limited electrophysiological studies of hPSC-derived T21 cortical neurons indicate that T21 neurons have reduced spontaneous synaptic activity compared to controls (Weick et al. 2013), although the phenotype has not been extensively reproduced

(Chen et al. 2014a). Synaptic transmission is also reduced, as shown in a single study using neurons genetically modified to express channel rhodopsin (Dong et al. 2020). Upon stimulation with blue light, control neurons exhibit long-term potentiation and long-term depression, while none were observed in T21 cultures, indicating aberrant plasticity in the T21 neurons. Chimeras provide a paradigm to model neuron dynamics in vivo and have been applied to DS with T21-derived neurons showing decreased neural network activity (Real et al. 2018).

Microglia may also contribute to the reduced synaptic phenotype and excitability observed in T21 neurons (Mizuno et al. 2018; Araujo et al. 2018). Microglia, a type of immune cell, are responsible for synaptic pruning postnatally by phagocytosing under-utilized synaptic connections. As microglia are formed from mesoderm, they are not present in directed neuronal differentiation of hPSCs. To study the effects of T21 microglia on synapse number, microglia were cocultured in cortical organoids and then transplanted into mice (Jin et al. 2022) to reveal increased synaptic pruning in both control and T21 organoids and the chimeric model. The increased phagocytic activity demonstrated by T21 microglia may further reduce an already aberrant number of synapses in the trisomic cortex.

Reduced synaptogenesis with altered astroglial and microglial function likely all contribute to the observed reduction in T21 cortical synapses; however, the relative contribution of each is currently unknown and an interesting topic for future studies using these models. Together these studies demonstrate T21 hPSC-derived cultures provide important insight into intrinsic electrophysiological properties, connectivity, and plasticity, and that T21 astrocytes play a significant role in altered maturation.

4.6 Mitochondrial Dysfunction and Oxidative Stress in hPSC Models Suggest Altered Metabolism in T21

Mitochondrial dysfunction and increased oxidative stress have been consistently observed in many DS cell types in vivo (Tan et al. 2023). Because cellular oxidative state and energy metabolism affect the balance between neural progenitor proliferation and differentiation, maturation, and cell functions, it is possible that altered metabolic processes contributes to T21 cortical phenotypes. Metabolic dysfunction in DS can be detected in the hPSC stage with significant increases in ROS and precocious mitochondrial network maturation in comparison to controls (Prutton et al. 2023). T21 NPCs have increases in O_2 consumption, energy utilization, and a glycolytic bias; however, they do not show the expected glycolysis to oxidative phosphorylation transition during differentiation as expected and observed in disomic NPCs. Impaired mitochondrial function plays a role in altered cortical development in DS and it starts very early in neural development (Martinez et al. 2024).

T21 affects mitochondrial features in hPSC-derived neurons including decreased mitochondrial basal respiration, ATP production, and maximal and spare respiration (Xu et al. 2022). Additionally, differences in mitochondrial morphology, which reflect changes in function, are observed in some T21 neurons, with decreased size and abnormal perinuclear aggregations. Other data indicate an increased number of mitochondria and decreased mitochondrial membrane potential in T21 neurons compared to controls (Murray et al. 2015; Weick et al. 2013). Dynamic hPSC-derived models provide a way to investigate how altered metabolism in T21 affects human neurogenesis.

4.7 Global Transcriptomic Changes in hPSC-Derived Cortical Models Suggest Altered Chromatin Structure and Dysregulated Signaling Pathways

Genomic and transcriptomic technologies applied to hPSC-derived models enable exploration of how T21 alters gene expression, chromatin state, and molecular pathways to drive phenotypes. Defining transcriptomic signatures via microarray, RNA sequencing (RNA-seq), and/or single cell or single nuclear sequencing (scSeq and snSeq, respectively) enables unbiased analysis of neural development and function. In T21 cells, upregulation of many HSA21-encoded genes is detected, as expected based on gene dosage, as well as dysregulation of genes across the genome with non-HSA21 genes both up- and downregulated compared to controls (De Toma et al. 2021). Gene ontology term (GO term) analysis to identify dysregulated developmental processes and pathways indicates common pathways are dysregulated across studies assessing T21 NPCs and neurons. These include processes of extracellular matrix, cell adhesion, morphogenesis, and neurodevelopment as well as mTOR, Shh, Wnt, and Notch signaling pathways (Hibaoui et al. 2014; Araujo et al. 2018; Huo et al. 2018; Gonzales et al. 2018; Ponroy Bally et al. 2020; Czermiński and Lawrence 2020; Giffin-Rao et al. 2022; Xu et al. 2019; Tang et al. 2021; Klein et al. 2022; Martinez et al. 2024). Comparisons of these transcriptomic data between hPSC-derived cells and publicly available databases of primary brain tissue transcriptomes will allow researchers to validate their findings by benchmarking their results to specific brain regions and developmental timepoints and taken together, may define dysregulation across time and space in vivo (Olmos-Serrano et al. 2016; Kang et al. 2011; He et al. 2023).

T21 hPSC models have significant changes in chromatin structure and topology. Techniques such as Hi-C and ATAC (Assay for Transposase-Accessible Chromatin) demonstrate that T21 results in spatial chromatin reorganization and differentially open chromatin regions across the genome, which correlates with transcriptomic data (Meharena et al. 2022; Tang et al. 2021; Ponroy Bally et al. 2020). Analyses of Laminar Associated Domains (LADS) and topologically associating domains (TADS) reveal that T21 causes chromosomal introversion, or a decrease in *trans*-chromosomal interactions and increases in *cis*-chromosomal interactions in NPCs (Meharena et al. 2022). Changes in open chromatin regions are useful in identifying alterations in chromatin structure where neurodevelopmentally significant transcription factors such as PAX6 and WNT7A bind, thus providing mechanistic insight into transcriptomic differences observed in developing T21 cortices. DNA methylation profiling partially explains these differences in chromatin accessibility, as one study describes 500 differentially methylated CpGs of which 151 are associated with genes enriched for DNA binding and chromatin structure (Laan et al. 2020). Interestingly, differential gene expression is organized into domains of up and down regulation throughout the genome, and these gene expression dysregulation domains correlate to known LADs (Letourneau et al. 2014). However, other data refute this hypothesis in favor of the natural clustering of coregulated genes (Ahlfors et al. 2019).

An abundance of information has been obtained from transcriptomic analysis from multiple hPSC cell lines and neuronal cell types, yet only one publication evaluates other mechanisms of post-transcriptional modifications such as alternative splicing and adenosine-to-inosine RNA editing (Gonzales et al. 2018). Splicing algorithms identify 117 and 36 differentially spliced transcripts in T21 iPSCs and cortical neurons, respectively. However, the study did not go beyond identification of differences, leaving the impact and evaluation of alternative splicing in other cell types and modified protein function open for investigation.

In summary, T21 hPSC-derived models of cortical development address the limitations of primary DS samples to study dysregulation of cellular processes and molecular pathways. These models also allow for the direct manipulation of candidate driver genes in specific cell types throughout differentiation to test specific hypotheses. Results from the growing number of studies reveal that T21 affects human cortical development in numerous ways and at many developmental stages (Fig. 3).

5 Investigating Alzheimer's Disease and Neurodegeneration in T21 hPSCs

DS has become a critical risk factor for early-onset AD (Fortea et al. 2020). As early as the 1940s, three individuals with DS were described by G.A. Jervis to have early senile dementia and AD-like neuropathologic features (Jervis 1948). Since then, an increasing number of studies have demonstrated the relationship between DS and AD (Leverenz and Raskind 1998; Wisniewski et al. 1985; Lott and Head 2019; Wiseman et al. 2015; Iulita et al. 2022). Down syndrome-associated Alzheimer's disease (DS-AD) is a genetic form of Alzheimer's disease (AD), and its development is associated with multiple HSA21 genes in DS. Almost all individuals with DS will develop Alzheimer's disease (AD) neuropathology by 40 years of age, and approximately two-thirds of individuals with DS develop clinical features of dementia by the age of 53 to 55 (Wiseman et al. 2015). DS has become a critical risk factor for early-onset AD (Fortea et al. 2020). Investigating the mechanism of AD in DS will offer insights into understanding the pathogenesis of both AD and DS and may discover potential therapeutic targets.

Decades of studies using transgenic mouse and patient primary tissues have provided a wealth of data for understanding the characteristics and pathological mechanisms of DS-AD, including key pathogenic genes (Ryoo et al. 2008; Chen et al. 2021; Wiseman et al. 2018; Hardy 2017; Zheng et al. 2022). Current studies have identified the HSA21 encoded genes that may promote AD (Hardy 2017; Zheng et al. 2022; Ryoo et al. 2008). For instance, the *APP* gene, located at 21q21.3, encodes the amyloid precursor protein and leads to amyloid plaque deposition and promotes the amyloidogenic pathway (Chen et al. 2021; Wiseman et al. 2018). More recently, the discovery of β2M-associated immune system abnormalities in DS has widened our understanding about mechanisms of cognitive impairment in DS (Gao et al. 2023). B2M can bind to Aβ and aggravate its neurotoxicity in AD patients (Zhao et al. 2023) and so may be critical in DS-AD as well. Recent advances in stem cell technology provide human-derived models with complete trisomy of HSA21 for DS-AD research (Gonzalez et al. 2018; Shi et al. 2012b).

5.1 APP, DYRK1A, and phospho-Tau

T21 hPSC-derived models enable detailed studies of DS-AD pathogenesis and drug screening. Overexpression of the *APP* gene in DS has been linked to increased brain amyloid levels and early-onset AD (Doran et al. 2017; Fortea et al. 2021; Rafii and Fortea 2023). Cortical neurons generated from T21 hPSCs develop characteristic pathological hallmarks of AD, including Aβ accumulation, tau protein expression, and tau hyperphosphorylation, and new candidate drugs have been identified through screening of this in vitro model, such as γ-secretase inhibitor, N-butylidenephthalide and N-acetylcysteine (Chang et al. 2015; Dashinimaev et al. 2017; Hu et al. 2018; Shi et al. 2012b; Toshikawa et al. 2021). Increased dosage of *APP* and the dual-specificity tyrosine phosphorylation regulated kinase 1A (DYRK1A) in T21 is considered necessary for DS-AD (Ryoo et al. 2008; Decourt et al. 2013; Wu et al. 2022a). By manipulating APP dosage through copy-number normalization of T21 hPSCs and APP-overexpressing hESCs, *APP* gene dosage was found to affect Aβ42/Aβ40 ratio, but not a range of tau-phosphorylation events (Ovchinnikov et al. 2018). Early amyloid plaque deposition was produced by the synergistic effect between APP and DYRK1A (Wegiel et al. 2011). Further, several studies used hPSCs-derived neurons to explore the role of the dosage sensitive HSA21 gene DYRK1A. DYRK1A phosphorylates Tau and APP at multiple sites, leading to microtubule instability and potentially affecting Aβ generation. Through inhibition or genetic correction of DYRK1A, pathological phenotypes, including elevated Tau phosphorylation, could be rescued in T21 neurons (Wu et al. 2022a). Furthermore, DYRK1A controls the intracellular distribution of APP by influencing axon transport function, and inhibition of DYRK1A to restore axonal transport of

APP may be a treatment approach for DS-AD (Fernandez Bessone et al. 2022). Taken together, altering *APP* gene dosage affects Aβ42/Aβ40 ratio, while APP and DYRK1A synergize to induce early amyloid plaque deposition. 2D cultured neurons derived from T21 hPSCs mimic DS-AD traits and aid in identifying pathogenesis and are useful in drug screening.

5.2 Amyloid Plaques and Neurofibrillary Tangles Are Recapitulated in T21 Brain Organoids

Over time, pathological features of AD including amyloid plaques and neurofibrillary tangles, have been reported to be recapitulated in 3D brain organoids derived from familial AD or T21 hiPSCs (Alić et al. 2021; Campbell et al. 2023; Fertan et al. 2024; Gonzalez et al. 2018; Wu et al. 2022b). In contrast, these pathological irregularities were absent in cerebral organoids derived from multiple control groups, including hiPSCs from those diagnosed with Creutzfeldt–Jakob disease, as well as from both mouse ESCs and mouse hiPSCs (Gonzalez et al. 2018; Zhao and Haddad 2022). Although most individuals with DS develop dementia symptoms at an early stage, there is still a small number of individuals who develop these symptoms later, or do not develop clinical dementia at all (Hithersay et al. 2019). Through the use of T21 hPSC cerebral organoids, BACE2 has been identified as a HSA21 encoded dose-sensitive AD-suppressor gene, and potentially explains the dementia delay in people with DS. BACE1 inhibitors inhibit Aβ-degrading activity of BACE2 in cerebral organoids, providing an additional explanation for the failure of some BACE1 inhibitor clinical trials (Alić et al. 2021). This research also highlights the potential of brain organoids as a drug discovering platform for DS-AD.

5.3 Inhibition of Type I Interferon Receptors Rescues DS-AD Microglia Pathological Phenotypes

Chimeric models have also been used in the study of DS-AD since the mouse brain provides appropriate physiological microenvironment for the long-term survival and maturation of human neurons. Because the improper functioning of microglia is a key factor in the development of AD, with several genes linked to this condition being predominantly or even exclusively active in microglia, much attention has been focused on microglia in T21 hiPSCs to study DS-AD. Researchers isolated soluble protein fractions from human brain samples diagnosed with DS-AD, which exhibited significant Tau protein abnormalities. These samples were matched in age and sex with control brain tissue. Mice with a chimeric constitution containing microglia from two different T21 hPSC lines were subsequently administered an injection of protein fractions from either the control or DS-AD tissue. Researchers observed a widespread presence of phosphorylated tau in the brains of those mice injected with DS-AD-derived Tau. Moreover, human DS-AD pathological Tau induces senescence in DS microglia. Inhibition of type I interferon (IFN-I) receptors (IFNARs) on HSA21 can rescue these DS microglia phenotypes. Chimeric brains, combining mouse and human elements, aid DS-AD research due to the supportive environment for human neurons. Through the use of chimeric brains, tau abnormalities and senescence in DS microglia can potentially be reversed via inhibiting specific interferon receptors (Table 1).

Table 1 Using stem cell-derived models to study DS-AD

Cell model derived from human PSCs	Reference	hPSCs lines		Related gene
		Control	T21	
2D cultured neurons	Shi et al. (2012b)	H9 (hESCs) BBHX (hiPSCs) CRL (hiPSCs)	SC-321 hESCs DS1-iPS4	
	Chang et al. (2015)	TW1 (hESCs)	allantoic fluid-derived trisomy 21 hiPSCs	
	Dashinimaev et al. (2017)	AFS-1 (hiPSCs) AFS-2 (hiPSCs) AFS-3 (hiPSCs)	AFS-DS1 (hiPSCs) AFS-DS2 (hiPSCs) AFS-DS3 (hiPSCs)	
	Hu et al. (2018)	NDC (hiPSCs) AD2--1 (hiPSCs)	PS1 Int4 (hiPSCs) APP^{Dp} (hiPSCs) Ts21 (hiPSCs) SFC808 (hiPSCs) Ts21.B (hiPSCs)	
	Toshikawa et al. (2021)	E-iPSCs (an isogenic euploid control clone established from the D-iPSCs)	D-iPSCs	
	Ovchinnikov et al. (2018)	Genea021 (hESCs) Genea022 (hESCs)	C11DS (hiPSCs)	*APP*
	Wu et al. (2022a)	Person #1 T21 (hiPSCs) Person #2 T21 (hiPSCs)	Person #1 T21-rev (hiPSCs) Person #2 T21-rev (hiPSCs)	*APP* *DYRK1A*
	Fernandez Bessone et al. (2022)	CVhiPSCs	/	*DYRK1A*

(continued)

Table 1 (continued)

Cell model derived from human PSCs	Reference	hPSCs lines		Related gene
		Control	T21	
Brain organoids	Gonzalez et al. (2018)	HC (hiPSC) fCJD-E200K (hiPSC) fCJD-Extra rep (hiPSC)	fAD (hiPSCs derived from familiar Alzheimer's disease) DS (hiPSCs derived from Down syndrome)	/
	Zhao and Haddad (2022)	Two isogenic control subclones isolated from DS1-iPS4 cells	Two DS hiPSC subclones isolated from DS1-iPS4 cells	/
	Alić et al. (2021)	D21C3 (hiPSC) D21C9 (hiPSC) D21C7 (hiPSC)	T21C6 (hiPSC) T21C13 (hiPSC) T21C5 (hiPSC) QM-DS1 (hiPSC) QM-DS2 (hiPSC) QM-DS3 (hiPSC) QM-DS4 (hiPSC) QM-DS5 (hiPSC) QM-DS6 (hiPSC) QM-DupAPP (hiPSC)	*BACE2*
Chimeric mouse brains	Jin et al. (2022)	Cont1 (hiPSC) Cont2 (hiPSC) Isogenic Di-DS3 (hiPSC)	DS1 (hiPSC) DS2 (hiPSC) Isogenic Tri-DS3 (hiPSC)	*IFN-I*

6 Using Stem Cell-Derived Models to Study Non-neurological Aspects of DS

In addition to the nervous system, the trisomy of HSA21 affects multiple organs in individuals, leading to a variety of diseases, including congenital heart defects (CHDs) and leukemia (Antonarakis 2017; Antonarakis et al. 2020; Esbensen 2010; Freeman et al. 1998). Human PSCs are also emerging to study the consequences of T21 in other organ systems.

6.1 Congenital Heart Defects

About half of individuals with DS are born with CHDs. Among them, 43% of CHD cases are atrioventricular septal defects (AVSDs), which are the most common forms of CHDs in DS and has severe effects on the quality of life of these individuals (Ferencz et al. 1989; Bergström et al. 2016). Current research, through the analysis of unique patient samples with partial T21 combined with CHD and DS mouse models with CHD symptoms including Ts65Dn, has uncovered some risk loci on HSA21. These include a 1.77 Mb area found in human samples between the *DSCAM* and *ZBTB21* genes (Korbel et al. 2009), and a 5 Mb area discovered in a Dp3Tyb mouse model (Lana-Elola et al. 2016). T21 hPSC models have focused on the specific mechanisms of how these genes influence disease phenotypes, but few reports exist. By differentiating T21 hESCs into cardiomyocytes, two genes encoded on HSA21, *ETS2* and *ERG*, were reported to be related to aberrant secondary heart field (SHF) development and abnormal electrophysiological phenotype in functional T21 cardiomyocytes (Bosman et al. 2015). Upregulated interferon signaling has also been reported to be involved in impaired cardiac differentiation (Chi et al. 2023); such upregulation can be caused by increased dosage of the HSA21 *IFNR* genes, which downregulates canonical Wnt signaling. Human PSC modeling has only begun to identify the mechanisms that underlie the effects of T21 on heart development and function.

6.2 Hematopoietic Abnormalities

Children with DS have a higher risk of developing hematopoietic abnormalities compared to children without DS, which can include acute leukemia and transient myeloproliferative disorder (TMD) (Tunstall et al. 2018). TMD affects approximately 10 to 15% of newborns with DS, characterized by the presence of circulating megakaryoblasts, increased white blood cell count, anemia, hepatomegaly, and splenomegaly. DS-TMD cells always carry mutations in *GATA1*, a hematopoietic transcription factor located on the X chromosome, which is associated with the clonal expansion of megakaryoblasts. These mutations result in the elimination of the full-length GATA1 protein, leaving the short isoform known as GATA1s. It is worth noting that *GATA1* mutations are rarely observed in leukemic cells without T21, indicating the significance of the interaction between T21 and GATA1s in the development of TMD in individuals with DS (Yoshida et al. 2013). Through combining patient-derived hPSCs, genome-editing technologies, and chromosome engineering techniques, several hPSCs harboring different combinations of HSA21 genes and *GATA1* isoforms were generated for exploring the synergistic interaction between GATA1 mutation and gene-dosage effects of T21 in hematopoiesis. A 4-Mb candidate region critical for hematopoietic defects in DS was identified

(Banno et al. 2016). Further work is needed to define the identity and functional significance of genes in this region.

Overall, the exploration of hPSCs for investigating peripheral system abnormalities in DS is still in its early stages. Additional techniques for differentiating and constructing peripheral lineage cell types, as well as developing organoid models for peripheral organs, are needed to advance the study of DS.

7 Summary/Future

The establishment of T21 hPSC models of DS have revolutionized the ability to study the impact of T21 on dynamic cellular processes of human cells. Critical insight has been gleaned from multiple cellular and molecular perturbations, shedding some light on the tissue hypo-cellularity phenotypes characteristic of DS. The ability to create hPSC models from numerous and diverse donors with DS will facilitate future studies to identify explanations for the variation among individuals with DS and establish models that better represent the variety of individuals with DS.

A wealth of information has emerged from the utilization of hPSC models to study neural development, function, and degeneration in DS. These tractable models can be genetically manipulated to investigate how HSA21 gene dosage affects neurodevelopment throughout the brain and identify molecular drivers. Future work has the potential to accelerate our understanding of the complexity of intellectual disability and subsequent search for therapeutics to help improve cognitive outcomes for individuals with DS.

Studies thus far have focused on development of the cerebral cortex as a proxy for neurodevelopment in general; however, other areas of the brain such as the cerebellum and brainstem are significantly impacted by T21 (Dierssen 2012). Human PSC differentiation protocols exist for other brain regions including motor, retinal, and spinal neurons, but have yet to be applied to modeling neurodevelopment in DS. The complex phenotypes associated with DS warrant investigation into these other brain areas and neural tissues, and will no doubt provide valuable insight into how neurogenic programs are altered in T21.

Analyses of other organ systems that are affected in DS with T21 hPSC are still sparse and thus represent an opportunity for future exploration. Studies can be extended to other affected organ systems through assessment of other differentiation trajectories that can be modeled with human iPSCs, including neural crest (craniofacial defects), cardiac (heart malformations), and pancreatic islets (diabetes). It will be important to discover whether common cellular and molecular mechanisms are impacted by T21 in different organs.

References

Ahlfors, H., Anyanwu, N., Pakanavicius, E., Dinischiotu, N., Lana-Elola, E., Watson-Scales, S., Tosh, J., Wiseman, F., Briscoe, J., Page, K., Fisher, E. M. C. & Tybulewicz, V. L. J. 2019. Gene Expression Dysregulation Domains Are Not A Specific Feature Of Down Syndrome. *Nat Commun,* **10,** 2489.

Alić, I., Goh, P. A., Murray, A., Portelius, E., Gkanatsiou, E., Gough, G., Mok, K. Y., Koschut, D., Brunmeir, R., Yeap, Y. J., O'Brien, N. L., Groet, J., Shao, X., Havlicek, S., Dunn, N. R., Kvartsberg, H., Brinkmalm, G., Hithersay, R., Startin, C., Hamburg, S., Phillips, M., Pervushin, K., Turmaine, M., Wallon, D., Rovelet-Lecrux, A., Soininen, H., Volpi, E., Martin, J. E., Foo, J. N., Becker, D. L., Rostagno, A., Ghiso, J., Krsnik, Ž., Šimić, G., Kostović, I., Mitrečić, D., Francis, P. T., Blennow, K., Strydom, A., Hardy, J., Zetterberg, H. & Nižetić, D. 2021. Patient-Specific Alzheimer-Like Pathology In Trisomy 21 Cerebral Organoids Reveals BACE2 As A Gene Dose-Sensitive AD Suppressor In Human Brain. *Mol Psychiatry,* **26,** 5766–5788.

Allen, N. J. & Lyons, D. A. 2018. Glia As Architects Of Central Nervous System Formation And Function. *Science,* 362, 181-–185.

Amiri, A., Coppola, G., Scuderi, S., Wu, F., Roychowdhury, T., Liu, F., Pochareddy, S., Shin, Y., Safi, A., Song, L., Zhu, Y., Sousa, A. M. M., Gerstein, M., Crawford, G. E., Sestan, N., Abyzov, A. & Vaccarino, F. M. 2018. Transcriptome And Epigenome Landscape Of Human Cortical Development Modeled In Organoids. *Science,* 362.

Anderson, S. A., Eisenstat, D. D., Shi, L. & Rubenstein, J. L. 1997. Interneuron Migration From Basal Forebrain To Neocortex: Dependence On Dlx Genes. *Science,* 278, 474–476.

Anderson, S. A., Kaznowski, C. E., Horn, C., Rubenstein, J. L. & Mcconnell, S. K. 2002. Distinct Origins Of Neocortical Projection Neurons And Interneurons In Vivo. *Cereb Cortex,* 12, 702–9.

Anderson, S. A., Marin, O., Horn, C., Jennings, K. & Rubenstein, J. L. 2001. Distinct Cortical Migrations From The Medial And Lateral Ganglionic Eminences. *Development,* 128, 353–363.

Antonarakis, S. E. 2017. Down Syndrome And The Complexity Of Genome Dosage Imbalance. *Nat Rev Genet,* 18, 147–163.

Antonarakis, S. E., Skotko, B. G., Rafii, M. S., Strydom, A., Pape, S. E., Bianchi, D. W., Sherman, S. L. & Reeves, R. H. 2020. Down Syndrome. *Nat Rev Dis Primers,* 6, 9.

Araujo, B. H. S., Kaid, C., De Souza, J. S., Gomes Da Silva, S., Goulart, E., Caires, L. C. J., Musso, C. M., Torres, L. B., Ferrasa, A., Herai, R., Zatz, M., Okamoto, O. K. & Cavalheiro, E. A. 2018. Down Syndrome IPSC-Derived Astrocytes Impair Neuronal Synaptogenesis And The MTOR Pathway In Vitro. *Molecular Neurobiology,* 55, 5962–5975.

Arber, C. & Li, M. 2013. Cortical Interneurons From Human Pluripotent Stem Cells: Prospects For Neurological And Psychiatric Disease. *Front Cell Neurosci,* 7, 10.

Bahn, S., Mimmack, M., Ryan, M., Caldwell, M. A., Jauniaux, E., Starkey, M., Svendsen, C. N. & Emson, P. 2002. Neuronal Target Genes Of The Neuron-Restrictive Silencer Factor In Neurospheres Derived From Fetuses With Down's Syndrome: A Gene Expression Study. *Lancet,* 359, 310–315.

Banno, K., Omori, S., Hirata, K., Nawa, N., Nakagawa, N., Nishimura, K., Ohtaka, M., Nakanishi, M., Sakuma, T., Yamamoto, T., Toki, T., Ito, E., Yamamoto, T., Kokubu, C., Takeda, J., Taniguchi, H., Arahori, H., Wada, K., Kitabatake, Y. & Ozono, K. 2016. Systematic Cellular Disease Models Reveal Synergistic Interaction Of Trisomy 21 And Gata1 Mutations In Hematopoietic Abnormalities. *Cell Rep,* 15, 1228–41.

Bansal, P., Banda, E. C., Glatt-Deeley, H. R., Stoddard, C. E., Linsley, J. W., Arora, N., Deleschaux, C., Ahern, D. T., Kondaveeti, Y., Massey, R. E., Nicouleau, M., Wang, S., Sabariego-Navarro, M., Dierssen, M., Finkbeiner, S. & Pinter, S. F. 2024. A Dynamic In Vitro Model Of Down Syndrome Neurogenesis With Trisomy 21 Gene Dosage Correction. *Sci Adv,* 10, Eadj0385.

Batista-Brito, R. & Fishell, G. 2009. The Developmental Integration Of Cortical Interneurons Into A Functional Network. *Curr Top Dev Biol,* 87, 81–118.

Becker, L. E., Armstrong, D. L. & Chan, F. 1986. Dendritic Atrophy In Children With Down's Syndrome. *Ann.Neurol.,* 20, 520–526.

Bergström, S., Carr, H., Petersson, G., Stephansson, O., Bonamy, A. K., Dahlström, A., Halvorsen, C. P. & Johansson, S. 2016. Trends In Congenital Heart Defects In Infants With Down Syndrome. *Pediatrics,* 138.

Bhatia, S. N. & Ingber, D. E. 2014. Microfluidic Organs-On-Chips. *Nat Biotechnol,* 32, 760–72.

Bhattacharyya, A., Mcmillan, E., Chen, S. I., Wallace, K. & Svendsen, C. N. 2009. A Critical Period In Cortical Interneuron Neurogenesis In Down Syndrome Revealed By Human Neural Progenitor Cells. *Dev Neurosci,* 31, 497–510.

Biancotti, J. C., Narwani, K., Buehler, N., Mandefro, B., Golan-Lev, T., Yanuka, O., Clark, A., Hill, D., Benvenisty, N. & Lavon, N. 2010. Human Embryonic Stem Cells As Models For Aneuploid Chromosomal Syndromes. *Stem Cells,* 28, 1530–40.

Bosman, A., Letourneau, A., Sartiani, L., Del Lungo, M., Ronzoni, F., Kuziakiv, R., Tohonen, V., Zucchelli, M., Santoni, F., Guipponi, M., Dumevska, B., Hovatta, O., Antonarakis, S. E. & Jaconi, M. E. 2015. Perturbations Of Heart Development And Function In Cardiomyocytes From Human Embryonic Stem Cells With Trisomy 21. *Stem Cells,* 33, 1434–46.

Breschi, A., Gingeras, T. R. & Guigó, R. 2017. Comparative Transcriptomics In Human And Mouse. *Nature Reviews Genetics,* 18, 425–440.

Briggs, J. A., Sun, J., Shepherd, J., Ovchinnikov, D. A., Chung, T.-L., Nayler, S. P., Kao, L.-P., Morrow, C. A., Thakar, N. Y., Soo, S.-Y., Peura, T., Grimmond, S. & Wolvetang, E. J. 2013. Integration-Free Induced Pluripotent Stem Cells Model Genetic And Neural Developmental Features Of Down Syndrome Etiology. *Stem Cells,* 31, 467–478.

Burridge, P. W., Matsa, E., Shukla, P., Lin, Z. C., Churko, J. M., Ebert, A. D., Lan, F., Diecke, S., Huber, B., Mordwinkin, N. M., Plews, J. R., Abilez, O. J., Cui, B., Gold, J. D. & Wu, J. C. 2014. Chemically Defined Generation Of Human Cardiomyocytes. *Nature Methods,* 11, 855–860.

Cadwell, C. R., Bhaduri, A., Mostajo-Radji, M. A., Keefe, M. G. & Nowakowski, T. J. 2019. Development And Arealization Of The Cerebral Cortex. *Neuron,* 103, 980–1004.

Camp, J. G., Badsha, F., Florio, M., Kanton, S., Gerber, T., Wilsch-Bräuninger, M., Lewitus, E., Sykes, A., Hevers, W., Lancaster, M., Knoblich, J. A., Lachmann, R., Pääbo, S., Huttner, W. B. & Treutlein, B. 2015. Human Cerebral Organoids Recapitulate Gene Expression Programs Of Fetal Neocortex Development. *Proc Natl Acad Sci U S A,* 112, 15672–7.

Campbell, N. B., Patel, Y., Moore, T. L., Medalla, M. & Zeldich, E. 2023. Extracellular Vesicle Treatment Alleviates Neurodevelopmental And Neurodegenerative Pathology In Cortical Spheroid Model Of Down Syndrome. *Int J Mol Sci,* 24.

Cao, S.-Y., Hu, Y., Chen, C., Yuan, F., Xu, M., Li, Q., Fang, K.-H., Chen, Y. & Liu, Y. 2017. Enhanced Derivation Of Human Pluripotent Stem Cell-Derived Cortical Glutamatergic Neurons By A Small Molecule. *Scientific Reports,* 7, 3282.

Carducci, F., Onorati, P., Condoluci, C., Di Gennaro, G., Quarato, P. P., Pierallini, A., Sara, M., Miano, S., Cornia, R. & Albertini, G. 2013. Whole-Brain Voxel-Based Morphometry Study Of Children And Adolescents With Down Syndrome. *Funct Neurol,* 28, 19–28.

Catalano, R. A. 1990. Down Syndrome. *Surv.Ophthalmol.,* 34, 385–398.

Chakrabarti, L., Best, T. K., Cramer, N. P., Carney, R. S., Isaac, J. T., Galdzicki, Z. & Haydar, T. F. 2010. OLIG1 And OLIG2 Triplication Causes Developmental Brain Defects In Down Syndrome. *Nat Neurosci,* 13, 927–34.

Chambers, S. M., Fasano, C. A., Papapetrou, E. P., Tomishima, M., Sadelain, M. & Studer, L. 2009. Highly Efficient Neural Conversion Of Human ES And IPS Cells By Dual Inhibition Of Smad Signaling. *Nature Biotechnology,* 27, 275–280.

Chang, C. Y., Chen, S. M., Lu, H. E., Lai, S. M., Lai, P. S., Shen, P. W., Chen, P. Y., Shen, C. I., Harn, H. J., Lin, S. Z., Hwang, S. M. & Su, H. L. 2015. N-Butylidenephthalide Attenuates Alzheimer's Disease-Like Cytopathy In Down Syndrome Induced Pluripotent Stem Cell-Derived Neurons. *Sci Rep,* 5, 8744.

Chen, C., Jiang, P., Xue, H., Peterson, S. E., Tran, H. T., Mccann, A. E., Parast, M. M., Li, S., Pleasure, D. E., Laurent, L. C., Loring, J. F., Liu, Y. & Deng, W. 2014a. Role Of Astroglia In Down's Syndrome Revealed By Patient-Derived Human-Induced Pluripotent Stem Cells. *Nature Communications,* 5, 4430.

Chen, K. G., Mallon, B. S., Mckay, R. D. & Robey, P. G. 2014b. Human Pluripotent Stem Cell Culture: Considerations For Maintenance, Expansion, And Therapeutics. *Cell Stem Cell,* 14, 13–26.

Chen, X. Q., Salehi, A., Pearn, M. L., Overk, C., Nguyen, P. D., Kleschevnikov, A. M., Maccecchini, M. & Mobley, W. C. 2021. Targeting Increased Levels Of APP In Down Syndrome: Posiphen-Mediated Reductions In APP And Its Products Reverse Endosomal Phenotypes In The Ts65Dn Mouse Model. *Alzheimers Dement,* 17, 271–292.

Chi, C., Knight, W. E., Riching, A. S., Zhang, Z., Tatavosian, R., Zhuang, Y., Moldovan, R., Rachubinski, A. L., Gao, D., Xu, H., Espinosa, J. M. & Song, K. 2023. Interferon Hyperactivity Impairs Cardiogenesis In Down Syndrome Via Downregulation Of Canonical WNT Signaling. *Iscience,* 26, 107012.

Chu, J. & Anderson, S. A. 2014. Development Of Cortical Interneurons. *Neuropsychopharmacology.*

Contestabile, A., Fila, T., Ceccarelli, C., Bonasoni, P., Bonapace, L., Santini, D., Bartesaghi, R. & Ciani, E. 2007. Cell Cycle Alteration And Decreased Cell Proliferation In The Hippocampal Dentate Gyrus And In The Neocortical Germinal Matrix Of Fetuses With Down Syndrome And In Ts65Dn Mice. *Hippocampus,* 17, 665–78.

Czerminski, J. T., King, O. D. & Lawrence, J. B. 2022. Large-Scale Organoid Study Suggests Effects Of Trisomy 21 On Early Fetal Neurodevelopment Are More Subtle Than Variability Between Isogenic Lines And Experiments. *Front Neurosci,* 16, 972201.

Czermiński, J. T. & Lawrence, J. B. 2020. Silencing Trisomy 21 With Xist In Neural Stem Cells Promotes Neuronal Differentiation. *Developmental Cell,* 52, 294–308.E3.

Dashinimaev, E. B., Artyuhov, A. S., Bolshakov, A. P., Vorotelyak, E. A. & Vasiliev, A. V. 2017. Neurons Derived From Induced Pluripotent Stem Cells Of Patients With Down Syndrome Reproduce Early Stages Of Alzheimer's Disease Type Pathology In Vitro. *J Alzheimers Dis,* 56, 835–847.

De Toma, I., Sierra, C. & Dierssen, M. 2021. Meta-Analysis Of Transcriptomic Data Reveals Clusters Of Consistently Deregulated Gene And Disease Ontologies In Down Syndrome. *Plos Comput Biol,* 17, E1009317.

Decourt, B., Mobley, W., Reiman, E., Shah, R. J. & Sabbagh, M. N. 2013. Recent Perspectives On APP, Secretases, Endosomal Pathways And How They Influence Alzheimer's Related Pathological Changes In Down Syndrome. *J Alzheimers Dis Parkinsonism,* Suppl 7, 002.

Delabar, J. M., Theophile, D., Rahmani, Z., Chettouh, Z., Blouin, J. L., Prieur, M., Noel, B. & Sinet, P. M. 1993. Molecular Mapping Of Twenty-Four Features Of Down Syndrome On Chromosome 21. *Eur J Hum Genet,* 1, 114–24.

Denas, O., Sandstrom, R., Cheng, Y., Beal, K., Herrero, J., Hardison, R. C. & Taylor, J. 2015. Genome-Wide Comparative Analysis Reveals Human-Mouse Regulatory Landscape And Evolution. *Bmc Genomics,* 16, 87.

Devlin, L. & Morrison, P. J. 2004. Mosaic Down's Syndrome Prevalence In A Complete Population Study. *Arch.Dis.Child,* 89, 1177–1178.

Dierssen, M. 2012. Down Syndrome: The Brain In Trisomic Mode. *Nat Rev Neurosci,* 13, 844–58.

Dong, Y., Xiong, M., Chen, Y., Tao, Y., Li, X., Bhattacharyya, A. & Zhang, S.-C. 2020. Plasticity Of Synaptic Transmission In Human Stem Cell-Derived Neural Networks. *Iscience,* 23, 100829.

Doran, E., Keator, D., Head, E., Phelan, M. J., Kim, R., Totoiu, M., Barrio, J. R., Small, G. W., Potkin, S. G. & Lott, I. T. 2017. Down Syndrome, Partial Trisomy 21, And Absence Of Alzheimer's Disease: The Role Of APP. *J Alzheimers Dis,* 56, 459-470.

Dumevska, B., Bosman, A., Mckernan, R., Main, H., Schmidt, U. & Peura, T. 2016. Derivation Of Trisomy 21 Affected Human Embryonic Stem Cell Line Genea021. *Stem Cell Research,* 16, 401-404.

Džaja, D., Hladnik, A., Bičanić, I., Baković, M. & Petanjek, Z. 2014. Neocortical Calretinin Neurons In Primates: Increase In Proportion And Microcircuitry Structure. *Front Neuroanat,* 8, 103.

Edmondson, R., **Broglie, J. J.**, Adcock, A. F. & Yang, L. 2014. Three-Dimensional Cell Culture Systems And Their Applications In Drug Discovery And Cell-Based Biosensors. *Assay Drug Dev Technol,* 12, 207–18.

Esbensen, A. J. 2010. Health Conditions Associated With Aging And End Of Life Of Adults With Down Syndrome. *Int Rev Res Ment Retard,* 39, 107–126.

Fang, Y. & Eglen, R. M. 2017. Three-Dimensional Cell Cultures In Drug Discovery And Development. *Slas Discov,* 22, 456–472.

Ferencz, C., Neill, C. A., Boughman, J. A., Rubin, J. D., Brenner, J. I. & Perry, L. W. 1989. Congenital Cardiovascular Malformations Associated With Chromosome Abnormalities: An Epidemiologic Study. *J Pediatr,* 114, 79–86.

Fernandez Bessone, I., Navarro, J., Martinez, E., Karmirian, K., Holubiec, M., Alloatti, M., Goto-Silva, L., Arnaiz Yepez, C., Martins-De-Souza, D., Nascimento, J. M., Bruno, L., Saez, T. M., Rehen, S. K. & Falzone, T. L. 2022. Dyrk1a Regulates The Bidirectional Axonal Transport Of APP In Human-Derived Neurons. *J Neurosci,* 42, 6344–58.

Fertan, E., Böken, D., Murray, A., Danial, J. S. H., Lam, J. Y. L., Wu, Y., Goh, P. A., Alić, I., Cheetham, M. R., Lobanova, E., Zhang, Y. P., Nižetić, D. & Klenerman, D. 2024. Cerebral Organoids With Chromosome 21 Trisomy Secrete Alzheimer's Disease-Related Soluble Aggregates Detectable By Single-Molecule-Fluorescence And Super-Resolution Microscopy. *Mol Psychiatry,* 29, 369–386.

Flames, N. & Marin, O. 2005. Developmental Mechanisms Underlying The Generation Of Cortical Interneuron Diversity. *Neuron,* 46, 377–381.

Foliaki, S. T., Schwarz, B., Groveman, B. R., Walters, R. O., Ferreira, N. C., Orrù, C. D., Smith, A., Wood, A., Schmit, O. M., Freitag, P., Yuan, J., Zou, W., Bosio, C. M., Carroll, J. A. & Haigh, C. L. 2021. Neuronal Excitatory-To-Inhibitory Balance Is Altered In Cerebral Organoid Models Of Genetic Neurological Diseases. *Molecular Brain,* 14, 156.

Fortea, J., Vilaplana, E., Carmona-Iragui, M., Benejam, B., Videla, L., Barroeta, I., Fernandez, S., Altuna, M., Pegueroles, J., Montal, V., Valldeneu, S., Gimenez, S., Gonzalez-Ortiz, S., Munoz, L., Estelles, T., Illan-Gala, I., Belbin, O., Camacho, V., Wilson, L. R., Annus, T., Osorio, R. S., Videla, S., Lehmann, S., Holland, A. J., Alcolea, D., Clarimon, J., Zaman, S. H., Blesa, R. & Lleo, A. 2020. Clinical And Biomarker Changes Of Alzheimer's Disease In Adults With Down Syndrome: A Cross-Sectional Study. *Lancet,* 395, 1988–1997.

Fortea, J., Zaman, S. H., Hartley, S., Rafii, M. S., Head, E. & Carmona-Iragui, M. 2021. Alzheimer's Disease Associated With Down Syndrome: A Genetic Form Of Dementia. *Lancet Neurol,* 20, 930–942.

Freeman, S. B., Taft, L. F., Dooley, K. J., Allran, K., Sherman, S. L., Hassold, T. J., Khoury, M. J. & Saker, D. M. 1998. Population-Based Study Of Congenital Heart Defects In Down Syndrome. *Am.J.Med.Genet.,* 80, 213–217.

Friederici, A. D. & Gierhan, S. M. 2013. The Language Network. *Curr Opin Neurobiol,* 23, 250–4.

Fusaki, N., Ban, H., Nishiyama, A., Saeki, K. & Hasegawa, M. 2009. Efficient Induction Of Transgene-Free Human Pluripotent Stem Cells Using A Vector Based On Sendai Virus, An Rna Virus That Does Not Integrate Into The Host Genome. *Proc Jpn Acad Ser B Phys Biol Sci,* 85, 348–62.

Ganguly, B. B. 2022. Genetics And Neurobiology Of Down Syndrome. Academic Press.

Gao, Y., Hong, Y., Huang, L., Zheng, S., Zhang, H., Wang, S., Yao, Y., Zhao, Y., Zhu, L., Xu, Q., Chai, X., Zeng, Y., Zeng, Y., Zheng, L., Zhou, Y., Luo, H., Zhang, X., Zhang, H., Zhou, Y., Fu, G., Sun, H., Huang, T. Y., Zheng, Q., Xu, H. & Wang, X. 2023. B2-Microglobulin Functions As An Endogenous NMDAR Antagonist To Impair Synaptic Function. *Cell,* 186, 1026-1038.E20.

Gardiner, K., Slavov, D., Bechtel, L. & Davisson, M. 2002. Annotation Of Human Chromosome 21 For Relevance To Down Syndrome: Gene Structure And Expression Analysis. *Genomics,* 79, 833-843.

Gelman, D. M., Marin & Rubenstein, J. L. R. 2012. The Generation Of Cortical Interneurons.

Giffin-Rao, Y., Sheng, J., Strand, B., Xu, K., Huang, L., Medo, M., Risgaard, K. A., Dantinne, S., Mohan, S., Keshan, A., Daley, R. A., Levesque, B., Amundson, L., Reese, R., Sousa, A. M. M.,

Tao, Y., Wang, D., Zhang, S.-C. & Bhattacharyya, A. 2022. Altered Patterning Of Trisomy 21 Interneuron Progenitors. *Stem Cell Reports,* 17, 1366-1379.
Golden, J. A. & Hyman, B. T. 1994. Development Of The Superior Temporal Neocortex Is Anomalous In Trisomy 21. *J Neuropathol Exp Neurol,* 53, 513-20.
Gonzales, P. K., Roberts, C. M., Fonte, V., Jacobsen, C., Stein, G. H. & Link, C. D. 2018. Transcriptome Analysis Of Genetically Matched Human Induced Pluripotent Stem Cells Disomic Or Trisomic For Chromosome 21. *Plos One,* 13, E0194581.
Gonzalez, C., Armijo, E., Bravo-Alegria, J., Becerra-Calixto, A., Mays, C. E. & Soto, C. 2018. Modeling Amyloid Beta And Tau Pathology In Human Cerebral Organoids. *Mol Psychiatry,* 23, 2363-2374.
Guidi, S., Bonasoni, P., Ceccarelli, C., Santini, D., Gualtieri, F., Ciani, E. & Bartesaghi, R. 2008. Neurogenesis Impairment And Increased Cell Death Reduce Total Neuron Number In The Hippocampal Region Of Fetuses With Down Syndrome. *Brain Pathol,* 18, 180-97.
Guidi, S., Ciani, E., Bonasoni, P., Santini, D. & Bartesaghi, R. 2011. Widespread Proliferation Impairment And Hypocellularity In The Cerebellum Of Fetuses With Down Syndrome. *Brain Pathol.,* 21, 361-373.
Guidi, S., Giacomini, A., Stagni, F., Emili, M., Uguagliati, B., Bonasoni, M. P. & Bartesaghi, R. 2018. Abnormal Development Of The Inferior Temporal Region In Fetuses With Down Syndrome. *Brain Pathol.*
Halevy, T., Biancotti, J. C., Yanuka, O., Golan-Lev, T. & Benvenisty, N. 2016. Molecular Characterization Of Down Syndrome Embryonic Stem Cells Reveals A Role For RUNX1 In Neural Differentiation. *Stem Cell Reports,* 7, 777-786.
Hansen, D. V., Lui, J. H., Parker, P. R. L. & Kriegstein, A. R. 2010. Neurogenic Radial GLIA In The Outer Subventricular Zone Of Human Neocortex. *Nature,* 464, 554-561.
Hardy, J. 2017. The Discovery Of Alzheimer-Causing Mutations In The App Gene And The Formulation Of The "Amyloid Cascade Hypothesis". *Febs J,* 284, 1040-1044.
Hasina, Z., Wang, N. & Wang, C. C. 2022. Developmental Neuropathology And Neurodegeneration Of Down Syndrome: Current Knowledge In Humans. *Front Cell Dev Biol,* 10, 877711.
Hattori, M., Fujiyama, A., Taylor, T. D., Watanabe, H., Yada, T., Park, H. S., Toyoda, A., Ishii, K., Totoki, Y., Choi, D. K., Soeda, E., Ohki, M., Takagi, T., Sakaki, Y., Taudien, S., Blechschmidt, K., Polley, A., Menzel, U., Delabar, J., Kumpf, K., Lehmann, R., Patterson, D., Reichwald, K., Rump, A., Schillhabel, M., Schudy, A., Zimmermann, W., Rosenthal, A., Kudoh, J., Schibuya, K., Kawasaki, K., Asakawa, S., Shintani, A., Sasaki, T., Nagamine, K., Mitsuyama, S., Antonarakis, S. E., Minoshima, S., Shimizu, N., Nordsiek, G., Hornischer, K., Brant, P., Scharfe, M., Schon, O., Desario, A., Reichelt, J., Kauer, G., Blocker, H., Ramser, J., Beck, A., Klages, S., Hennig, S., Riesselmann, L., Dagand, E., Haaf, T., Wehrmeyer, S., Borzym, K., Gardiner, K., Nizetic, D., Francis, F., Lehrach, H., Reinhardt, R. & Yaspo, M. L. 2000. The Dna Sequence Of Human Chromosome 21. *Nature,* 405, 311–319.
He, C., Kalafut, N. C., Sandoval, S. O., Risgaard, R., Sirois, C. L., Yang, C., Khullar, S., Suzuki, M., Huang, X., Chang, Q., Zhao, X., Sousa, A. M. M. & Wang, D. 2023. Boma, A Machine-Learning Framework For Comparative Gene Expression Analysis Across Brains And Organoids. *Cell Rep Methods,* 3, 100409.
Heinke, D., Isenburg, J. L., Stallings, E. B., Short, T. D., Le, M., Fisher, S., Shan, X., Kirby, R. S., Nguyen, H. H., Nestoridi, E., Nembhard, W. N., Romitti, P. A., Salemi, J. L. & Lupo, P. J. 2021. Prevalence Of Structural Birth Defects Among Infants With Down Syndrome, 2013-2017: A US Population-Based Study. *Birth Defects Res,* 113, 189–202.
Hibaoui, Y. & Feki, A. 2015. Concise Review: Methods And Cell Types Used To Generate Down Syndrome Induced Pluripotent Stem Cells. *J Clin Med,* 4, 696-714.
Hibaoui, Y., Grad, I., Letourneau, A., Sailani, M. R., Dahoun, S., Santoni, F. A., Gimelli, S., Guipponi, M., Pelte, M. F., Béna, F., Antonarakis, S. E. & Feki, A. 2014. Modelling And Rescuing Neurodevelopmental Defect Of <Span Style="Font-Variant:Small-Caps;">D</Span> Own Syndrome Using Induced Pluripotent Stem Cells From Monozygotic Twins Discordant For Trisomy 21. *Embo Molecular Medicine,* 6, 259-277.

Hithersay, R., Startin, C. M., Hamburg, S., Mok, K. Y., Hardy, J., Fisher, E. M. C., Tybulewicz, V. L. J., Nizetic, D. & Strydom, A. 2019. Association Of Dementia With Mortality Among Adults With Down Syndrome Older Than 35 Years. *Jama Neurol,* 76, 152-160.

Hu, N. W., Corbett, G. T., Moore, S., Klyubin, I., O'Malley, T. T., Walsh, D. M., Livesey, F. J. & Rowan, M. J. 2018. Extracellular Forms Of Aβ And Tau From Ipsc Models Of Alzheimer's Disease Disrupt Synaptic Plasticity. *Cell Rep,* 23, 1932–1938.

Huo, H.-Q., Qu, Z.-Y., Yuan, F., Ma, L., Yao, L., Xu, M., Hu, Y., Ji, J., Bhattacharyya, A., Zhang, S.-C. & Liu, Y. 2018. Modeling Down Syndrome With Patient Ipscs Reveals Cellular And Migration Deficits Of Gabaergic Neurons. *Stem Cell Reports,* 10, 1251–1266.

Itskovitz-Eldor, J., Schuldiner, M., Karsenti, D., Eden, A., Yanuka, O., Amit, M., Soreq, H. & Benvenisty, N. 2000. Differentiation Of Human Embryonic Stem Cells Into Embryoid Bodies Compromising The Three Embryonic Germ Layers. *Mol Med,* 6, 88–95.

Iulita, M. F., Garzón Chavez, D., Klitgaard Christensen, M., Valle Tamayo, N., Plana-Ripoll, O., Rasmussen, S. A., Roqué Figuls, M., Alcolea, D., Videla, L., Barroeta, I., Benejam, B., Altuna, M., Padilla, C., Pegueroles, J., Fernandez, S., Belbin, O., Carmona-Iragui, M., Blesa, R., Lleó, A., Bejanin, A. & Fortea, J. 2022. Association Of Alzheimer Disease With Life Expectancy In People With Down Syndrome. *Jama Netw Open,* 5, E2212910.

Jervis, G. A. 1948. Early Senile Dementia In Mongoloid Idiocy. *Am J Psychiatry,* 105, 102–6.

Jiang, J., Jing, Y., Cost, G. J., Chiang, J.-C., Kolpa, H. J., Cotton, A. M., Carone, D. M., Carone, B. R., Shivak, D. A., Guschin, D. Y., Pearl, J. R., Rebar, E. J., Byron, M., Gregory, P. D., Brown, C. J., Urnov, F. D., Hall, L. L. & Lawrence, J. B. 2013. Translating Dosage Compensation To Trisomy 21. *Nature,* 500, 296–300.

Jin, M., Xu, R., Wang, L., Alam, M. M., Ma, Z., Zhu, S., Martini, A. C., Jadali, A., Bernabucci, M., Xie, P., Kwan, K. Y., Pang, Z. P., Head, E., Liu, Y., Hart, R. P. & Jiang, P. 2022. Type-I-Interferon Signaling Drives Microglial Dysfunction And Senescence In Human Ipsc Models Of Down Syndrome And Alzheimer's Disease. *Cell Stem Cell,* 29, 1135–1153.E8.

Jung, R. E. & Haier, R. J. 2007. The Parieto-Frontal Integration Theory (P-Fit) Of Intelligence: Converging Neuroimaging Evidence. *Behav Brain Sci,* 30, 135–54; Discussion 154-87.

Kang, H. J., Kawasawa, Y. I., Cheng, F., Zhu, Y., Xu, X., Li, M., Sousa, A. M., Pletikos, M., Meyer, K. A., Sedmak, G., Guennel, T., Shin, Y., Johnson, M. B., Krsnik, Z., Mayer, S., Fertuzinhos, S., Umlauf, S., Lisgo, S. N., Vortmeyer, A., Weinberger, D. R., Mane, S., Hyde, T. M., Huttner, A., Reimers, M., Kleinman, J. E. & Sestan, N. 2011. Spatio-Temporal Transcriptome Of The Human Brain. *Nature,* 478, 483–9.

Kawatani, K., Nambara, T., Nawa, N., Yoshimatsu, H., Kusakabe, H., Hirata, K., Tanave, A., Sumiyama, K., Banno, K., Taniguchi, H., Arahori, H., Ozono, K. & Kitabatake, Y. 2021. A Human Isogenic IPSC-Derived Cell Line Panel Identifies Major Regulators Of Aberrant Astrocyte Proliferation In Down Syndrome. *Communications Biology,* 4, 730.

Kennedy, M., D'souza, S. L., Lynch-Kattman, M., Schwantz, S. & Keller, G. 2007. Development Of The Hemangioblast Defines The Onset Of Hematopoiesis In Human ES Cell Differentiation Cultures. *Blood,* 109, 2679–87.

Kepecs, A. & Fishell, G. 2014. Interneuron Cell Types Are Fit To Function. *Nature,* 505, 318–26.

Kessaris, N., Magno, L., Rubin, A. N. & Oliveira, M. G. 2014. Genetic Programs Controlling Cortical Interneuron Fate. *Curr Opin Neurobiol,* 26, 79–87.

Kim, J., Koo, B. K. & Knoblich, J. A. 2020. Human Organoids: Model Systems For Human Biology And Medicine. *Nat Rev Mol Cell Biol,* 21, 571–584.

Klein, J. A., Li, Z., Rampam, S., Cardini, J., Ayoub, A., Shaw, P., Rachubinski, A. L., Espinosa, J. M., Zeldich, E. & Haydar, T. F. 2022. Sonic Hedgehog Pathway Modulation Normalizes Expression Of Olig2 In Rostrally Patterned NPCs With Trisomy 21. *Frontiers In Cellular Neuroscience,* 15, 794675.

Korbel, J. O., Tirosh-Wagner, T., Urban, A. E., Chen, X. N., Kasowski, M., Dai, L., Grubert, F., Erdman, C., Gao, M. C., Lange, K., Sobel, E. M., Barlow, G. M., Aylsworth, A. S., Carpenter, N. J., Clark, R. D., Cohen, M. Y., Doran, E., Falik-Zaccai, T., Lewin, S. O., Lott, I. T., Mcgillivray, B. C., Moeschler, J. B., Pettenati, M. J., Pueschel, S. M., Rao, K. W., Shaffer,

L. G., Shohat, M., Van Riper, A. J., Warburton, D., Weissman, S., Gerstein, M. B., Snyder, M. & Korenberg, J. R. 2009. The Genetic Architecture Of Down Syndrome Phenotypes Revealed By High-Resolution Analysis Of Human Segmental Trisomies. *Proc Natl Acad Sci U S A,* 106, 12031–6.

Korenberg, J. R., Kawashima, H., Pulst, S. M., Ikeuchi, T., Ogasawara, N., Yamamoto, K., Schonberg, S. A., West, R., Allen, L., Magenis, E. & Et Al. 1990. Molecular Definition Of A Region Of Chromosome 21 That Causes Features Of The Down Syndrome Phenotype. *Am J Hum Genet,* 47, 236–46.

Kriegstein, A. & Parnavelas, J. G. 2006. Progress In Corticogenesis. *Cereb Cortex,* 16 Suppl 1, I1–2.

Kriegstein, A. R. & Noctor, S. C. 2004. Patterns Of Neuronal Migration In The Embryonic Cortex. *Trends Neurosci,* 27, 392–9.

Kwan, K. Y., Lam, M. M., Johnson, M. B., Dube, U., Shim, S., Rašin, M. R., Sousa, A. M., Fertuzinhos, S., Chen, J. G., Arellano, J. I., Chan, D. W., Pletikos, M., Vasung, L., Rowitch, D. H., Huang, E. J., Schwartz, M. L., Willemsen, R., Oostra, B. A., Rakic, P., Heffer, M., Kostović, I., Judaš, M. & Sestan, N. 2012. Species-Dependent Posttranscriptional Regulation Of NOS1 By FMRP In The Developing Cerebral Cortex. *Cell,* 149, 899–911.

Laan, L., Klar, J., Sobol, M., Hoeber, J., Shahsavani, M., Kele, M., Fatima, A., Zakaria, M., Annerén, G., Falk, A., Schuster, J. & Dahl, N. 2020. Dna Methylation Changes In Down Syndrome Derived Neural Ipscs Uncover Co-Dysregulation Of ZNF And HOX3 Families Of Transcription Factors. *Clinical Epigenetics,* 12, 9.

Lana-Elola, E., Watson-Scales, S., Slender, A., Gibbins, D., Martineau, A., Douglas, C., Mohun, T., Fisher, E. M. & Tybulewicz, V. L. 2016. Genetic Dissection Of Down Syndrome-Associated Congenital Heart Defects Using A New Mouse Mapping Panel. *Elife,* 5.

Lejeune, J. T. R. G. M. 1959. Le Mogolisme, Premier Exemple D'aberration Autosomique Humaine. *Ann Genet,* 1, 41–49.

Letourneau, A., Santoni, F. A., Bonilla, X., Sailani, M. R., Gonzalez, D., Kind, J., Chevalier, C., Thurman, R., Sandstrom, R. S., Hibaoui, Y., Garieri, M., Popadin, K., Falconnet, E., Gagnebin, M., Gehrig, C., Vannier, A., Guipponi, M., Farinelli, L., Robyr, D., Migliavacca, E., Borel, C., Deutsch, S., Feki, A., Stamatoyannopoulos, J. A., Herault, Y., Van Steensel, B., Guigo, R. & Antonarakis, S. E. 2014. Domains Of Genome-Wide Gene Expression Dysregulation In Down's Syndrome. *Nature,* 508, 345–350.

Leverenz, J. B. & Raskind, M. A. 1998. Early Amyloid Deposition In The Medial Temporal Lobe Of Young Down Syndrome Patients: A Regional Quantitative Analysis. *Exp Neurol,* 150, 296–304.

Levine, S. & Grabel, L. 2017. The Contribution Of Human/Non-Human Animal Chimeras To Stem Cell Research. *Stem Cell Research,* 24, 128–134.

Li, Z., Klein, J. A., Rampam, S., Kurzion, R., Campbell, N. B., Patel, Y., Haydar, T. F. & Zeldich, E. 2022. Asynchronous Excitatory Neuron Development In An Isogenic Cortical Spheroid Model Of Down Syndrome. *Frontiers In Neuroscience,* 16, 932384.

Lott, I. T. 2012. Neurological Phenotypes For Down Syndrome Across The Life Span. *Prog.Brain Res.,* 197, 101–121.

Lott, I. T. & Head, E. 2019. Dementia In Down Syndrome: Unique Insights For Alzheimer Disease Research. *Nat Rev Neurol,* 15, 135–147.

Lu, H.-E., Yang, Y.-C., Chen, S.-M., Su, H.-L., Huang, P.-C., Tsai, M.-S., Wang, T.-H., Tseng, C.-P. & Hwang, S.-M. 2013. Modeling Neurogenesis Impairment In Down Syndrome With Induced Pluripotent Stem Cells From Trisomy 21 Amniotic Fluid Cells. *Experimental Cell Research,* 319, 498–505.

Ludwig, T. E., Kujak, A., Rauti, A., Andrzejewski, S., Langbehn, S., Mayfield, J., Fuller, J., Yashiro, Y., Hara, Y. & Bhattacharyya, A. 2018. 20 Years Of Human Pluripotent Stem Cell Research: It All Started With Five Lines. *Cell Stem Cell,* 23, 644–648.

Maffioletti, S. M., Gerli, M. F., Ragazzi, M., Dastidar, S., Benedetti, S., Loperfido, M., Vandendriessche, T., Chuah, M. K. & Tedesco, F. S. 2015. Efficient Derivation And Inducible

Differentiation Of Expandable Skeletal Myogenic Cells From Human ES And Patient-Specific IPS Cells. *Nat Protoc,* 10, 941-58.

Marin, O. 2013. Cellular And Molecular Mechanisms Controlling The Migration Of Neocortical Interneurons. *Eur.J.Neurosci.*

Marin-Padilla, M. 1972. Structural Abnormalities Of The Cerebral Cortex In Human Chromosomal Aberrations: A Golgi Study. *Brain Res,* 44, 625-9.

Martinez, J. L., Piciw, J. G., Crockett, M., Sorci, I. A., Makwana, N., Sirois, C. L., Giffin-Rao, Y. & Bhattacharyya, A. 2024. Transcriptional Consequences Of Trisomy 21 On Neural Induction. *Front Cell Neurosci,* 18, 1341141.

Masaki, H. & Nakauchi, H. 2017. Interspecies Chimeras For Human Stem Cell Research. *Development,* 144, 2544–2547.

Mascetti, V. L. & Pedersen, R. A. 2016. Contributions Of Mammalian Chimeras To Pluripotent Stem Cell Research. *Cell Stem Cell,* 19, 163–175.

Meharena, H. S., Marco, A., Dileep, V., Lockshin, E. R., Akatsu, G. Y., Mullahoo, J., Watson, L. A., Ko, T., Guerin, L. N., Abdurrob, F., Rengarajan, S., Papanastasiou, M., Jaffe, J. D. & Tsai, L.-H. 2022. Down-Syndrome-Induced Senescence Disrupts The Nuclear Architecture Of Neural Progenitors. *Cell Stem Cell,* 29, 116–130.E7.

Menghini, D., Costanzo, F. & Vicari, S. 2011. Relationship Between Brain And Cognitive Processes In Down Syndrome. *Behav Genet,* 41, 381–93.

Mito, T. & Becker, L. E. 1993. Developmental Changes Of S-100 Protein And Glial Fibrillary Acidic Protein In The Brain In Down Syndrome. *Exp Neurol,* 120, 170–6.

Mizuno, G. O., Wang, Y., Shi, G., Wang, Y., Sun, J., Papadopoulos, S., Broussard, G. J., Unger, E. K., Deng, W., Weick, J., Bhattacharyya, A., Chen, C. Y., Yu, G., Looger, L. L. & Tian, L. 2018. Aberrant Calcium Signaling In Astrocytes Inhibits Neuronal Excitability In A Human Down Syndrome Stem Cell Model. *Cell Rep,* 24, 355–365.

Montoliu-Gaya, L., Strydom, A., Blennow, K., Zetterberg, H. & Ashton, N. J. 2021. Blood Biomarkers For Alzheimer's Disease In Down Syndrome. *J Clin Med,* 10.

Moyer, A. J., Gardiner, K. & Reeves, R. H. 2021. All Creatures Great And Small: New Approaches For Understanding Down Syndrome Genetics. *Trends Genet,* 37, 444–459.

Murray, A., Letourneau, A., Canzonetta, C., Stathaki, E., Gimelli, S., Sloan-Bena, F., Abrehart, R., Goh, P., Lim, S., Baldo, C., Dagna-Bricarelli, F., Hannan, S., Mortensen, M., Ballard, D., Syndercombe Court, D., Fusaki, N., Hasegawa, M., Smart, T. G., Bishop, C., Antonarakis, S. E., Groet, J. & Nizetic, D. 2015. Brief Report: Isogenic Induced Pluripotent Stem Cell Lines From An Adult With Mosaic Down Syndrome Model Accelerated Neuronal Ageing And Neurodegeneration. *Stem Cells,* 33, 2077–2084.

Noctor, S. C., Martinez-Cerdeno, V. & Kriegstein, A. R. 2007. Neural Stem And Progenitor Cells In Cortical Development. *Novartis Found Symp,* 288, 59–73; Discussion 73–8, 96–8.

Olmos-Serrano, J. L., Kang, H. J., Tyler, W. A., Silbereis, J. C., Cheng, F., Zhu, Y., Pletikos, M., Jankovic-Rapan, L., Cramer, N. P., Galdzicki, Z., Goodliffe, J., Peters, A., Sethares, C., Delalle, I., Golden, J. A., Haydar, T. F. & Sestan, N. 2016. Down Syndrome Developmental Brain Transcriptome Reveals Defective Oligodendrocyte Differentiation And Myelination. *Neuron,* 89, 1208–1222.

Ovchinnikov, D. A., Korn, O., Virshup, I., Wells, C. A. & Wolvetang, E. J. 2018. The Impact Of App On Alzheimer-Like Pathogenesis And Gene Expression In Down Syndrome IPSC-Derived Neurons. *Stem Cell Reports,* 11, 32–42.

Papavassiliou, P., Charalsawadi, C., Rafferty, K. & Jackson-Cook, C. 2015. Mosaicism For Trisomy 21: A Review. *Am J Med Genet A,* 167a, 26–39.

Papavassiliou, P., York, T. P., Gursoy, N., Hill, G., Nicely, L. V., Sundaram, U., Mcclain, A., Aggen, S. H., Eaves, L., Riley, B. & Jackson-Cook, C. 2009. The Phenotype Of Persons Having Mosaicism For Trisomy 21/Down Syndrome Reflects The Percentage Of Trisomic Cells Present In Different Tissues. *Am J Med Genet A,* 149a, 573–83.

Park, I. H., Arora, N., Huo, H., Maherali, N., Ahfeldt, T., Shimamura, A., Lensch, M. W., Cowan, C., Hochedlinger, K. & Daley, G. Q. 2008a. Disease-Specific Induced Pluripotent Stem Cells. *Cell,* 134**,** 877–886.

Park, I. H., Zhao, R., West, J. A., Yabuuchi, A., Huo, H., Ince, T. A., Lerou, P. H., Lensch, M. W. & Daley, G. Q. 2008b. Reprogramming Of Human Somatic Cells To Pluripotency With Defined Factors. *Nature,* 451**,** 141–146.

Parker, S. E., Mai, C. T., Canfield, M. A., Rickard, R., Wang, Y., Meyer, R. E., Anderson, P., Mason, C. A., Collins, J. S., Kirby, R. S. & Correa, A. 2010. Updated National Birth Prevalence Estimates For Selected Birth Defects In The United States, 2004-2006. *Birth Defects Res.A Clin.Mol.Teratol.,* 88**,** 1008–1016.

Parnavelas, J. G., Anderson, S. A., Lavdas, A. A., Grigoriou, M., Pachnis, V. & Rubenstein, J. L. 2000. The Contribution Of The Ganglionic Eminence To The Neuronal Cell Types Of The Cerebral Cortex. *Novartis Found Symp,* 228**,** 129–39; Discussion 139–47.

Paşca, A. M., Sloan, S. A., Clarke, L. E., Tian, Y., Makinson, C. D., Huber, N., Kim, C. H., Park, J.-Y., O'rourke, N. A., Nguyen, K. D., Smith, S. J., Huguenard, J. R., Geschwind, D. H., Barres, B. A. & Paşca, S. P. 2015. Functional Cortical Neurons And Astrocytes From Human Pluripotent Stem Cells In 3D Culture. *Nature Methods,* 12**,** 671–678.

Petit, T. L., Leboutillier, J. C., Alfano, D. P. & Becker, L. E. 1984. Synaptic Development In The Human Fetus: A Morphometric Analysis Of Normal And Down's Syndrome Neocortex. *Exp Neurol,* 83**,** 13–23.

Ponroy Bally, B., Farmer, W. T., Jones, E. V., Jessa, S., Kacerovsky, J. B., Mayran, A., Peng, H., Lefebvre, J. L., Drouin, J., Hayer, A., Ernst, C. & Murai, K. K. 2020. Human IPSC-Derived Down Syndrome Astrocytes Display Genome-Wide Perturbations In Gene Expression, An Altered Adhesion Profile, And Increased Cellular Dynamics. *Human Molecular Genetics,* 29**,** 785–802.

Presson, A. P., Partyka, G., Jensen, K. M., Devine, O. J., Rasmussen, S. A., Mccabe, L. L. & Mccabe, E. R. 2013. Current Estimate Of Down Syndrome Population Prevalence In The United States. *J Pediatr,* 163**,** 1163–8.

Prutton, K. M., Marentette, J. O., Leifheit, B. A., Esquer, H., Labarbera, D. V., Anderson, C. C., Maclean, K. N. & Roede, J. R. 2022. Oxidative Stress As A Candidate Mechanism For Accelerated Neuroectodermal Differentiation Due To Trisomy 21. *Free Radical Biology And Medicine,* 186**,** 32–42.

Prutton, K. M., Marentette, J. O., Maclean, K. N. & Roede, J. R. 2023. Characterization Of Mitochondrial And Metabolic Alterations Induced By Trisomy 21 During Neural Differentiation. *Free Radical Biology And Medicine,* 196**,** 11–21.

Pueschel, S. M. 1990. Clinical Aspects Of Down Syndrome From Infancy To Adulthood. *Am.J.Med.Genet.Suppl,* 7**,** 52–56.

Purpura, D. P. 1975. Normal And Aberrant Neuronal Development In The Cerebral Cortex Of Human Fetus And Young Infant. *Ucla Forum Med Sci***,** 141–69.

Rafii, M. S. & Fortea, J. 2023. Down Syndrome In A New Era For Alzheimer Disease. *Jama.*

Real, R., Peter, M., Trabalza, A., Khan, S., Smith, M. A., Dopp, J., Barnes, S. J., Momoh, A., Strano, A., Volpi, E., Knott, G., Livesey, F. J. & De Paola, V. 2018. In Vivo Modeling Of Human Neuron Dynamics And Down Syndrome. *Science,* 362.

Rehder, H. 1981. Pathology Of Trisomy 21--With Particular Reference To Persistent Common Atrioventricular Canal Of The Heart. *Hum Genet Suppl,* 2**,** 57–73.

Reiche, L., Kury, P. & Gottle, P. 2019. Aberrant Oligodendrogenesis In Down Syndrome: Shift In Gliogenesis? *Cells,* 8.

Risgaard, K. A., Sorci, I. A., Mohan, S. & Bhattacharyya, A. 2022. Meta-Analysis Of Down Syndrome Cortical Development Reveals Underdeveloped State Of The Science. *Front Cell Neurosci,* 16**,** 915272.

Ross, M. H., Galaburda, A. M. & Kemper, T. L. 1984. Down's Syndrome: Is There A Decreased Population Of Neurons? *Neurology,* 34**,** 909–16.

Rueda, N., Florez, J. & Martinez-Cue, C. 2013. Apoptosis In Down's Syndrome: Lessons From Studies Of Human And Mouse Models. *Apoptosis,* 18, 121–34.

Ryoo, S. R., Cho, H. J., Lee, H. W., Jeong, H. K., Radnaabazar, C., Kim, Y. S., Kim, M. J., Son, M. Y., Seo, H., Chung, S. H. & Song, W. J. 2008. Dual-Specificity Tyrosine(Y)-Phosphorylation Regulated Kinase 1A-Mediated Phosphorylation Of Amyloid Precursor Protein: Evidence For A Functional Link Between Down Syndrome And Alzheimer's Disease. *J Neurochem,* 104, 1333–44.

Salari, S., Adashi, E. Y., Keller, L., Johnson, T. R. B. & Smith, G. D. 2023. Human Embryos Donated For Human Embryonic Stem Cell Derivation. *Fertil Steril,* 119, 3–10.

Sandoval, S. O., Cappuccio, G., Kruth, K., Osenberg, S., Khalil, S. M., Méndez-Albelo, N. M., Padmanabhan, K., Wang, D., Niciu, M. J., Bhattacharyya, A., Stein, J. L., Sousa, A. M. M., Waxman, E. A., Buttermore, E. D., Whye, D., Sirois, C. L., Williams, A., Maletic-Savatic, M. & Zhao, X. 2024. Rigor And Reproducibility In Human Brain Organoid Research: Where We Are And Where We Need To Go. *Stem Cell Reports,* 19, 796–816.

Shahbazi, M. N., Wang, T., Tao, X., Weatherbee, B. A. T., Sun, L., Zhan, Y., Keller, L., Smith, G. D., Pellicer, A., Scott, R. T., Jr., Seli, E. & Zernicka-Goetz, M. 2020. Developmental Potential Of Aneuploid Human Embryos Cultured Beyond Implantation. *Nat Commun,* 11, 3987.

Shi, Y., Kirwan, P. & Livesey, F. J. 2012a. Directed Differentiation Of Human Pluripotent Stem Cells To Cerebral Cortex Neurons And Neural Networks. *Nature Protocols,* 7, 1836-1846.

Shi, Y., Kirwan, P., Smith, J., Maclean, G., Orkin, S. H. & Livesey, F. J. 2012b. A Human Stem Cell Model Of Early Alzheimer's Disease Pathology In Down Syndrome. *Science Translational Medicine,* 4.

Shin, M., Besser, L. M., Kucik, J. E., Lu, C., Siffel, C. & Correa, A. 2009. Prevalence Of Down Syndrome Among Children And Adolescents In 10 Regions Of The United States. *Pediatrics,* 124, 1565–71.

Sidman, R. L. & Rakic, P. 1973. Neuronal Migration, With Special Reference To Developing Human Brain: A Review. *Brain Res.,* 62, 1–35.

Silverman, W. 2007. Down Syndrome: Cognitive Phenotype. *Ment Retard Dev Disabil Res Rev,* 13, 228–36.

Sloan, S. A., Andersen, J., Pasca, A. M., Birey, F. & Pasca, S. P. 2018. Generation And Assembly Of Human Brain Region-Specific Three-Dimensional Cultures. *Nature Protocols,* 13, 2062–2085.

Stagni, F., Giacomini, A., Emili, M., Guidi, S. & Bartesaghi, R. 2018. Neurogenesis Impairment: An Early Developmental Defect In Down Syndrome. *Free Radic Biol Med,* 114, 15–32.

Stagni, F., Giacomini, A., Emili, M., Uguagliati, B., Bonasoni, M. P., Bartesaghi, R. & Guidi, S. 2019. Subicular Hypotrophy In Fetuses With Down Syndrome And In The Ts65Dn Model Of Down Syndrome. *Brain Pathol,* 29, 366–379.

Stagni, F., Giacomini, A., Emili, M., Uguagliati, B., Bonasoni, M. P., Bartesaghi, R. & Guidi, S. 2020. Neuroanatomical Alterations In Higher-Order Thalamic Nuclei Of Fetuses With Down Syndrome. *Clin Neurol Neurosurg,* 194, 105870.

Takahashi, K., Tanabe, K., Ohnuki, M., Narita, M., Ichisaka, T., Tomoda, K. & Yamanaka, S. 2007. Induction Of Pluripotent Stem Cells From Adult Human Fibroblasts By Defined Factors. *Cell,* 131, 861–872.

Tan, K. L., Lee, H. C., Cheah, P. S. & Ling, K. H. 2023. Mitochondrial Dysfunction In Down Syndrome: From Pathology To Therapy. *Neuroscience,* 511, 1–12.

Tang, X.-Y., Xu, L., Wang, J., Hong, Y., Wang, Y., Zhu, Q., Wang, D., Zhang, X.-Y., Liu, C.-Y., Fang, K.-H., Han, X., Wang, S., Wang, X., Xu, M., Bhattacharyya, A., Guo, X., Lin, M. & Liu, Y. 2021. DSCAM/PAK1 Pathway Suppression Reverses Neurogenesis Deficits In IPSC-Derived Cerebral Organoids From Patients With Down Syndrome. *Journal Of Clinical Investigation,* 131, E135763.

Thomson, J. A., Itskovitz-Eldor, J., Shapiro, S. S., Waknitz, M. A., Swiergiel, J. J., Marshall, V. S. & Jones, J. M. 1998. Embryonic Stem Cell Lines Derived From Human Blastocysts. *Science,* 282, 1145–1147.

Tokusumi, T., Iida, A., Hirata, T., Kato, A., Nagai, Y. & Hasegawa, M. 2002. Recombinant Sendai Viruses Expressing Different Levels Of A Foreign Reporter Gene. *Virus Res,* 86, 33–8.

Toshikawa, H., Ikenaka, A., Li, L., Nishinaka-Arai, Y., Niwa, A., Ashida, A., Kazuki, Y., Nakahata, T., Tamai, H., Russell, D. W. & Saito, M. K. 2021. N-Acetylcysteine Prevents Amyloid-B Secretion In Neurons Derived From Human Pluripotent Stem Cells With Trisomy 21. *Sci Rep,* 11, 17377.

Trounson, A. 2006. The Production And Directed Differentiation Of Human Embryonic Stem Cells. *Endocrine Reviews,* 27, 208–219.

Tunstall, O., Bhatnagar, N., James, B., Norton, A., O'Marcaigh, A. S., Watts, T., Greenough, A., Vyas, P., Roberts, I. & Wright, M. 2018. Guidelines For The Investigation And Management Of Transient Leukaemia Of Down Syndrome. *Br J Haematol,* 182, 200–211.

Van Den Heuvel, M. P. & Sporns, O. 2013. Network Hubs In The Human Brain. *Trends Cogn Sci,* 17, 683–96.

Wegiel, J., Gong, C. X. & Hwang, Y. W. 2011. The Role Of DYRK1A In Neurodegenerative Diseases. *FEBS J,* 278, 236–45.

Weick, J. P., Held, D. L., Bonadurer, G. F., Doers, M. E., Liu, Y., Maguire, C., Clark, A., Knackert, J. A., Molinarolo, K., Musser, M., Yao, L., Yin, Y., Lu, J., Zhang, X., Zhang, S.-C. & Bhattacharyya, A. 2013. Deficits In Human Trisomy 21 IPSCs And Neurons. *Proceedings Of The National Academy Of Sciences,* 110, 9962–9967.

Weitzdoerfer, R., Dierssen, M., Fountoulakis, M. & Lubec, G. 2001. Fetal Life In Down Syndrome Starts With Normal Neuronal Density But Impaired Dendritic Spines And Synaptosomal Structure. *J Neural Transm Suppl,* 59–70.

Wiseman, F. K., Al-Janabi, T., Hardy, J., Karmiloff-Smith, A., Nizetic, D., Tybulewicz, V. L., Fisher, E. M. & Strydom, A. 2015. A Genetic Cause Of Alzheimer Disease: Mechanistic Insights From Down Syndrome. *Nat Rev Neurosci,* 16, 564–74.

Wiseman, F. K., Pulford, L. J., Barkus, C., Liao, F., Portelius, E., Webb, R., Chávez-Gutiérrez, L., Cleverley, K., Noy, S., Sheppard, O., Collins, T., Powell, C., Sarell, C. J., Rickman, M., Choong, X., Tosh, J. L., Siganporia, C., Whittaker, H. T., Stewart, F., Szaruga, M., Murphy, M. P., Blennow, K., De Strooper, B., Zetterberg, H., Bannerman, D., Holtzman, D. M., Tybulewicz, V. L. J., Fisher, E. M. C., Consortium, L. D. S. & Consortium, L. 2018. Trisomy Of Human Chromosome 21 Enhances Amyloid-B Deposition Independently Of An Extra Copy Of App. *Brain,* 141, 2457–2474.

Wisniewski, K. E., Dalton, A. J., Mclachlan, C., Wen, G. Y. & Wisniewski, H. M. 1985. Alzheimer's Disease In Down's Syndrome: Clinicopathologic Studies. *Neurology,* 35, 957–61.

Wonders, C. P. & Anderson, S. A. 2006. The Origin And Specification Of Cortical Interneurons. *Nat Rev Neurosci,* 7, 687–96.

Wu, C. I., Vinton, E. A., Pearse, R. V., 2nd, Heo, K., Aylward, A. J., Hsieh, Y. C., Bi, Y., Adeleye, S., Fancher, S., Duong, D. M., Seyfried, N. T., Schwarz, T. L. & Young-Pearse, T. L. 2022a. App And DYRK1A Regulate Axonal And Synaptic Vesicle Protein Networks And Mediate Alzheimer's Pathology In Trisomy 21 Neurons. *Mol Psychiatry,* 27, 1970–1989.

Wu, Q., Liu, J., Fang, A., Li, R., Bai, Y., Kriegstein, A. R. & Wang, X. 2014. The Dynamics Of Neuronal Migration. *Adv Exp Med Biol,* 800, 25–36.

Wu, Y., West, N. R., Bhattacharyya, A. & Wiseman, F. K. 2022b. Cell Models For Down Syndrome-Alzheimer's Disease Research. *Neuronal Signal,* 6, NS20210054.

Xu, L., Huo, H.-Q., Lu, K.-Q., Tang, X.-Y., Hong, Y., Han, X., Fu, Z.-X., Fang, K.-H., Xu, M., Guo, X. & Liu, Y. 2022. Abnormal Mitochondria In Down Syndrome IPSC-Derived Gabaergic Interneurons And Organoids. *Biochimica Et Biophysica Acta (BBA) - Molecular Basis Of Disease,* 1868, 166388.

Xu, R., Brawner, A. T., Li, S., Liu, J.-J., Kim, H., Xue, H., Pang, Z. P., Kim, W.-Y., Hart, R. P., Liu, Y. & Jiang, P. 2019. OLIG2 Drives Abnormal Neurodevelopmental Phenotypes In Human IPSC-Based Organoid And Chimeric Mouse Models Of Down Syndrome. *Cell Stem Cell,* 24, 908–926.E8.

Yoshida, K., Toki, T., Okuno, Y., Kanezaki, R., Shiraishi, Y., Sato-Otsubo, A., Sanada, M., Park, M. J., Terui, K., Suzuki, H., Kon, A., Nagata, Y., Sato, Y., Wang, R., Shiba, N., Chiba, K., Tanaka, H., Hama, A., Muramatsu, H., Hasegawa, D., Nakamura, K., Kanegane, H., Tsukamoto, K., Adachi, S., Kawakami, K., Kato, K., Nishimura, R., Izraeli, S., Hayashi, Y., Miyano, S., Kojima, S., Ito, E. & Ogawa, S. 2013. The Landscape Of Somatic Mutations In Down Syndrome-Related Myeloid Disorders. *Nat Genet,* 45, 1293–9.

Yu, J., Vodyanik, M. A., Smuga-Otto, K., Antosiewicz-Bourget, J., Frane, J. L., Tian, S., Nie, J., Jonsdottir, G. A., Ruotti, V., Stewart, R., Slukvin, I. I. & Thomson, J. A. 2007. Induced Pluripotent Stem Cell Lines Derived From Human Somatic Cells. *Science,* 318, 1917–1920.

Zdaniuk, G., Wierzba-Bobrowicz, T., Szpak, G. M. & Stępień, T. 2011. Astroglia Disturbances During Development Of The Central Nervous System In Fetuses With Down's Syndrome. *Folia Neuropathol,* 49, 109–14.

Zhang, S.-C., Wernig, M., Duncan, I. D., Brüstle, O. & Thomson, J. A. 2001. In Vitro Differentiation Of Transplantable Neural Precursors From Human Embryonic Stem Cells. *Nature Biotechnology,* 19, 1129-1133.

Zhang, W., Jiang, J., Xu, Z., Yan, H., Tang, B., Liu, C., Chen, C. & Meng, Q. 2023. Microglia-Containing Human Brain Organoids For The Study Of Brain Development And Pathology. *Mol Psychiatry,* 28, 96–107.

Zhao, H. H. & Haddad, G. G. 2022. Alzheimer's Disease Like Neuropathology In Down Syndrome Cortical Organoids. *Front Cell Neurosci,* 16, 1050432.

Zhao, Y., Zheng, Q., Hong, Y., Gao, Y., Hu, J., Lang, M., Zhang, H., Zhou, Y., Luo, H., Zhang, X., Sun, H., Yan, X. X., Huang, T. Y., Wang, Y. J., Xu, H., Liu, C. & Wang, X. 2023. B(2)-Microglobulin Coaggregates With Aβ And Contributes To Amyloid Pathology And Cognitive Deficits In Alzheimer's Disease Model Mice. *Nat Neurosci,* 26, 1170–1184.

Zheng, Q., Song, B., Li, G., Cai, F., Wu, M., Zhao, Y., Jiang, L., Guo, T., Shen, M., Hou, H., Zhou, Y., Zhao, Y., Di, A., Zhang, L., Zeng, F., Zhang, X. F., Luo, H., Zhang, X., Zhang, H., Zeng, Z., Huang, T. Y., Dong, C., Qing, H., Zhang, Y., Zhang, Q., Wang, X., Wu, Y., Xu, H., Song, W. & Wang, X. 2022. USP5 Inhibition Ameliorates Alzheimer's Pathology Through The Regulation Of App Processing And Aβ Generation. *J Clin Invest,* 132.

Afterword

How can one study Down syndrome? In this volume we discuss commonly used model systems for understanding the complex consequences on development and function of slightly overexpressing the several hundred genes on human chromosome 21 (HSA21). We anticipate that these descriptions will be helpful for investigators entering the field of Down syndrome research. But which system(s) should be used for which application?

The answer depends on the specific question being asked and by the *comparative phenotypes* that can be measured in a non-human system. Many fundamental properties of interactions between a small number of genes are likely to be conserved across large evolutionary distances. Studies of addition or subtraction of single genes can be accomplished in all of the systems described here, but studies of multiple gene interactions—even for two genes—is a different matter. Given 158 HSA21 protein coding genes conserved with rodents, a simple (!) subtraction of each gene (murine or human) from the TcMAC21 trisomic model would require establishment of 158 strains of mice; creating transgenic mice with overrepresentation of each possible pairwise combination on an otherwise euploid background would require 12,482 strains. This is clearly not a viable approach. However, a genetic screen in which the HSA21 orthologs in *C. elegans*, *Drosophila*, or zebrafish are targeted to increase transcription is a possible approach to combinatorial gene overexpression, potentially of many genes. In fact, given a strong selection and relevant and readily evaluated phenotype, a CRSPR-based screen could be carried out in human iPSCs to find not only trisomic genes that contribute to the phenotype but also disomic modifiers throughout the genome. Zebrafish offer opportunities for considering multiple gene interactions, and as a vertebrate species, presents some developmental outcomes more similar to mammals. They may present more challenges for conducting comprehensive unbiased screens, but results of large-scale screens could identify a limited number of target candidates for assessment in other systems. As a generalization, the number of genes that can be queried is related to the method of delivery and the complexity of phenotype assessment.

B. Ye, R. H. Reeves (eds.), *Genetic Models of Down Syndrome*,
https://doi.org/10.1007/978-3-031-78611-2

What aspect of Down syndrome do you want to study? A mammal would certainly be the appropriate model type if the goal is to provide preclinical data for a clinical trial. Understanding why trisomy 21 is protective against many forms of solid cancers in adults, but susceptible to specific leukemias in childhood, likely requires a mammalian model. Taking the next step to find all possible partners of candidate genes or refine understanding of mechanisms might proceed more efficaciously in other model systems. Heart development is accessible in fish and in some cases, even in flies. Human iPSC-derived organoids might be employed for specific questions before moving to more time-consuming models. Understanding the impact of a gene or genes on development of specific types of neurons might be carried out more rapidly and broadly in zebrafish, flies, or worms depending on the target.

The most frequently cited "common" characteristic of Down syndrome is intellectual disability (ID), but "ID" *is not a phenotype in any scientifically meaningful sense*. Cognitive abilities in people with Down syndrome range from individuals who are nonverbal and noninteractive to those with IQs in the near normal range. Are the processes that affect their brain function identical across this range? Obviously not. Further, the many possible effects of trisomy 21 on brain development and function are not the same as those in any other syndrome, disease, or trauma in *Homo sapiens*, let alone in other species. Despite many advances in understanding how the brain works, we are very far from a complete understanding. Simplifying assumptions are essential to progress in this area, but equating any variance in the outcomes of very different etiologies (e.g. by unspecified use of terms like "ID" to describe many uniquely different functions) immediately obfuscates understanding. Of the common impacts of trisomy 21 on the life and opportunities of people in which it occurs, minimizing the impact on cognitive potential must be among the most important goals. It can only be addressed by clear and consistent definition of the issues being addressed.

Another near-universal feature of trisomy 21 is the impact on the craniofacial skeleton. Retrusion of the midface skeleton and mandible underlies the characteristic appearance of Down syndrome, further modified by variations in musculature and likely innervation of facial muscles. Precise metrics are available for characterization of facial appearance. Despite the readily recognizable impact on features, children with trisomy 21 look much more like their siblings than they do like other children with Down syndrome. The degree of impact varies significantly among individuals. Many additional aspects of bone development and metabolism have been identified using models to study axial skeleton.

The early onset of Alzheimer's disease (AD) histopathology occurs in all individuals with trisomy 21 and the accompanying progressive dementia might affect all of those who live long enough. Subtle differences in histopathology appear to distinguish AD in Down syndrome from that in people with two copies of HSA21, but in both cases, the overexpression of the *APP* gene is a major (or the) precipitating cause. However, the age of onset and rate of progression are variable in Down syndrome as in the euploid population and the basis for this is not well understood.

Modification of APP expression and removal of the products of the protein are major targets for therapy.

A number of features are more common, but not universal, in those with trisomy 21. In other words, trisomy 21 is a risk factor but is not sufficient to cause these outcomes; additional genetic, environmental, or stochastic contributions must be involved. Congenital heart disease is seen in 40–45% of newborns with Down syndrome, or from the opposite perspective more than half of babies with trisomy 21 have a normal heart. Hirschsprung disease is more frequent in Down syndrome than in the euploid population. Substantial anecdotal information and a few reviewed scientific studies suggest a lower incidence of atherosclerotic disease. Autism may affect 15% or more of those with trisomy 21.

To design experiments and select a model system, one must ultimately address the question: *What is Down syndrome?* From a clinical perspective, this syndrome is a collection of features commonly, but *not universally*, seen more frequently in individuals with trisomy 21 than in the euploid population. From a developmental perspective, one might consider Down syndrome to be a condition in which regulatory mechanisms that normally limit the range of variability are expanded allowing more divergence in structure or function. Genetically, the research community has considered for many years whether Down syndrome is a collection of single gene dosage effects or the product of complex genetic interactions involving many or most HSA21 genes and the resulting impact on expression of thousands of disomic genes. The reductionist concept is clearly more readily approachable from an experimental standpoint, but has led to misleading statements to the effect that dosage of a single gene is responsible for major phenotypes or even the entire syndrome.

All of these are considerations in choosing an experimental system. But the primary consideration for experiments leading to an understanding of the effects of trisomy 21 rests with caregivers and affected individuals. *What aspects of Down syndrome provide the most significant barriers to realization of an individual's potential?* Interaction with the families, caregivers, and individuals with Down syndrome is key. It is not for researchers to tell those living with Down syndrome what they want, or to assume they know what is most important. Whether doing a mass unbiased genome-wide gene expression screen in worms, differentiating derived iPSC in organoid cultures, or refining measures of learning and memory to support more precise understanding of clinical outcomes, research is justified to the degree it can provide support to the population of those with Down syndrome.

For most researchers, the reason for efforts to improve cognition is to provide more opportunities for personal growth and development. A significant number of families feel that any change that would affect personality (through reduction of barriers to cognitive function) is inappropriate. I am frequently reminded of a father addressing a parent group in Atlanta at a meeting where the Roche clinical trial of basmisanil to improve learning and memory was being discussed. This father gave a moving description of how much he saw himself in his son who had Down syndrome, the red hair, love of sports, and stubborn personality that they shared. He gave a quite beautiful portrayal of his love for this child as he was. But he added, "When I see him struggling with math homework near tears because he just can't understand it, would I do something to make this easier for him? Damn right I would."

Learning and memory barriers, Alzheimer's disease, congenital heart disease; approaches to prophylaxis for cancer and atherosclerosis for the euploid population; these are prominent reasons why we seek to understand the consequences of trisomy 21. The model systems described here and used extensively in research are a useful component of that ongoing effort.

Roger H. Reeves

Index

A

Animal model, 3, 72, 81, 89, 189, 201, 235

C

Cre/loxP-mediated chromosome engineering, 46–59

D

Dosage imbalance, 6, 7, 23–25, 51, 55, 57, 71–74, 76, 79, 81
Down syndrome (DS), 2, 50, 72, 87, 151, 187, 238, 269
Down syndrome critical region 1 (DSCR1), 168, 169, 173, 174, 176
Drosophila, 112, 124, 151–178, 190, 195, 202, 269
Drug discovery, 214–217, 240
Dscam, 152, 157, 159, 162–165, 169–171, 174, 175, 202, 204, 254
Dyrk1A, 13, 16, 24, 53, 54, 93, 122, 123, 127, 166–168, 172–173, 175–176, 189, 194, 198, 202, 210, 211, 214, 215, 218, 250–252

E

Environmental effects, 6, 35, 36
Extra chromosome, 3–6, 9, 21, 24, 27, 31, 36, 51, 54–59, 72–75, 77, 83, 97, 177, 243

G

Gene dosage, 5, 6, 24, 26, 51, 54–59, 71, 73, 74, 78, 80–83, 110, 177, 248, 250, 251, 254, 255, 271
Genetic background, 6, 7, 32–33, 35, 36, 73, 77, 116, 238
Genetic models, 127, 204, 212, 213
Genetic screening, 161

H

Heart development, 157, 169, 171, 193, 201, 202, 254, 270
Human, 2, 48, 71, 89, 151, 187, 234, 269
Human chromosome 21 (Hsa21), 2, 50, 71, 87, 151, 188, 234, 269

L

Learning, 22, 27, 31, 82, 83, 94, 102, 113, 118, 157, 166, 168, 171, 176, 210–212, 271, 272

M

Mammal, 93, 94, 102, 110, 112–114, 124, 152, 162, 163, 167, 197, 199–201, 204, 205, 217, 269, 270
Modeling, 47–51, 59, 76, 81, 83, 109, 122, 127, 128, 201, 238, 240, 241, 254, 255
Mouse, 2, 46, 71, 87, 162, 188, 234
Mouse models, 2, 48, 72, 87, 162, 188, 244

B. Ye, R. H. Reeves (eds.), *Genetic Models of Down Syndrome*,
https://doi.org/10.1007/978-3-031-78611-2

N
Nematode, 88
Neural, 100, 111, 126, 156, 166, 172, 175, 191, 198, 203, 206–208, 210, 212, 235, 239–245, 247, 248, 255
Neurodevelopment, 124, 194, 213, 248, 255

O
Optical transparency, 188, 197, 199, 207, 216
Organ, 72, 74, 82, 91, 113–114, 116, 122, 152–157, 159, 188, 197, 204, 235–236, 239, 240, 253, 255
Overexpression, 31, 53, 80, 97, 98, 107, 110, 115, 122–128, 152, 157, 159, 162, 163, 165–177, 189, 191–195, 198–200, 202, 209, 210, 213, 214, 234, 245, 250, 269, 270

P
Phenotype variability, 4, 5, 37
Post-mortem, 235–236, 246

R
Rat, 71–77, 79, 80, 82, 83, 168, 207, 217
Rodent models, 71–83, 216

S
Segmental duplication, 24–28, 196
Segmental trisomy, 6–23, 25, 31, 32, 53, 56, 169
Single-cell RNA-seq, 103, 191
Stem cells, 30, 72, 199, 214, 236–241, 250
Synteny, 76, 110, 128, 189, 191, 196, 216, 217

T
Therapeutic treatment, 175–177
Tissue, 2, 29, 30, 72, 73, 75, 78, 79, 96, 103, 104, 108, 113, 114, 120, 155, 159, 160, 166, 175, 177, 190, 194, 195, 201, 217, 235–236, 239, 240, 244, 246, 248, 250, 251, 255
Traits, 2, 4–6, 21, 26, 29, 32, 33, 35, 37, 87–99, 251
Transchromosomic, 6, 9, 29–31, 36, 78, 83
Transgenic line, 195, 199, 200, 203, 208–209, 211, 217
Translocation, 6–8, 19–21, 23–24, 46–48, 52, 55, 72, 196, 234
Trisomy, 2, 52, 71, 87, 152, 187, 234, 270
Trisomy 21 (Ts21), 2, 3, 6, 33, 34, 37, 52, 53, 58, 72, 74–76, 80, 82, 83, 87, 88, 157, 171, 177, 189, 199, 201, 234–249, 252, 270–272

W
Whole-brain activity, 206–207, 218

Z
Zebrafish, 89, 187–218, 269, 270

The manufacturer's authorised representative in the EU is Springer Nature Customer Service Centre GmbH, Europaplatz 3, 69115 Heidelberg, Germany. If you have any concerns regarding our products, please contact ProductSafety@springernature.com

Printed and bound by CPI Group (UK) Ltd, Croydon, CR0 4YY
30/09/2026
02225600-0005